$$s_p^2 = \frac{(N_1 - 1)s_1^2 + (N_2 - 1)s_2^2}{N_1 + N_2 - 2} \qquad (8.2)$$

When $\sigma_1^2 \neq \sigma_2^2$, $\quad s_{\bar{x}_1 - \bar{x}_2} = \sqrt{\dfrac{s_1^2}{N_1} + \dfrac{s_2^2}{N_2}} \qquad (8.2)$

$$\text{Satterwaite } df = \frac{(s_{\bar{x}_1}^2 + s_{\bar{x}_2}^2)^2}{\dfrac{(s_{\bar{x}_1}^2)^2}{N_1 - 1} + \dfrac{(s_{\bar{x}_2}^2)^2}{N_2 - 1}} \qquad (9.3)$$

where $s_{\bar{x}_1}^2 = \dfrac{s_1^2}{N_1}$ and $s_{\bar{x}_2}^2 = \dfrac{s_2^2}{N_2}$

TWO RELATED MEANS

$$t = \frac{\bar{D} - \mu_D}{s_{\bar{D}}}, \qquad df = N_D - 1, \quad \text{and} \quad s_{\bar{D}} = \frac{s_D}{\sqrt{N_D}} \qquad (8.3)$$

$$\text{where } s_D = \sqrt{\frac{\Sigma D^2 - \dfrac{(\Sigma D)^2}{N_D}}{N_D - 1}} \qquad (8.3)$$

TWO SEPARATE VARIANCES

With unequal Ns, $\quad F = \dfrac{\text{larger } s^2}{\text{smaller } s^2} \quad$ for df_1 and $df_2 \qquad (9.2)$

With equal Ns, $\quad F_{\max} = \dfrac{\text{largest } s^2}{\text{smallest } s^2} \qquad (9.2)$

for $k = 2$ and $df = N_{\text{per batch}} - 1$

k SEPARATE VARIANCES

With unequal Ns,

$$\chi^2 = \frac{2.3026\,[(A)(E - C)]}{1 + \dfrac{1}{3(k - 1)}\left(D - \dfrac{1}{A}\right)} \qquad (9.4)$$

where A, B, C, and D are table sums and

$$E = \log_{10}\left(\frac{B}{A}\right), df = k - 1$$

With equal Ns,

$$F_{\max} = \frac{\text{largest } s^2}{\text{smallest } s^2} \qquad (9.4)$$

for k = number of s^2s in the set and $df = N_{\text{per batch}} - 1$

TWO RELATED VARIANCES

$$t = \frac{(s_1^2 - s_2^2)\sqrt{(N - 2)}}{\sqrt{4s_1^2 s_2^2 (1 - r_{12}^2)}} \quad \text{and} \quad df = N - 2 \qquad (9.5)$$

T5-BZZ-503

NUMBERS
A PRIMER OF DATA ANALYSIS

NUMBERS
A PRIMER OF DATA ANALYSIS

EDWARD L. WIKE
The University of Kansas

Charles E. Merrill Publishing Company
A Bell & Howell Company
Columbus Toronto London Sydney

Published by
Charles E. Merrill Publishing Company
A Bell & Howell Company
Columbus OH 43216

This book was set in Century Schoolbook and Frutiger.
Production Editor: Megan Rowe
Text Designer: Cynthia Brunk
Cover Designer: Cathy Watterson
Cover Photographer: Larry Hammill

The computer output appearing throughout the text is from *Minitab Reference Manual* by Thomas A. Ryan Jr., Brian L. Joiner, and Barbara F. Ryan, 1981. Copyright 1981 by Thomas A. Ryan Jr. Reproduced by the Academic Computer Center, The University of Kansas. Reprinted by permission.

"Who likes short shorts? The hemline theorists" (p. 152) by Steve Rosen, is reprinted by permission of the Kansas City Star, copyright 1981; "Rise in child health problems seen," "One-third of workers in survey admit stealing," "Factories operating at highest capacity since February 1982," "Economic gaps still separate races, study says," "Fed expected to revise money-supply targets," "Researchers say smokers' children ill more than those of non-smokers," "Unemployment declines for third straight month," and "Residents to weigh arsenic risk from Tacoma, Wash., smelter" (pp. 153–158) are reprinted by permission of Associated Press Newsfeatures.

Copyright © 1985 by Bell & Howell Company. All rights reserved. No part of this book may be reproduced in any form, electronic or mechanical, including photocopy, recording, or any information storage and retrieval system, without permission in writing from the publisher.

Library of Congress Catalog Card Number: 84–61672
International Standard Book Number: 0–675–20309–0
Printed in the United States of America
1 2 3 4 5 6 7 8 9 10—89 88 87 86 85

CONTENTS

CHAPTER ONE
NUMBERS NUMBERS NUMBERS   2

1.1   The world of numbers   4
1.2   Kinds of numbers   4
1.3   The tasks of statistics   7
1.4   The data analytic orientation   8
1.5   Mental health and statistics   10
1.6   Outline of the book   12
       Summary   12

CHAPTER TWO
PUTTING NUMBERS INTO PILES
FOR FUN AND PROFIT   14

2.1   Numbers from research are disorganized   16
2.2   It helps to put numbers into several piles   16
2.3   Breaking up piles into smaller piles   17
2.4   From piles of numbers to pictures of numbers   21
2.5   What do pictures of numbers tell us?   22
2.6   The compared-to-what problem—percentiles   26
       Summary   31

Problems 32
Addendum: Using computers in statistics 33

CHAPTER THREE
BEING MORE EXACT ABOUT CENTERS 36

3.1 Why do we need more exact information about centers? 38
3.2 Common types of centers—mean, median, and mode 38
3.3 Calculating the mean 40
3.4 Finding the median 43
3.5 The mode—hello and goodbye 46
3.6 The mean versus the median 47
3.7 New types of centers 48
Summary 49
Problems 51
Appendix: Finding the mean from a frequency distribution 51

CHAPTER FOUR
NAILING DOWN SPREADS 54

4.1 Why spread is the fundamental concept in statistics 56
4.2 Some spreads—standard deviation, semi-interquartile range, and range 56
4.3 Key concepts—sum of squares, standard deviation, and variance 57
4.4 Getting standard deviations and variances 59
4.5 Getting the semi-interquartile range 62
4.6 A quick-and dirty spread—the range 65
4.7 Whens and whys for the three spreads 65
4.8 Summarizing batches of numbers 67
4.9 The compared-to-what problem again—the standard-score approach 70
4.10 The normal distribution and Z scores 73
Summary 76
Problems 78
Appendix: Calculating s from a frequency distribution 79
Addendum: Using computers in statistics 80

CHAPTER FIVE
FLIPPIN' COINS AND BUYIN' BEERS   82

5.1 The suds scene   84
5.2 Putting the scene into a statistical context   84
5.3 Testing hypotheses with the binomial distribution   87
5.4 Testing hypotheses with the normal distribution   97
5.5 Testing hypotheses about centers, spreads, and other statistics   99
     Summary   100
     Problems   102
     Addendum: Using computers in statistics   103

CHAPTER SIX
RHO RHO RHO THE BOAT   106

6.1 So you have two numbers for every case   108
6.2 Finding an index to summarize a relationship   113
6.3 An example—selling ice cream and liking kids   116
6.4 Is our relationship just a chance affair?   118
6.5 Suppose you have a pair of ranks for every case?   122
     Summary   126
     Problems   127
     Addendum: Using computers in statistics   128

CHAPTER SEVEN
STILL RHOING   130

7.1 Why it's a good idea to square r and multiply by 100   132
7.2 Spreads and r 133
7.3 The Crunchie Ice Cream Bar example revisited   135
7.4 Why correlation can be "an instrument of the devil"   144
     Summary   146
     Problems   147
     Addendum: Using computers in statistics   148

NUMBERS IN THE NEWS   151

CHAPTER EIGHT
COMPARING CENTERS   160

8.1   Testing a single mean   162
8.2 t for two—testing the means of two separate batches   170
8.3 t for twins—testing means for two related batches   176
8.4   Testing hypotheses versus confidence limits   181
      Summary   184
      Problems   186
      Addendum: Using computers in statistics   187

CHAPTER NINE
COMPARING SPREADS   190

9.1   Why spreads should be compared   192
9.2   Testing spreads from two separate batches   194
9.3   The t-for-two test revisited   199
9.4   Testing spreads from k separate batches   201
9.5   Comparing spreads from two related batches   205
      Summary   207
      Problems   208
      Addendum: Using computers in statistics   210

CHAPTER TEN
A PIE WITH A FEW SLICES—TESTING k MEANS   212

10.1 The case of k separate batches—a simple analysis of variance (ANOVA)   214
10.2 Genuine Swiss Army Knives and Singapore Slings—another example   218
10.3 What to do after an F test?   222
10.4 Seminasty problems   226
10.5 The case of k related batches—a treatments-by-subjects or repeated-measures ANOVA   231
10.6 A strange example—the Washington (DC) connection   231
10.7 New complications   235
      Summary   237
      Problems   239
      Addendum: Using computers in statistics   241

CHAPTER ELEVEN
A PIE WITH MORE SLICES—COMPARING MEANS FROM A FACTORIAL EXPERIMENT 244

11.1 What is a factorial design and what is so great about it? 246
11.2 A mildly exciting example—a close shave in Singapore 248
11.3 After a two-factor ANOVA—what next? 253
11.4 "Dead rats belong in the Results Section" 264
       Summary 268
       Problems 270
       Addendum: Using computers in statistics 272

CHAPTER TWELVE
MEETING RANKS AND SIGNS AT HIGH NOON 274

12.1 Statistics (nonparametric) that do not take so much for granted 276
12.2 Testing the center of a single batch 277
12.3 Comparing centers for two separate batches—Big T for two 281
12.4 Comparing centers for two related batches—a tired example 284
12.5 Comparing the centers of k separate batches of ranks—the Singapore story one more time 286
12.6 Comparing centers for k related batches—a reanalysis of the secret mission numbers 288
       Summary 291
       Problems 292
       Addendum: Using computers in statistics 293

CHAPTER THIRTEEN
MEETING FREAKS AT HIGH NOON 296

13.1 Bring on the freaks—in a single batch 298
13.2 More freaks—in two separate batches 300
13.3 More freaks—in k separate batches 303
13.4 More freaks—in two related batches 306

13.5 An ending—freaks in k related batches   308
Summary   311
Problems   312
Addendum: Using computers in statistics   313

CHAPTER FOURTEEN
NONTRIVIAL ODDS AND ENDS   316

14.1 The return of the outliers   318
14.2 Spreading it around some more   321
14.3 Re-expression sometimes helps   325
14.4 Power—a critical but overlooked problem   327
14.5 Doing data analysis on a computer   331
14.6 The data-analytic orientation revisited   332
14.7 Something blue   333
Summary   333
Problems   335
Addendum: Using computers in statistics   36

TABLES   340

ANSWERS TO SELECTED PROBLEMS   363

REFERENCES   371

GLOSSARY OF KEY TERMS   377

INDEX   391

PREFACE

A friend visited a large book warehouse in a midwestern city. "It was weird," he related. "There were thousands of books and most of them were stat books. It was kind of a graveyard for stat books." Considering the body count in the past and the number of elementary statistics texts currently available, why should anyone write another book? The answer is simple. Aspiring authors of statistics books are deluded. They share a common delusion that *they* can write a book that is different and better. I am in this deluded group.

What is wrong with elementary statistics books? First, they aren't written, they are cloned. As a result, they are remarkably similar to one another. I have tried to break the genetic template by including different topics and some changes in emphasis. Second, these books are dull reading because they are frequently written in a dead language—academese. Again, I have attempted to "lighten up" in places and depart from the dead-language model. Third, these textbooks are often unsuited for the mixed bag of students enrolled in statistics. What do students in nursing, physical therapy, personnel administration, education, sociology, psychology, and who-knows-what have in common? What are we trying to teach them in the beginning statistics course? How do we meet the divergent needs of the draftees ("I'm only here because this course is required") and the volunteers ("I'm here because I'm going to graduate school")?

My thesis is that we live in a world of numbers, and that these numbers have a profound impact on our lives. Research of all kinds generates numbers. The *Numbers in the News* section beginning on page 151 bears evidence of that. The aim of applied statistics is to make sense of numbers. This is a how-to-do-it book on applied statistics. I hope, however, that an intuitive presentation will promote some understanding of statistical thinking. (A fuller understanding can only be gained, I believe, by completing several statistics courses and by encounters with real data.) But what about the students who will go beyond a beginning course? To challenge them and prepare them for later courses, the book contains some topics that are not usually presented in beginning statistics books.

The general outline of the book is as follows. Chapters 2–4 are devoted to exploratory data analysis (descriptive statistics). We don't look at our numbers enough! Even simple procedures like sorting numbers from the smallest to the largest can reveal valuable information. Here the idea that spread or variability is *the* fundamental concept in statistics is first presented.

Chapters 5–13 are concerned with confirmatory data analysis (statistical inference). Chapter 5 introduces inference in the context of simple binomial problems. The treatment of probability is cursory. Considerable data analysis can be done by relating test statistics to theoretical distributions without a deep understanding of probability. Correlation and regression are the subjects of Chapters 6 and 7. While these are important topics in their own right, they are introduced early in the book to lay the foundation for an accounted-for variance interpretation of statistics.

Chapters 8 and 10 are devoted to tests of means. In these chapters and throughout the book, statistical tests are organized into (a) tests for a single batch, (b) tests for two separate and related batches, and (c) tests for k separate and related batches. Chapter 9, which discusses comparing variances, is novel for an elementary statistics textbook. It argues that comparing variances is just as legitimate and important as comparing means.

Analysis of variance (ANOVA) for two-factor designs is described in Chapter 11. The meaning and analysis of two-factor interactions receives more attention than is typical. In addition, the chapter features a section on ANOVA with disproportionality.

Chapters 12 and 13 consider nonparametric tests. The first chapter presents tests for ranks and signs; the second focuses upon frequency analysis. Both chapters stress analytical procedures to follow after a significant overall test.

The final chapter is devoted to a series of diverse topics, including outlier analysis, the importance of spreads and a form-robust test for spreads, the role of re-expression in data analysis, an introduction to power analysis, comments on computer analysis of data, and a re-emphasis on the data analytic orientation.

Some features of the book are
- chapter outlines
- detailed chapter summaries
- helpful suggestions for statistics students
- instructions for calculator analyses of data
- displays of computer analysis of data
- chapter problems
- glossary of terms

Topics receiving special emphasis include
- exploratory data analysis
- the data analytic orientation
- power and Type 2 errors
- tests of variances
- the assessment of normality
- the accounted-for variance interpretation of statistics
- protected tests for multiple comparisons
- the analysis of interactions in ANOVA
- outlier analysis
- re-expression in data analysis

As always, the instructor must decide which topics to omit or add depending upon his or her goals and interests, the length of the course, and the mix of the students. For a single quarter or a semester course, some of the following topics might be excluded: Section 3.7—new types of centers; Section 4.8—summarizing batches of numbers; Chapter 9—testing variances; Section 10.7—the assumptions for a repeated-measures ANOVA; Section 11.3—the analysis of interactions; Section 11.4—ANOVA with disproportionality; Chapter 14—odds and ends. A two-quarter or two-semester course could supplement the book with lectures on (a) multiple correlation and regression, (b) ANOVA with more than two factors and factorial designs involving within factors, and (c) a fuller treatment of power analysis.

Every book is a social product. I am indebted to Joan M. Bussell and Deb Swihart for their patience and skill in turning my crude handwriting into a beautifully typed manuscript. At Charles E. Merrill Publishing I wish to thank Marilyn R. Freedman for her thoughtful editorial guidance and Megan Rowe for her careful editing. The manuscript benefitted greatly from the suggestions provided by Lyle Bourne, University of Colorado; Leslie J. Caplan, Ball State University; Anders Ericsson, University of Colorado; Philip S. Gallo Jr., San Diego State University; Michael Lupfer, Memphis State University; Marion MacDonald, University of Massachusetts; David A. Nordlie, Bemidji State University; Joe Rodgers, University of Oklahoma; Ed Shirkey, University of Central Florida; Benjamin Wallace, Cleveland State University; and Howard M. Weiss, Purdue University. I also wish to thank David Vaughn of McDonnell-Douglas for checking the manuscript for accuracy; and Leslie E. Fisher, Cleveland State University, who with Benjamin Wallace wrote the student manual. Two colleagues, Edwin Martin and David S. Holmes, supplied useful examples and suggestions. My research collaborator, James D. Church of the Department of Mathematics, was of great assistance in clarifying certain issues and commenting upon the glossary. My greatest debt is to Joan M. Bussell for her constant emotional support.

I am grateful to the Literary Executor of the late Sir Ronald A. Fisher, F.R.S., to Dr. Frank Yates, F.R.S., and to Longman Group Ltd, London for permission to reprint Tables IIi and VII from their book *Statistical Tables for Biological, Agricultural and Medical Research* (6th Edition, 1974). Finally, I wish to thank the following for permission to reprint various materials: American Statistical Association (Tables I and K); Biometrika Trustees (Tables A, E, F, and J); *Journal of Mathematical Statistics* (Table C); American Cyanamid Company (Tables G and H); Academic Press and Dr. Joan Welkowitz (Tables L and M); Associated Press Newsfeatures (newspaper articles); *The Kansas City Star* (a newspaper article); and MINITAB Project and Dr. Barbara F. Ryan (examples of MINITAB output).

NUMBERS
A PRIMER OF
DATA ANALYSIS

CHAPTER OUTLINE

1.1 THE WORLD OF NUMBERS

1.2 KINDS OF NUMBERS

1.3 THE TASKS OF STATISTICS

1.4 THE DATA ANALYTIC ORIENTATION

1.5 MENTAL HEALTH AND STATISTICS

1.6 OUTLINE OF THE BOOK

CHAPTER ONE
NUMBERS
NUMBERS
NUMBERS

THE WORLD OF NUMBERS

1.1

We live in a world of numbers. If you are not convinced of that, pick up any newspaper or magazine. A host of terms like social indicators, indexing, consumer price index, unemployment rates, economic indicators, prime rate, MPG, the Dow-Jones average, and so on confronts us daily. Furthermore, these terms reduce to numbers, which have a profound impact on our lives.

Most research generates numbers. We hang numbers on things. By numbering we gain knowledge about what is out there and, we hope, in research we learn what is related to what. In the simplest form of research, an investigator asks questions. The investigator then collects data relevant to the questions in the appropriate setting. "Collecting data" means getting numbers. Then, analyzing the numbers provides answers to the questions. Reread the last four simple sentences. Do not let their simplicity minimize the power inherent in research.

KINDS OF NUMBERS

1.2

Consider the following statements:

1. Yesterday, the bloodied, tattered jersey of old Number 13, legendary Detroit Destroyers quarterback Bobby Joe Jack, was placed in the Hall of Fame.
2. Veronica Diedre Veronica of Dead Rock, Virginia, last night became Number 1: Miss America. Veronica, who entertained the audience with her tap dancing, also swept the field in the swimsuit competition. Number 2 was Avis Tulong, Miss Louisiana, who won the talent competition with her exotic bird calls.
3. Tomorrow's expected high in Phoenix will be 91°.
4. The overall length of the 1983 Cadillac DeVille is 221 inches.

A physicist, N.R. Campbell (1938), argued that these numbers—13, 1, 91, and 221—exemplify four levels of measurement. The 13 denotes a class—quarterbacks—so it is only meaningful to talk about how many quarterbacks—the frequency—fall in that class. Campbell termed this level of measurement *nominal*. In the Miss America example we are dealing with a ranking. Veronica was ranked 1, Tulong was 2, and so on. This level of measurement is labeled *ordinal*. The ranks only indicate that Veronica is more beautiful or talented or whatever than Tulong, not *how much*. *Interval* level measurements, such as Fahrenheit temperature, further refine measurement. Interval measurement has a scale with equal intervals and an arbitrary zero point. *Length* is an example of a *ratio* scale with a "true" zero point and equality of ratios. Equality of ratios means that 12 inches is two times 6 inches (a 2 to 1 ratio), 24 inches is twice 12 inches, and so on. In short, Campbell proposed a hierarchy of measurement ranging from the simplest level—nominal—to the most complex—ratio.

Why consider levels of measurement? A psychophysicist, the late S.S. Stevens (1951) proposed that: (a) the level of measurement defines the types of arithmetic operations (e.g., multiplication) that can be performed on numbers and (b) the level of measurement dictates the kinds of statistics that can be applied to numbers. Stevens contended that to perform certain widely used statistical procedures, such as analysis of variance, t tests, and the Pearson product-moment r (which will be described in later chapters), the numbers must be measured on at least an interval scale. Some authors of statistics books quickly accepted Stevens' proscriptions (e.g., Siegel, 1956) and others (e.g., Blalock, 1979) still do despite repeated criticisms of Stevens' position (see Gaito, 1980, for a review.).

All statistical procedures depend upon assumptions, but these assumptions *do not* include assumptions about the level of measurement. Campbell's four levels of measurement do not constitute an exhaustive classification. For example, Cohen and Cohen (1975) contend that rating scales that have a zero point but intervals of questionable equality fail to fit into Campbell's scheme. Furthermore, Kirk asserts

> Unfortunately, ratio measurement and even interval measurement are rare in the behavioral sciences and education. At best our measurement falls somewhere between the ordinal and interval levels. (1978, p. 16)

If Kirk's assessment is correct and Stevens' statistical proscriptions are accepted, then much of the statistical analysis being done in the behavioral sciences would be invalid. I do not believe that the analyses are invalid, simply because measurement level is not among the assumptions for the statistical techniques.

Most statistical analysis is performed today using computers. How do statistical packages for computers deal with different kinds of numbers and levels of measurement? First, information is divided into three classes: F, I, and A. F stands for *fixed* or *real* numbers like 3.734. I stands for *integers* like 1, 2, 3, . . . , N. A stands for *alphanumeric* information like married or single. F numbers are punched into computer cards or typed into terminals. I numbers are used to produce repetitive operations in programming, but in data analysis they are simply converted to F numbers. Alphanumeric data can be nicely double converted. We can code married persons as 2s and singles as 1s or vice versa (it makes no difference as long as the coding is consistent). Then, the integers are converted to real numbers with 1 becoming 1.000 . . . , 2 becoming 2.000 . . . , and so on. In short, computer number crunching converts the numbers into real numbers. Second, although computer packages consider assumptions for various statistical techniques and include ways to evaluate them, the assumptions do not dictate that numbers be measured on a certain scale. Third, the computational procedures applied in statistical packages do not conform to Stevens' proscriptions. For example, analysis of variance may be done by methods (regression) in which some variables are *qualitative* (e.g., group codes) rather than being measured on an interval scale. In summary, computer packages for data analysis have not accepted Stevens' proscriptions for levels of measurement and permissible statistics.

This book will largely reject Stevens' position and follow the advice of a wise and inventive statistician, John W. Tukey: "Campbellian measurement has scared us far too long" (1969, p. 87). So let's stop being scared. In deference to Campbell and Stevens, we will recognize that there are different kinds of numbers—*frequencies, ranks,* and *real numbers*—but we will not be concerned with interval versus ratio measurement. And going with the computer flow, we will enter these different kinds of numbers into computers and calculators, let these wonderful gadgets perform whatever operations are programmed, and hope that the output is simple and beautiful.

This section on kinds of numbers ends with three examples of frequencies, ranks, and real numbers.

1. A panel of 100 respondents is asked: "In the coming election do you intend to vote for candidate A, for candidate B, or are you undecided?" Of the respondents, 52 say they intend to vote for A, 41 for B, and 7 are undecided. The numbers 52, 41, and 7 are *frequencies.*

2. The aggressiveness of 11 preschool children during a free-play period is assessed. Aggressiveness is defined for an observer, who then assigns 1 to the most aggressive child, 2 to the next most aggressive child, and so on. The numbers 1, 2, ..., 11 are *ranks*.
3. A weight-loss program requires participants to report every Monday morning for a "weigh-in" to monitor their weight changes. The resulting weights (197 lb, 264 lb, etc.) are *real numbers*.

THE TASKS OF STATISTICS
1.3

Statistics is a part of mathematics with two branches. One branch, *theoretical statistics,* is concerned with solving abstract problems. These statisticians are interested in distributions, derivations, and proofs. The other branch, *applied statistics,* is devoted to solving practical problems. These statisticians are intrigued with numbers. There isn't a brick wall between theoretical and applied statistics and, as in walks of life, some people do not know where they are. We, however, do know where we are. We are in the applied branch. We won't derive or prove anything. We will borrow techniques from theoretical statistics, but we won't try to play theoretical statisticians' games. Does this mean that this book is a cookbook? Definitely! Many authors in the prefaces of elementary statistics books proudly say two things: (a) students need not know anything to use this book and (b) this book is not a cookbook. With this book, if you don't know a little arithmetic and algebra, you will encounter grave difficulties. Further, if you find cookbooks offensive, then you should enroll in a statistics course that focuses on theoretical statistics.

Applied statistics is divided into *descriptive* and *inferential* statistics. The descriptive approach aims to take a batch of numbers and describe it succinctly and meaningfully. For example, what is the center of the batch? How do the numbers spread out? If we turned the numbers into a picture, what would it look like? In inferential statistics, the task is to go beyond a batch of numbers—a *sample*—and make statements about the *population* from which the numbers came. We have *some* cases (a sample) and we want to talk about *all* cases (a population). A population is a complete aggregation of cases, "things," or numbers. A sample is a subset of a population.

Tukey calls the descriptive/inferential dichotomy "exploratory vs. confirmatory" statistics. The word *exploratory* is well-chosen. It implies that we should feel free to explore numbers, to try to look at them in novel ways, to explore them by picturing them in different ways, and so on. He believes that we don't explore numbers enough and do not do it creatively enough before we move into the *confirmatory* mode. Even with planned experiments, Tukey asserts: ". . . restricting one's self to the planned analysis—failing to accompany it with exploration—loses sight of the most interesting results too frquently to be comfortable" (1977, p. 8). Tukey (1969) once gave a speech entitled "Analyzing data: Sanctification or detective work." Tukey pointed out that exploratory data analysis means doing detective work on numbers and pictures of numbers. It is a process of looking for *clues* in numbers and pictures of what is below the surface. He regards confirmatory analysis, on the other hand, as "judicial or quasi-judicial" in nature. Tukey's emphasis on exploratory analysis is wise advice for helping to achieve the goal of applied statistics: *making sense of numbers*.

THE DATA ANALYTIC ORIENTATION

1.4

Earlier, a simple portrait of research showed an investigator asking questions, collecting numbers, analyzing them and thereby answering the questions. Often researchers do not merely ask questions. Instead they test *hypotheses,* the implications derived from theories. The goal, of course, is to arrive at decisions regarding these hypotheses and to verify or falsify the portions of the theory that generated the hypotheses. But whether answering questions or testing hypotheses, we are still confronted by numbers.

Previously, we said that we would borrow statistical techniques from the theoretical statisticians to help cope with numbers. Now we come to a problem. The techniques borrowed from the theoretical statisticians are very rigorous *when* the underlying assumptions regarding the numbers are met. Often the numbers obtained in research, however, do not fulfill the assumptions of the statistical techniques. How do we solve this problem?

Tukey has come to our rescue. In an important paper, he observed

> The physical sciences are used to "praying over" their data, examining the same data from a variety of points of view. This process

has been very rewarding, and has led to many extremely valuable insights. Without this sort of flexibility progress in physical sciences would have been much slower. (1962, p. 46)

Again, Tukey suggests exploring numbers as a detective would. And besides using statistics as a tool, you should apply your knowledge and the experience of your discipline to the numbers. A rat runs down a runway in X, a fraction of a minute. When you convert the number to speed of running, you are astounded to learn that the rat ran 973 miles per hour. Don't believe it! Your X is a bad number. Keep your hand firmly on the tiller of reality.

Secondly, Tukey suggests viewing the outcomes of data analyses with caution:

Flexibility in analysis is often to be had honestly at a price of a willingness not to demand that what has already been observed shall establish or prove, what analysis *suggests*. In the physical sciences generally the results of praying over the data are thought of as something to be put to further tests in another experiment as indications rather than conclusions. (1962, p. 46)

The key words here are *prove, indications,* and *conclusions.* Given that numbers in behaviorial sciences often violate the assumptions required by statistical methods and given the exploratory attack waged upon such numbers, we do not perform a single experiment and then prove things or draw definitive conclusions. Instead, the products of our analyses are *indications.* An indication is properly *more provisional* than a proof or a conclusion. An indication is a tentative pointing-at-something that has emerged from our analytical endeavors. If a single experiment or a study only permits indications, then how are conclusions ever achieved? Indications become conclusions when they turn up time and time again in experiments. In other words, conclusions are the product of successful experimental replications. You can feel happy over the outcome of a single experiment, but don't bet your polyester leisure suit on it.

Another approach to the problem of numbers resulting from research not always fitting the assumptions for statistical techniques is to study the effects of violations of assumptions upon the performance of statistical tests. These so-called *Monte Carlo studies,* involving computer simulations of violations of assumptions, offer evidence about the consequences of departures from assumptions and help to decide what statistical technique might be optimal under particular conditions. We will draw upon some of these Monte Carlo investigations later in the book.

MENTAL HEALTH AND STATISTICS*

1.5   Students cope with a lot of complex material. For example, often they work well with enormously complicated theories. But some students become lost when confronted by numbers. Then add a few symbols and formulas and the fear becomes panic. Are their reactions a result of math phobia? Are the students afraid that they will make an error? Numbers are unavoidable—they're like taxes, doing laundry, and buying groceries. You cannot escape numbers, so you must learn to live with them. Symbols are not enemies, they are friends. Symbols will help you deal successfully with numbers. Combine some symbols and you have another helper—a formula. Formulas are the recipes in our cookbook. Some students never write down formulas and then can't believe it when their answers are wrong.

Consider two approaches to errors:

1. I am the greatest and I could never make an error.
2. I might make an error and that would be the end of the world as we know it.

Both of these viewpoints are unrealistic. If you swim, you are going to get wet. If you work with numbers, you are going to make errors. Accept that as a fact. It follows that you should be paranoid about the accuracy of your work. One task is to do a problem. Another task is to detect any errors that you may have made and to eradicate those errors. Sometimes in the course of calculating there are indications that something is amiss (examples of this will appear elsewhere in the book). When this happens, backtrack and search for errors. Always ask yourself: *Does my answer make any sense?* In pursuing this question, don't hesitate to study your batch of numbers.

The *arithmetic mean,* for example, is defined as the sum of a batch of numbers divided by the count or the number of numbers in the batch. Consider these numbers: 5, 6, 8, 8, 9, 10. The arithmetic mean is reported to be 11.7. That is an impossible answer. How could the arithmetic mean be greater than the largest number, 10? It ought to be a number somewhere in the center of the batch. Here, the error is obvious. But with a large batch of unordered numbers, an error in the arithmetic mean is less detectable. In summary, errors are inevitable. Therefore, evaluating answers,

(*or, Why students crack up when facing numbers and symbols)

looking for errors, and checking calculations are absolutely essential aspects of data analysis.

We have been talking about the stuff—fears of numbers, symbols, and formulas and making errors—of which bad dreams are made. Is there any good news? Yes, there is. You are extremely fortunate to be taking statistics at this time. Why? Because for about $20 you can buy a calculator that will become invaluable to you. Don't be dazzled by the beauty of the thing. It is what it can do that counts. Therefore, go directly to the calculator's manual. Minimally you want a calculator that will tell you (a) what the sum of a batch of number is, (b) what the sum of a batch of squared numbers is, and (c) how many numbers are in the batch. You can do many of the calculations that are necessary in statistics with these quantities. If your calculator has statistical functions like calculation of the *arithmetic mean* ($\overline{X}$) and the *standard deviation* (s), that is even better. Often calculators called "scientific calculators" will be suitable for statistical analysis. Choose your calculator carefully, study the manual, and master it. After a while you will discover that the calculator is a treasured tool.

Finally, it will probably surprise you to learn that a statistics book is not like a paper back novel. You must fight statistics books. Study the examples with care. Be picky about symbols and formulas. Tackle the problems. Look for errors. Do your answers make sense? Check and recheck you work. If your instructor's lectures baffle you, don't be afraid to ask questions.

In brief, to succeed in your statistics course, you must take an *active approach*. You should be warned, however, that statistics is very simple in the beginning and then rapidly becomes complex. Do not stop working after your initial success. On the other hand, if you encounter difficulty in the beginning, you should seriously think about acquiring more background in mathematics before taking a statistics course.

Finally, some of you may have access to a computer. Today, elegant and powerful software packages are available for doing statistics. If you do not encounter these packages in your beginning course, you will in more advanced courses. Samples of the output from one computer package will appear throughout the book.

A word of warning is in order regarding calculators and computers. While they are powerful helpmates, they are also idiots. They will do exactly what you tell them to do. So it is essential to issue the correct commands. If you don't enter the right numbers and do the right things to those numbers, your answers will be wrong. Again, asking if your answer makes sense and checking and rechecking are still absolutely necessary. "It doesn't help to run the data from a bad experiment through a computer—it's not a washing machine" (Wike, 1971, p. 6). A computer will not wash wrong numbers, either.

1.6 OUTLINE OF THE BOOK

Chapters 2–4 will be devoted to descriptive statistics. These chapters will discuss (a) organizing numbers into classes, (b) converting numbers into pictures, (c) finding single numbers that represent the center of a batch of numbers, and (d) obtaining single numbers that summarize the spread for a batch of numbers.

Inferential statistics begins in Chapter 5 with the analysis of frequencies from experiments in which one of two outcomes is possible for each trial. In Chapters 6 and 7 the relationship between two variables and predicting from one variable for the other will be examined as well as the relationship between two variables when the numbers are ranks.

Chapters 8, 10, and 11 will focus upon comparing centers ranging from a single mean to the means from a factorial experiment. Chapter 9 will consider comparisons of two or more spreads. This subject receives more attention than is customary in elementary statistics texts.

Chapters 12 and 13 will discuss nonparametric statistics. These statistics, whose assumptions are less restrictive, permit the analysis of numbers in the form of ranks, signs, and frequencies. The final chapter will include brief treatments of some diverse topics: outlier analysis, spreads, re-expression of numbers, power analysis, data analysis on a computer, and data analytic orientation.

This is not a textbook on theoretical statistics. Rather, it is a how-to-do-it book. The "it" is applied statistics, the goal of which is to make sense of numbers.

SUMMARY

We live in a world of numbers, numbers that have a profound impact upon our lives. Understanding these numbers helps us learn about the world. In research, for example, analysis of the numbers collected produces answers. Numbers are measured at four

☐ SUMMARY

levels: nominal, ordinal, interval, and ratio. Assumptions for behavioral science statistics do not specify level of measurement, however. Further, computer packages solve measurement level problems by reducing integer and alphanumeric data to real numbers. Instead of the traditional measurement hierarchy, this book will consider the analysis of three kinds of numbers: frequencies, ranks, and real numbers.

Of the two branches of statistics, theoretical and applied, this book deals with the latter branch, split into descriptive and inferential (exploratory and confirmatory) statistics. The orientation advocated is data analytic, emphasizing exploring numbers, being cautious about the outcomes of analysis (indications, not proofs or conclusions), and seeking firm conclusions through replication of experiments. The basic task of applied statistics is *to make sense of numbers*.

Finally, since errors in analysis are inescapable, questioning answers, seeking errors, checking, and rechecking are advisable. An active approach to statistics will promote success. Calculators and computers are powerful tools in data analysis.

CHAPTER OUTLINE

2.1 NUMBERS FROM RESEARCH ARE DISORGANIZED

2.2 IT HELPS TO PUT NUMBERS INTO SEVERAL PILES

2.3 BREAKING UP PILES INTO SMALLER PILES

2.4 FROM PILES OF NUMBERS TO PICTURES OF NUMBERS

2.5 WHAT DO PICTURES OF NUMBERS TELL US?

2.6 THE COMPARED-TO-WHAT PROBLEM—PERCENTILES

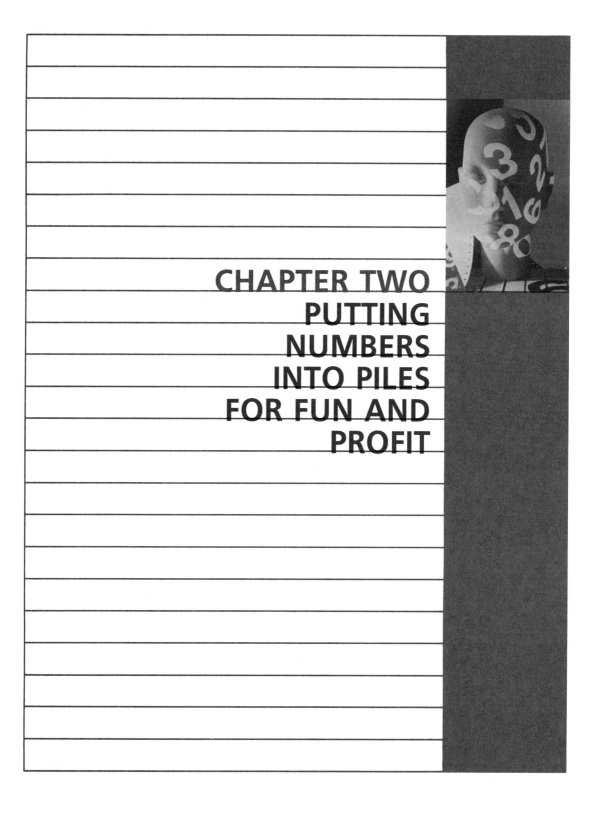

CHAPTER TWO
PUTTING NUMBERS INTO PILES FOR FUN AND PROFIT

NUMBERS FROM RESEARCH ARE DISORGANIZED

2.1 One warm day last fall the following scenario was unfolding. The students were in class, but as evidenced by their glazed-over eyes and nodding heads, they weren't really there. While I was there, yakking away and making chalk dust, I was elsewhere too. I was thinking about things other than statistics (e.g., "Am I too old to become a licensed plumber?" "Those new little Japanese diesel pickup trucks really look great. If I bought one, would that bring eternal happiness?").

To break out of this scenario, wonderful as it was, I suggested to my sleepy group that we see how good they were at estimating an interval of time. Granted that was not a very exciting thing for them to do, but it gave me a chance to play with the buttons on my digital watch, which is exciting for me. So they were instructed to guess the time interval in seconds that elapsed between my saying "start" and "stop," to write the estimate on a piece of paper, and also to indicate their sex. After a period devoted to remastery of the buttons on my watch, we did it. And what did we get? A large, disorganized pile of scraps of paper. The point is, when you do research, you obtain numbers that are a mess. Here are the actual numbers:[1] 70, 105, *85,* 40, 90, 120, *75, 60,* 140, *120,* 180, 65, 80, 50, 57, *70,* 90, 40, *90,* 140, 82, 90, 87, *85, 94, 100,* 107, *95,* 105, 68, 80, 94, *65,* 70, 92, *90,* 75, 100, 90, *85,* 150, 60, *127, 75,* 85, *80,* 120, 120, *105,* 120, *105,* 50, 75, 90, *100,* 82, 83, *45,* 90, *110,* 68, 82, 100, *94,* 90, 92, 82, 105, *80, 130,* 80, 75, *105,* 135, 180, *93,* 95, *93, 93,* 132, 65, 180, 120, *105,* 150, *93,* 150, *137,* 109, 90, 140, 70, *40,* 100, 80, *80, 74, 180,* 120, 170, *77, 50,* 90, *90,* 70, 90, 120, 65, *60, 140,* 180. A careful study of these numbers leads to two observations: (a) we have a fairly large batch of mixed-up numbers, and (b) the students seem to make more even-numbered estimates than odd-numbered ones.

IT HELPS TO PUT NUMBERS INTO SEVERAL PILES

2.2 We are engaged in data analysis. To analyze means to separate something into its parts. Therefore, we will begin to explore the batch of numbers by separating them into two piles. When we sort the pieces of paper into two piles according to the sex of the

---

1. Numbers in regular type were reported by females; those in italics, males.

estimator, we obtain the following numbers for the female students: 70, 105, 40, 90, 120, 140, 180, 65, 80, 50, 57, 90, 40, 140, 82, 90, 87, 107, 105, 68, 80, 94, 70, 92, 75, 100, 90, 150, 60, 85, 120, 120, 120, 50, 75, 90, 82, 83, 90, 68, 82, 100, 90, 92, 82, 105, 80, 75, 135, 180, 95, 132, 65, 180, 120, 150, 150, 109, 90, 140, 70, 100, 80, 120, 170, 90, 70, 90, 120, 65, 180.

Some ways of doing things are better than others. Instead of listing the numbers in a *string*, why not put them in a *matrix?* Doing this for the male students produces

```
 85  90  65  80  94  93  (40) 50
 75  85  90 105  80  93   80  90
 60  94  85 100 130 105   74  60
120 100 127  45 105  93 (180) 140
 70  95  75 110  93 137   77  105
```

Why is the matrix better than the string? It is better because we can count the numbers more quickly (there are 7 columns of 5 plus 4, or 39 males). Chapter 1 stressed the importance of looking for errors, checking, and rechecking. A recheck of all numbers for accuracy results here in a count of 111. A recheck of the women's numbers produces a count of 71. We have lost a male subject in the shuffle (111 − 71 = 40, not 39)—can you find the missing number?

Why is dividing the batch of numbers into fewer piles helpful? First, we might want to compare the women's time estimates with the men's. Second, we have fewer numbers to consider—40 or 71 versus 111.

BREAKING UP PILES INTO SMALLER PILES

2.3

Breaking our large batch of numbers into a couple of smaller piles is a limited (but helpful) procedure. Why not do some more of it? Let's take the smaller pile, the men's estimates (including the lost case), and break the numbers into more piles by *sorting* the numbers from the smallest number to the largest. This procedure yields

```
40  65  77  85  93   94  (105) 127
45  70  80  85  93   95   105  130
50  74  80  90  93  100   105  137
60  75  80  90  93  100   110  140
60  75  85  90  94  105   120  180
```

What does this arrangement tell us? Quite a bit, including (a) the smallest number is 40, (b) the largest number is 180, (c) the *range* of the numbers (the largest number to the smallest) is 140, (d) the *center* of the numbers is around 90, and (e) one number, 180, deviates quite a bit from its neighbor, 140. Sorting is useful for examining a batch of numbers.

Another way of sorting is to put a batch of numbers into a limited number of *ordered piles* or *classes*. If too few piles are used, then information is lost. If too many are used, the work becomes cumbersome. In consulting books for an answer to the question of how many piles, you will find different answers like 6–20, 8–15, 10–20, and so on. We will adopt *8–15* piles or classes as our convention, but in the exploratory spirit we won't be rigid. If a little fewer or more classes are convenient, we won't hesitate to depart from the 8–15 rule. Again, let us examine the men's time estimates. Suppose we want about eight piles. What size of class would yield about eight classes? We can get an approximate idea of the *size of the class—i*—by dividing the range of numbers by the number of classes desired: $i \cong$ range/number of classes. Recalling from the sort above that the range equals the largest number minus the smallest, we have $i \cong (180 - 40)/8 \cong 17.5$. Since it is conventional to employ i-values like 1, 2, 3, 5, or some multiple of 5, let us try an $i = 20$. It is also conventional to start each class with a multiple of the class size. We could start with 0–19. That would be a waste of time because there are no numbers of that size. Accordingly, for the lowest class we will use 40–59. The next class will be 60–79, and so on. This yields the set of classes in column one of Table 2.1.

The next step is tallying—sorting the numbers into the classes. The number 85 goes in the 80–99 class. Accordingly, we make a

TABLE 2.1
Frequency and cumulative frequency distributions of the men's time estimates

Class	Tally	f	cf
180–199	I	1	40
160–179		0	39
140–159	I	1	39
120–139	IIII	4	38
100–119	IIII II	7	34
80–99	IIII IIII IIII I	16	27
60–79	IIII III	8	11
40–59	III	3	3

tally next to that class. We tally 75 in the 60–79 class, and so on. In column three, labeled f, the tallies have been converted to *frequencies* for each class. Finally, it is beneficial to add another column, *cumulative frequencies (cf)*. The cumulative frequency for any class is the frequency for the given class plus the sum of the frequencies for all classes below it. Thus, for the 60–79 class f is 8 and for the 40–59 class f is 3, so the cumulative frequency for the 60–79 class is 8 + 3 = 11. Notice that for the highest class, 180–199, cf = 40, or the total number of males. It is always wise to perform this check.

The classes and their frequencies form a distribution—specifically, a *frequency distribution*. At a glance it shows how many time estimates fall within each class. The classes and their cumulative frequencies constitute another distribution termed a *cumulative frequency distribution*. It reveals how many time estimates fall in a class *and* in all classes below it.

Before leaving frequency distributions we need to consider a few important details. In the example our class interval size, i, was 20. The stated limits of the lowest class were 40–59. Since 59 − 40 = 19, and 19 ≠ 20, how do we resolve this problem? The answer is that the *real limits* of a class extend from one-half unit below to one-half unit above the *stated limits*. The reason for the real limits is that a number like 40 includes a range of values from 39.5 to 40.5, 59 includes 58.5–59.5, and so on. In other words, the real limits of the 40–59 class are 39.5–59.5. Note that 59.5 − 39.5 = 20 = i. A time estimate of 39.54 sec would fall in the 40–59 class; a time estimate of 39.43 would not fall in the 40–59 class. The *lower real limit* of a class is labeled *LRL*, and the *upper real limit* is labeled *URL*.

Where is the midpoint or center of a class? It is the lower real limit plus half of the class interval. For the 40–59 class it is 39.5 + (20/2) = 49.5. Is that a reasonable answer? Well, the *LRL* = 39.5, the midpoint = 49.5, and the *URL* = 59.5 (i.e., 49.5 + 10). So 49.5 is halfway between 39.5 and 59.5. To repeat, the formula for the midpoint of a class is midpoint = *LRL* + (i/2). Another way to find the midpoint is to calculate the number halfway between the *stated* class limits of 40–59. In this case, 59 − 40 = 19, 19/2 = 9.5, and 40 + 9.5 = 49.5. Observe that we obtain the same answer, 49.5, as before and that 40 + 9.5 = 49.5 and 49.5 + 9.5 = 59. You may wonder, "Will this be on the test?" Everything will be on the test. Later in the book the *LRL, URL,* and midpoint will be needed to make graphs and compute certain statistics.

The following is a summary of the rules for constructing a frequency distribution and a cumulative frequency distribution:

> 1. Select a class interval size, i, that will produce about 8–15 classes. The i-value can be estimated by dividing the range of scores by the number of desired classes. It is conventional to employ i-values like 1, 2, 3, 5, or a multiple of 5.
> 2. Start each class with a multiple of i and the lowest class at the bottom.
> 3. After tallying the numbers into the classes, convert the class tallies into frequencies (f) for each class.
> 4. Successively accumulate the frequencies in the classes beginning with the lowest class to obtain a cumulative frequency (cf) for each class. The cumulative frequency is the frequency for a class plus all frequencies in the classes below.
> 5. Make sure that the cumulative frequency for the highest class is equal to N, the batch count.

Tukey (1977) proposed a new type of frequency distribution that he called a *stem-and-leaf*. When we made a frequency distribution, we set up the classes and then by tallying found the *frequency* of numbers falling within each class. In the stem-and-leaf distribution the classes (stems) are set up and then the *numbers* (leaves) within each class are listed.[2] This is shown for the time estimates of the male students in Table 2.2. To the basic stem-and-leaf distribution we have added frequency and cumulative frequency columns to ensure that all 40 numbers were included. Why compile a stem-and-leaf distribution? The major advantage of the stem-and-leaf over frequency distributions is that the former still lists the numbers. In a frequency distribution, unless you have the original data, all that you know is *how many* numbers fell where. In a stem-and-leaf you know *which* numbers fell where. Knowing exactly what the numbers are is valuable because they can be subjected to further analysis. For example, doing a sort of the numbers and calculating certain statistics is easy using a stem-and-leaf distribution.

2. Tukey's treatment of stems and leaves is shown in the Addendum to this chapter. For comparison with a typical frequency distribution, the classes have been used as stems and the actual estimates as leaves.

TABLE 2.2
A stem-and-leaf distribution of the men's time estimates

Stem	Leaf	f	cf
180–199	180	1	40
160–179		0	39
140–159	140	1	39
120–139	120, 127, 130, 137	4	38
100–119	100, 100, 105, 105, 105, 105, 110	7	34
80–99	80, 80, 80, 85, 85, 85, 90, 90, 90, 93, 93, 93, 93, 94, 94, 95	16	27
60–79	60, 60, 65, 70, 74, 75, 75, 77	8	11
40–59	40, 45, 50	3	3

FROM PILES OF NUMBERS TO PICTURES OF NUMBERS

If a batch of numbers has been sorted and put into a frequency or a stem-and-leaf distribution, it is convenient to examine; easier still is a picture of numbers. We can convert a frequency distribution (or a stem-and-leaf distribution, which is readily transformed into a frequency distribution) into two kinds of pictures: a *histogram* (a type of bar graph) and *frequency polygon*. A histogram displays the classes in equal intervals on the horizontal, or X axis. The vertical, or Y axis shows frequency. The vertical axis is generally about two-thirds the height of the horizontal axis. A bar corresponds to the frequency for each class. Figure 2.1 is a histogram showing the men's time estimates.

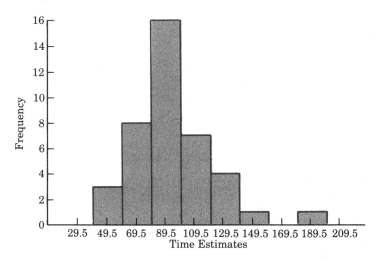

FIGURE 2.1
A histogram of the men's time estimates

FIGURE 2.2
A frequency polygon of the men's time estimates

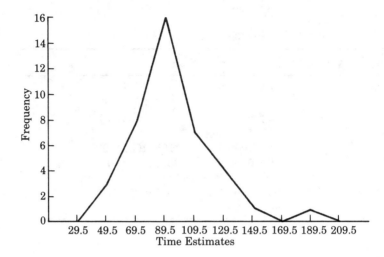

The other common way to illustrate distributions is a frequency polygon. Again, the classes are laid out on the horizontal axis and the frequency scale on the vertical axis. Instead of erecting bars corresponding to the frequencies of the classes, a dot representing the frequency for a class is plotted at the midpoint of each class. Then the dots are connected by a straight line. Figure 2.2 shows a frequency polygon of the men's time estimates. Notice that an extra class has been added at each end of the distribution and the straight line has been extended to the baseline. Which picture is better? Most of the time it probably does not matter. Guilford (1956), however, argued that frequency polygons reveal the contour of the data better and are especially handy when two distributions are plotted on the same baseline.

WHAT DO PICTURES OF NUMBERS TELL US?

2.5

A histogram or a frequency polygon provides visual information about (a) shape, (b) center, (c) spread, and (d) outliers. "Shape" is not readily definable. At the very least, we are concerned with the form of the histogram or frequency polygon. A further complication is that when we look at the form of a distribution we are implicitly comparing it with some standard form, such as a *normal distribution*. A normal distribution is a symmetrical, bell-shaped curve.

While we can not define shape precisely, it is possible by looking at a histogram to get hints about two dimensions of shape—*kurtosis* and *skewness*. Kurtosis refers to the degree of peakedness of the distribution relative to the normal curve. If a distribution is too pointed in comparison to a normal distribution, it is termed *leptokurtic*. If it is too flat, it is termed *platykurtic*. If it has the proper degree of peakedness, it is referred to as *mesokurtic*. Kurtosis is not easy to assess visually except when departures from normal are gross. Figure 2.3 depicts different degrees of kurtosis.

Skewness refers to the degree of symmetry of a distribution. A simple definition of symmetry is as follows: if a distribution is folded in the middle and nothing hangs out, it is symmetrical. Figures 2.4, 2.5, and 2.6 are all symmetrical. Figures 2.7 and 2.8 fail the fold test. When a distribution is folded and something hangs out, generally that something is the *tail*. Figures 2.7 and 2.8 are called *asymmetrical* or *skewed* distributions. In Figure 2.7 the tail is on the left—toward smaller numbers—and it is *negatively skewed*. In Figure 2.8 the tail is on the right—toward larger numbers—and it is *positively skewed*. How about the men's time estimates in Figures 2.1 and 2.2? Are they symmetrical or asymmetrical? If the latter, are they skewed positively or negatively?

Why do we care about shape? First, presenting information about skewness or kurtosis is more descriptive of a batch of numbers than simply saying "there they are." Second, in performing more complicated statistical analyses, we will often assume that our numbers conform approximately to a certain shape. This matter will be considered in greater detail later.

One of the goals of descriptive statistics or exploratory data analysis is to simplify a batch of numbers. If we can reduce the

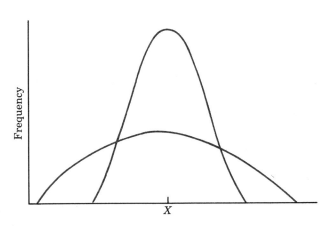

FIGURE 2.3
Distributions with the same mean but differing in kurtosis

FIGURE 2.4
A symmetrical distribution

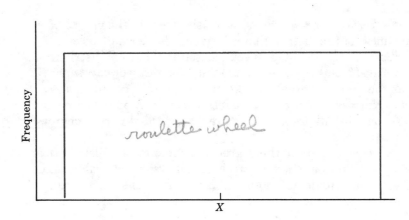

FIGURE 2.5
A symmetrical distribution

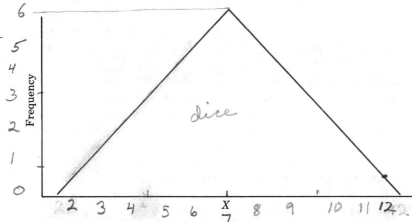

FIGURE 2.6
A symmetrical distribution

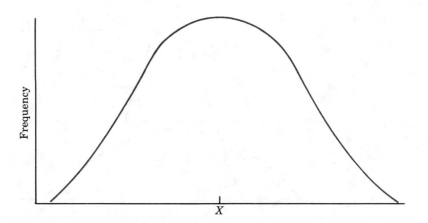

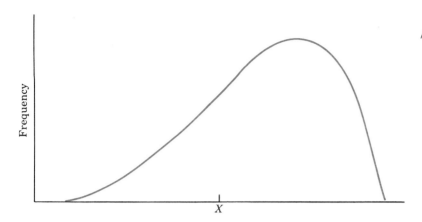

FIGURE 2.7
A negatively-skewed distribution

batch to a few numbers that summarize the whole batch, that goal will be realized. What kind of summarizing numbers do we want? We are generally concerned with the *center* of the batch of numbers. In everyday life we refer to this notion as the "average." We can eyeball a histogram or frequency polygon to see where the numbers "peak." By comparing this peak to its class interval or the midpoint of that class interval, we have a rough estimate for the center. For example, the men's time estimates (Figure 2.1) peak on the 80–99 class, whose midpoint is 89.5.

Another summarizing number that is less intuitive than the center of a set of numbers is the *spread* of the numbers. How widely scattered are they? Are the numbers tightly packed around the center (Figure 2.9) or do they spread out from the cen-

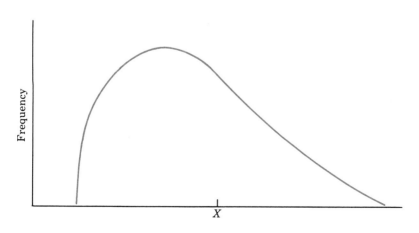

FIGURE 2.8
A positively skewed distribution

FIGURE 2.9
A distribution with a small spread

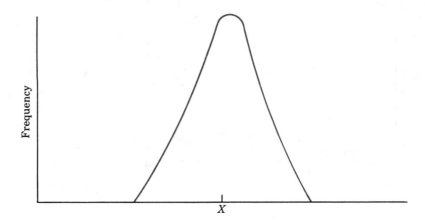

ter more (Figure 2.10)? Notice that the centers in Figures 2.9 and 2.10 are the same, but the distributions differ in spread. Even at the descriptive level the notion of spread is essential to summarize batches of numbers. There is also a link between spread and symmetry. In a symmetrical distribution the numbers spread equally from the center; in a skewed distribution the numbers do not spread equally from the center. Figure 2.1 is an example of unequal spreads. In the next two chapters we will depart from the eyeball approach and consider quantitative indices of centers and spreads.

Finally, looking at a histogram or frequency polygon may disclose a potential outlier or outliers. An *outlier* is a number either too large or too small in comparison with the batch of numbers. How large is "too large" or how small is "too small"? What does an outlier mean? Is it an error in observation? Is it the re-

FIGURE 2.10
A distribution with the same mean as in Figure 2.9 but a wider spread

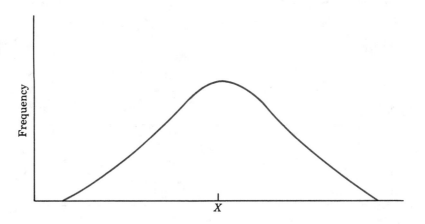

sult of the experimental treatment? What does one do with such a discrepant number? One possibility is to delete it (trim). Another is to include it in the batch (retain). Or we might try to modify it by transforming our numbers to a different scale (re-expression) or by replacing it by the nearest nonoutlying number (Windsorizing). We do not want to trim a "real" number, because that will distort our numbers. On the other hand, we do not want to retain an erroneous number, because certain quantitative indices of centers and spreads can be distorted greatly by extreme numbers. Outliers are often revealed by "holes" in the tail of a frequency distribution. A hole is a class in which $f = 0$ (see Figure 2.1 for a distribution with a hole in the right tail). Outliers can be masked in frequency distributions when too few classes are imposed upon a batch of numbers.

THE COMPARED-TO-WHAT PROBLEM—PERCENTILES

2.6

Besides enabling us to go from numbers to pictures, frequency distributions are valuable for solving what is called here the *compared-to-what problem*. Suppose a student scored 17 on a statistics test. Or suppose a student received 27 on a math quiz and 43 on an English theme. What do these various numbers mean? Is 17 a "good" score? Did the person do better in math or English? By seeing where these scores fell in the statistics, math, and English frequency distributions we could crudely answer these questions. However, we can achieve more precise answers by transforming the numbers to *percentiles*. By converting 17, 27, and 43 to percentiles with a potential range of 0–100 we are "measuring" the numbers with a common scale. A *percentile* is a point in a distribution of numbers at or below which a specified percentage of the numbers fall. If 17 on the statistics test corresponds to the 60th percentile, it would mean that 60% of the students in the statistics class scored 17 or lower. Converting a number to a percentile is one problem. The converse problem is converting a percentile to a number. The number corresponding to the 60th percentile is 17.

Calling upon the men's time estimates once more, we will first do a number-to-a-percentile problem. Consider this question: What percentage of the men estimated the time interval at 80 seconds or fewer? The formula for solving this problem is

$$\text{percentile} = B + \left(\frac{X - LRL}{i}\right)K$$

where B is the percent of numbers falling below the critical class, X is the number, LRL is the lower real limit of the critical class, i is the class size, and K is the percent of numbers falling *in* the critical class.

The only mysterious phrase is the critical class—it is the class in which the *number*, $X = 80$, falls. Table 2.3 is the frequency distribution with a cumulative frequency column included and the critical class boxed.

An orderly way to attack this problem is to determine the values for the symbols in the formula, put the values into the formula, and then do the arithmetic. Be an obsessive-compulsive—do it that way! If B is the percent of numbers falling below the critical class of 80–99, and the cumulative frequency for the 60–79 class is 11, then the *proportion* of cases below the critical class is *this cf/N* or $11/40 = 0.275$. Since we want the *percentage* of cases below the critical class, however,

$$B = \frac{cf}{N}(100) = \frac{11}{40}(100) = 27.5$$

X is our number (80); LRL is the lower real limit of the critical class 80–99 (79.5); i is the class size (20, i.e., $i = 99.5 - 79.5 = 20$); the frequency in the critical class is 16. The proportion of cases in the critical class is *this f/N* $= 16/40 = 0.4$. Again, to produce the percentage of cases in the critical class:

$$K = \frac{f}{N}(100)$$

$$= \frac{16}{40}(100) = 40$$

TABLE 2.3
Distribution of the men's time estimates (critical class in box)

Class	f	cf
180–199	1	40
160–179	0	39
140–159	1	39
120–139	4	38
100–119	7	34
80–99	16	27
60–79	8	11
40–59	3	3

SEC. 2.6 □ THE COMPARED-TO-WHAT PROBLEM—PERCENTILES

We have the pieces, now we will put them together:

$$\text{percentile} = B + \left(\frac{X - LRL}{i}\right)K$$

$$= 27.5 + \left(\frac{80 - 79.5}{20}\right)40$$

$$= 27.5 + 1 = 28.5$$

We can solve the equation using a calculator. First, we will compute the terms beyond the + sign (remembering to deal with the inner parentheses before the outer ones): 80 ⊟ 79.5 ⊟ 0.5 ⊡ 20 ⊟ 0.025 ⊠ 40 ⊟ 1.0 Next, B is added to this result: ⊞ 27.5 ⊟ 28.5.

Is 28.5 a sensible answer? Since 27.5% of the numbers fell below the critical class and our X is at the bottom of the critical class, the answer appears to be reasonable. Again, it means that 28.5% of the numbers fell at or below a time estimate of 80 seconds. Or, stated another way, an estimate of 80 seconds corresponds to the 28.5th percentile.

Back to the compared-to-what problem: if 17 on the statistics test corresponded to the 60th percentile, then the student's score has been specified more precisely. We know that 60% of the students scored 17 or lower. If the scores for math and English were transformed into percentiles, then we can compare the two percentiles to see how the student performed in the two classes as well as the student's standing with respect to other students in each class.

The converse problem—converting a percentile to a number—is our next concern. A formula for this procedure may be obtained by solving the percentile formula above for X. The resulting formula is

$$X = LRL + \frac{i(\text{percentile} - B)}{K}$$

where all terms have the same definitions as before except for the critical class. The critical class is the class in which the *percentile* falls.

To check the previous answer, let us find the X corresponding to the 28.5th percentile. Since 27.5% of the cases fell at or below the 60–79 class, the critical class is, as before, the next class, 80–99. Solving for X, we have

$$X = LRL + \frac{i(\text{percentile} - B)}{K}$$

$$= 79.5 + \frac{20(28.5 - 27.5)}{40}$$

$$= 79.5 + 0.5 = 80$$

A time estimate of 80 seconds corresponds to the 28.5th percentile, and the 28.5th percentile is 80 seconds.

How about a new question? What number, X, corresponds to the 50th percentile? Half or 50% of the cases ($N/2$) is 20. The critical class is the one in which the 20th case falls. Looking at the cumulative frequency column, the critical class is still the 80–99 class. Because all other values are the same, except for the percentile value, the solution is

$$X = LRL + \frac{i(\text{percentile} - B)}{K}$$

$$= 79.5 + \frac{20(50 - 27.5)}{40}$$

$$= 79.5 + 11.25 = 90.75$$

How about the 75th percentile? The ($3N/4$) or 30th case falls in the 100–119 class. Accordingly, we have

$$X = LRL + \frac{i(\text{percentile} - B)}{K}$$

$$= 99.5 + \frac{20(75 - 67.5)}{17.5}$$

$$= 99.5 + 8.57 = 108.07$$

Where did the numbers 67.5 and 17.5 come from? Recall again that B is the percentage of cases below the critical class. Therefore,

$$B = \frac{cf}{N}(100)$$

$$= \frac{27}{40}(100) = 67.5$$

And K is the percent of cases in the critical class, or

$$K = \frac{f}{N}(100)$$

$$= \frac{7}{40}(100) = 17.5$$

Do the answers for the last two examples make any sense? Why?

In summary, one solution to the compared-to-what problem is converting numbers to percentiles. This procedure, relating a number to a distribution of numbers, provides more precise information regarding the position of the number in the batch. Conversely, if a percentile value is known, it is possible to determine the number corresponding to the percentile. In addition, converting numbers from different distributions to percentiles enables us to compare the numbers.

SUMMARY

Research results in a lot of disorganized numbers. Rearranging the numbers into fewer piles can help one organize and comprehend a batch of numbers. Sorting numbers from the smallest to the largest provides information about high and low numbers, the range of the numbers (largest to smallest), the center of the numbers, and possible outliers (extreme numbers). Putting a batch of numbers into a limited number (8–15) of ordered classes results in a frequency distribution. Dividing the range of numbers by the desired number of classes yields an estimate of the class size, i. The real limits of a class extend from one-half unit below to one-half unit above the stated class limits. The midpoint of a class is found by adding half of the class interval to the lower real limit (LRL) of the class. If the frequencies in the classes are accumulated from the lowest to the highest class, a cumulative frequency distribution results. Another type of distribution—a stem-and-leaf—displays the actual numbers in each class instead of the frequencies.

Frequency distributions can be converted to pictures—histograms (a type of bar graph) and frequency polygons. These pictures provide visual information regarding the shape, center, and spread of a distribution as well as indications of possible outliers. Assessing shape implies a comparison with a standard shape, often a normal distribution. A normal distribution is a symmetrical, bell-shaped curve. Two dimensions of shape are kurtosis (degree of peakedness) and skewness (degree of symmetry). In a positively skewed distribution the tail is on the right; in a negatively skewed distribution the tail is on the left.

Centers and spreads are numbers that reduce and summarize a batch of numbers. Center refers to where the numbers peak. Spread refers to how the numbers scatter about the center.

Outliers are troublesome. They pose problems of detection and interpretation. These extreme numbers, when retained, can profoundly affect quantitative indices of centers and spreads.

In compared-to-what problems a number from a test or class is described more precisely by transforming it into a percentile. Numbers from different tests or classes can be compared by converting them to a common percentile scale. A percentile is a number at or below which a specified percentage of the numbers fall. The converse problem is converting a percentile to a number.

PROBLEMS

1. Here are 40 scores from a first examination in statistics. Sort the scores from smallest to largest.

17	85	33	86	23	93	35	57
2	39	1	53	40	18	91	23
64	47	10	37	81	92	24	6
97	9	93	90	39	59	92	33
77	44	68	22	82	63	47	56

2. Using the sorted scores in Problem 1, find: (a) the lowest score, (b) the highest score, (c) the range of scores, and (d) the approximate center of the scores. Do any scores seem much larger or much smaller than the other scores in the batch?

3. Sort the 71 time estimates for the female students (see Section 2.2). What is the range of their estimates?

4. Using an $i = 15$, construct a frequency distribution of the women's time estimates. Include an additional column for cumulative frequency values. Start your distribution with the 30–44 class at the bottom.

5. What is the lower real limit (LRL) of the 30–44 class? The upper real limit (URL)? The midpoint?

6. Using the same set of classes construct a stem-and-leaf distribution of the women's time estimates. Add columns for f and cf. Why is a stem-and-leaf distribution more informative than a frequency distribution?

7. From the frequency distribution in Problem 4, make (a) a histogram and (b) a frequency polygon. Graph paper is helpful in constructing these figures.

8. Examine the histogram or frequency polygon from Problem 7. Does the distribution appear to be symmetrical? If not, how is it skewed? Are there holes in the distribution (i.e., classes in which $f = 0$)? Do some scores depart greatly from the rest of the batch? What is the approximate center of the distribution?

9. One woman estimated the time interval to be 83 seconds. What percentile does this estimate correspond to?
10. What estimate corresponds to the 50th percentile? What does the 50th percentile mean?
11. Check your answer in Problem 10 by finding the percentile corresponding to the estimate.

ADDENDUM: USING COMPUTERS IN STATISTICS

Statistics can be calculated rapidly with a computer. In addition, statistical packages have potential for a variety of complex analyses. We will demonstrate these capabilities using some sample output from MINITAB, an interactive statistical package designed by Ryan, Joiner, and Ryan (1981). First, the time estimates for the men are put into memory by a "SET" command (1). Then a histogram (2) and stem-and-leaf display (3) are output upon command. Note that the MINITAB program for the histogram produces eight classes and an $i = 20$ but different midpoints from those we used. Column 1 of the stem-and-leaf consists of the

```
--SET IN C1
--85 75 60 120 70 90 85 94 100 95 65 90 85 127 75 80 105 105
--100 45 110 94 80 130 105 93 93 93 105 93 137 40 80 74 180          (1)
--77 50 90 60 140

--HISTOGRAM C1                          --STEM-AND-LEAF C1

  C1                                    STEAM-AND-LEAF DISPLAY OF C1
                                        LEAF DIGIT UNIT = 1.0000
  MIDDLE OF    NUMBER OF                1 2 REPRESENTS 12.
  INTERVAL     OBSERVATIONS
     40.         2  **                     2    4  05
     60.         4  ****                   3    5  0
     80.        11  ***********            6    6  005
    100.        16  **************** (2)  11    7  04557
    120.         3  ***                   17    8  000555                (3)
    140.         3  ***                  (10)   9  0003333445
    160.         0                        13   10  005555
    180.         1  *                      7   11  0
                                           6   12  07
                                           4   13  07
                                           2   14  0
                                                HI 180
```

cumulative frequencies above and below the median class (10). The stems are in column 2, (for example, 4) and the column 3 values are the leaves. Thus, 05 in conjunction with the stem 4 denotes the numbers 40 and 45.

CHAPTER OUTLINE

3.1 WHY DO WE NEED MORE EXACT INFORMATION ABOUT CENTERS?

3.2 COMMON TYPES OF CENTERS—MEAN, MEDIAN, AND MODE

3.3 CALCULATING THE MEAN

3.4 FINDING THE MEDIAN

3.5 THE MODE—HELLO AND GOODBYE

3.6 THE MEAN VERSUS THE MEDIAN

3.7 NEW TYPES OF CENTERS

CHAPTER THREE
BEING MORE EXACT ABOUT CENTERS

WHY DO WE NEED MORE EXACT INFORMATION ABOUT CENTERS?

3.1

In Chapter 2, we said we want to simplify batches of numbers by reducing them from many numbers to a few that summarize the batches. Measures for the *center* and *spread* of a distribution of numbers were found to be particularly valuable for this task. Rough estimates of the center and spread of distribution can be made by visual inspection. We looked for the middle of the array of the ordered time estimates for the men and at the peak of the frequency distribution for rough estimates of the center.

While you should never forsake your eyeball, you can profitably go beyond it. Centers can be computed more accurately and reliably than with the eyeball method. Why do we want more exact numbers? First, more exact numbers tell us more accurately what is out there. Compare "there are a lot of unemployed people" with "the current unemployment rate is 10.2%." Second, in everyday life and science we compare and relate many things. We can do a superior job of comparing and relating when we have more accurate numbers. It is difficult to compare "there are a lot of unemployed people" with "there are many unemployed people."

COMMON TYPES OF CENTERS—MEAN, MEDIAN, AND MODE

3.2

The three most common types of centers are the *arithmetic mean*, *median*, and *mode*. We will define these terms somewhat intuitively, and then turn to the problem of calculating them. The sample *mean* ($\overline{X}$) represents the balance point of a batch of numbers arranged in a frequency distribution. As in a teeter-totter, the "weights" of the numbers are equal on both sides of the mean or balance point. The *median* (*Mdn*) is also a balance point. But, in the case of the median, the balance point divides a batch of ordered numbers into two equal piles. Let N denote the count or number of numbers in the batch. The median is a number that splits the batch into two piles, each containing $N/2$ numbers. Figures 3.1 and 3.2 depict the mean and median, respectively. In Figure 3.1 the weights in each bar are products of a number's distance from the mean and the frequency. If a distribution is symmetrical, the mean and median will be the same number, but if the distribution is asymmetrical or skewed, they will not. Lastly, we will mention the "poor relation" of the center set—the *mode*.

The mode is the most frequently occurring number in a batch of numbers. If there is one such number, the distribution is *unimodal*. If there are two such numbers that are nonadjacent, the distribution is *bimodal*.

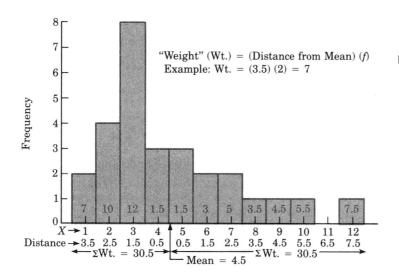

FIGURE 3.1
The mean as a balance point for the "weights" in a distribution

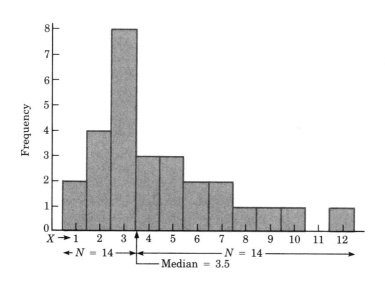

FIGURE 3.2
The median as a balance point for the number of cases in a distribution

CALCULATING THE MEAN, $\overline{X}$

3.3

To obtain $\overline{X}$ we need to introduce two new symbols: Σ and X. Σ (a capital sigma) is an *arithmetic operator*. It directs you to add some numbers. X is a *variable*. What is a variable? A variable is something that varies or takes on different *values*. So far our variable, X, has been an estimate of time. The numbers the students recorded are the values for this variable. Note that X is a concept and, as such, it can denote income, reaction time, degree of introversion, or whatever variable an investigator is studying. Concepts are potent tools because they are capable of encompassing many phenomena.

We need to describe the summation operator, Σ, more fully. How do we know what numbers to add? This can be specified exactly by attaching limits to Σ. For example, $\Sigma_{i=1}^{N}$ means sum from $i = 1$, the first number, through N, the last number. If $N = 6$, $\Sigma_{i=1}^{6}$ means sum from the first number through the sixth number. By adding X, we specify which variable to operate on. So a complete instruction to add all men's time estimates would be: $\Sigma_{i=1}^{N} X_i$. Using the limits and subscripts is tedious so we will usually shorten the instruction to ΣX. But remember that ΣX means $\Sigma_{i=1}^{N} X_i$. Since $\overline{X}$ is the sum of the numbers in a batch divided by the count or number of numbers, N, the formula for the mean is

$$\overline{X} = \frac{\Sigma X}{N}$$

We can calculate the mean of the men's time estimates from the sorted batch in Chapter 2 with the help of our calculator. Before beginning, determine from your calculator's manual whether or not the calculator has to be placed in a statistics mode of operation.

If your calculator has an $\boxed{\overline{X}}$ key, the procedure in step 3 can be shortened to
3. Read $\overline{X}$: $\boxed{\overline{X}}$ 91.875

It is still advisable to include step 4 as a check to ensure that 40 numbers were entered in memory.

Is the answer for the mean reasonable? In Chapter 2 we observed that the center of the men's sorted time estimates was about 90 and that the midpoint of the class with the peak in the frequency distribution was 89.5. Accordingly, our answer of 91.88 appears to be reasonable. How could we check it further? We could calculate it again. The problem with checking by repeating

1. List the formula: $\overline{X} = \dfrac{\Sigma X}{N}$
2. Enter the numbers in
   memory and sum:
   40 [M+]
   45 [M+]
   50 [M+]
   . .
   . .
   . .
   180 [M+]
3. Recall ΣX: [ΣX] 3675
4. Recall N: [N] 40
5. List formula 1: $\overline{X} = \dfrac{\Sigma X}{N}$
6. Substitute and calculate: 3675 [÷] 40 [=] 91.875
7. Round even: 91.88

is that sometimes the same error is repeated. But it is better than not checking. Another way to check the answer is to assess the accuracy of the ΣX by simply adding the numbers: 40 [+] 45 [+] 50 [+] ... 180 [+]. If the answer is 3675, that is reassuring. If the answer is not 3675, then you will have to resolve the discrepancy. In entering a repeated number in memory, calculators differ. In some calculators, the numbers must be repeatedly entered: 80 [M+] 80 [M+] 80 [M+].[1] In other calculators repeating [M+] will work: 80 [M+] [M+] [M+]. Some calculators display the number being stored in memory, while others display the count of the number being entered. Another matter to check is how to clear the memory registers of the calculator before you begin calculating. Again, study your calculator's manual, practice with your calculator, and discover its potential. And do not be frightened by all those functions in your calculator. Only a few functions will require mastery for this course.

In calculating the mean we have used a calculator with an *algebraic* logic. The clue to this logic is the presence of an [=] key. Other companies produce calculators with a *reverse* or *RPN* logic.

1. The presence of an [M+] key is no guarantee that the calculator will output ΣX and ΣX^2, since some inexpensive calculators have these keys. Examine the manual for a calculator carefully before you purchase it.

CH. 3 □ BEING MORE EXACT ABOUT CENTERS

Suppose you want to multiply 10 by 3. In the algebraic system it is 10 ⌧X⌫ 3 ⌧=⌫ 30. In the reverse system it is 10 ⌧ENTER⌫ 3 ⌧X⌫ 30. Now we will recalculate the $\overline{X}$ by a calculator with reverse logic.

1. List the formula: $\overline{X} = \dfrac{\Sigma X}{N}$

2. Enter number in memory and sum:
   40 ⌧Σ+⌫
   45 ⌧Σ+⌫
   50 ⌧Σ+⌫
   . .
   . .
   . .
   180 ⌧Σ+⌫

3. Recall ΣX: ⌧RCL⌫ ⌧3⌫ 3675
4. Recall N: ⌧RCL⌫ ⌧2⌫ 40

5. List formula 1: $\overline{X} = \dfrac{\Sigma X}{N}$

6. Substitute and calculate: 3675 ⌧ENTER⌫ 40 ⌧÷⌫ 91.875
7. Round even: 91.88

Again, if your calculator has a ⌧X̄⌫ key, the procedure in step 3 can be shortened to
3. Read $\overline{X}$:   ⌧X̄⌫ 91.875

And again, it is beneficial to count the number of values that the mean is based upon:
4. Obtain N: ⌧RCL⌫ ⌧2⌫ 40

As you may have guessed, the ⌧RCL⌫ key stands for recall and ⌧2⌫ and ⌧3⌫ are memory registers. In other calculators different registers may store N and ΣX; for example, ⌧7⌫ and ⌧8⌫, and so on. Consult your calculator manual to determine which quantities are stored in the various memory registers.

We have calculated the mean from the raw scores for the time estimates. It is also possible to calculate a mean from a frequency distribution. Since this latter, antiquated method should be applied when only a frequency distribution is available, it is presented in the appendix to Chapter 3.

Imagine the following situation. Suppose that on the same day that I asked my class to estimate a time interval, introductory statistics teachers all over the country did the same exercise. Imagine further that all the resulting numbers were summed and then divided by the total N. What would we have? We would have a *parameter*, a *mean* of the time estimates for the total *population* of statistics students. If two things have the same name, much confusion may result. Thus, the mean of the population is termed a parameter, while the *mean* of our *sample* is termed a *statistic*. Once more to avoid confusion different symbols are employed. The population mean is labeled μ (for the Greek letter mu), and the sample (our class) mean is, as you now know, labeled $\overline{X}$. It is customary to use Greek letters for parameters and Roman letters for statistics. There are also different formulas for μ and $\overline{X}$:

$$\mu = \frac{\sum_{i=1}^{N_p} X_i}{N_p}$$

where N_P is the count for whole population, and

$$\overline{X} = \frac{\sum_{i=1}^{N} X_i}{N}$$

where N is the sample count. Knowing μ is the goal, but we usually only know $\overline{X}$. The $\overline{X}$ is called upon to provide an *estimate* of μ. How well does $\overline{X}$ do? That depends in part upon the size of N and the extent to which the sample is *representative* of the population.

FINDING THE MEDIAN

3.4

Now let us turn our attention to the second center measure, the median. Remember that a *median* is a number that splits an *ordered* batch of numbers into two equal-sized piles. In other words, the median is a number that divides the batch of N numbers into two piles of N/2 numbers. The numbers must be ordered before locating the number that divides the batch.

There are three ways to obtain a median: (a) from an ordered batch, (b) from a frequency distribution, and (c) from a graph. To find the median of a batch of numbers we must first sort the numbers from small to large. If N is *odd*, the median is the middle score. Here are 11 sorted numbers: 2, 3, 8, 9, 11, 12, 13, 15, 19, 22, 27. The middle number is the sixth. Why? Because it splits the numbers into two piles of five each. The sixth number is 12, which is the median of the batch. When N is *even*, the median is halfway between the two middle numbers. Adding another number to the batch yields an even N: 2, 3, 8, 9, 11, 12, 13, 15, 19, 22, 27, 31. The median is now halfway between the sixth and seventh numbers. The median, therefore, is: $(12 + 13)/2 = 12.5$. Note that the median partitions the batch into two piles with six numbers each. More generally, the median is the $(N + 1)/2$th number. If $N = 12$, the $(N + 1)/2$th number is the 6.5th number, which means halfway between the sixth and seventh numbers.

Returning to the men's time estimates in which $N = 40$, the $(N + 1)/2$th number is 20.5. So the median is halfway between the 20th and 21st number. The 20th number is 90, and the 21st number is 93; therefore, the $Mdn = (90 + 93)/2 = 91.5$. Finding the median this way from a small ordered batch is quick. With a large batch of numbers, sorting becomes unwieldly unless it is done by a computer. When hand sorting, only half of the numbers plus one beyond need to be ordered when the only desired statistic is the median. Again, it should be pointed out that even with a large batch of numbers a median can be obtained easily from a stem-and-leaf distribution. By sorting the leaves within the stems and counting, the $(N + 1)/2$th number is readily discernible. In Chapter 2 sorting was shown to be useful for revealing some items of eyeball information. Sorting a batch of numbers also quickly leads to the median, a *quantitative* measure of the center of a batch.

The second way to obtain a median is from a frequency distribution. In Section 2.5 we calculated the 50th percentile of men's time estimates and found it to be 90.75. What is the 50th percentile? It is the point in a distribution at or below which 50% of the numbers fall. If 50% of the numbers are at or below the point, then 50% are above it. Since 50% is equal to $N/2$ cases (and recalling the definition of the median given earlier), *the median is the 50th percentile*. Accordingly, we found the median from a frequency distribution by following the *percentile-to-a-number* method as described in Chapter 2. Since the median is the number corresponding to the 50th percentile, we calculated X. Therefore, the appropriate formula for the median is

$$X = LRL + \frac{i(\text{percentile} - B)}{K}$$

With the sorting-and-counting method the $Mdn = 91.5$, but with the percentile procedure the $Mdn = 90.75$. Why do these two medians differ? Because different batches of numbers are being analyzed. In the sorting-and-counting method the batch consists of the actual numbers. In the percentile procedure the frequencies in the classes serve as the data base. In the percentile procedure it is assumed that the numbers are evenly distributed within each class. What method should be used when? With only a frequency distribution, you must employ the percentile method. If you know the numbers in the batch, then you should employ the sorting-and-counting method. This contradiction between a $Mdn = 91.5$ versus $Mdn = 90.75$ does not exhaust the puzzles regarding the median. See Stavig (1978) for a discussion of these matters.

A third method of obtaining a median is graphic and employs percentiles. In this method we need to construct a new type of graph—a *cumulative percentage polygon,* or *ogive.* In a frequency polygon we plotted the class frequencies at the midpoints of the classes and connected the dots with straight lines. In a cumulative percentage polygon we plot the cumulative percentages for the classes (y-axis) at the upper real limits (URLs) of the classes (x-axis). Table 3.1 is the cumulative frequency distribution for the men's time estimates. To this distribution columns have been added for the URLs and cumulative percentages.

Class	cf	URL	cf%
180–199	40	199.5	100.0
160–179	39	179.5	97.5
140–159	39	159.5	97.5
120–139	38	139.5	95.0
100–119	34	119.5	85.0
80–99	27	99.5	67.5
60–79	11	79.5	27.5
40–59	3	59.5	7.5

TABLE 3.1
Cumulative percentage distribution of the men's time estimates

Recall again that the *URL* for a class is the stated upper limit, *UL,* plus one-half. Thus, for the lowest class, the $URL = 59 + 0.5 = 59.5$, and so on. The cumulative percentage, *cf%,* is

$$cf\% = \frac{cf}{N}(100)$$

For the lowest class

$$cf\% = \frac{3}{40}(100) = 7.5$$

FIGURE 3.3
A cumulative percentage polygon for the men's time estimates

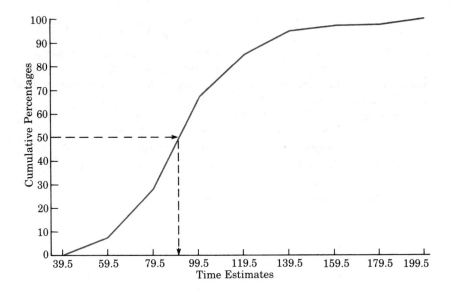

Repeating this process produces the fourth column of data. Note that the cumulative frequency percentage for the highest class is 100.0—that is as it should be. Next, we plot the cumulative frequency percentages against the *URLs* and connect the dots with straight lines (if our curve were too "bumpy," we might smooth it with a French curve). Figure 3.3, then, is the completed cumulative percentage polygon (brought down to 0% by the addition of a nonexistent class at the lower end). It is possible to extract the number corresponding to any desired percentile from the ogive. To obtain the median, first draw a horizontal line from the 50% point on the cumulative frequency percentages vertical axis to the intersection with the polygon. Second, drop a vertical line from the intersection to the baseline scale. Third, read the number, the median, on the baseline scale. The median, found in this manner, is approximately 91. From a carefully drawn and smoothed ogive it is possible to find the median as well as other percentiles. To anticipate a future topic, think about this question: How would you find the 25th and 75th percentiles from Figure 3.3?

THE MODE—HELLO AND GOODBYE

3.5

We won't spend much time on the third center measure—the *mode*. Why? Because it has only limited applicability. It does a feeble job in descriptive statistics and does not apply to inferential statistics. Although the definition of the mode is clear—it is the most frequently occurring number in a batch—it is assailed by nagging problems. Suppose that two adjacent numbers like 90 and 91 have the highest frequency in the batch. What is the mode? Are there two modes? Should we say the mode is $(90 + 91)/2 = 90.5$? What if 90 occurs seven times and 91 occurs six times? What is the mode? Suppose we have a *rectangular distribution,* termed a *uniform distribution* by statisticians, in which all numbers occur equally often. What is the mode? To summarize: The mode is not entirely satisfactory as a descriptive statistic and plays no role in inferential statistics.

THE MEAN VERSUS THE MEDIAN

3.6

We have just questioned the utility of the mode. What are the strengths and weaknesses of the mean and median? In the case of the mean, on the positive side (a) it is the most widely used measure of the center, (b) it is part of elaborate inferential procedures, (c) it is the most reliable center, (d) the sum of the deviations from the mean—$\Sigma(X - \overline{X})$—is 0, and (e) the sum of the squared deviations from the mean—$\Sigma(X - \overline{X})^2$—is minimal.

On the negative side, outliers greatly distort the mean. Consider the numbers 6, 7, 8, 8, 9, 57. Their mean is 15.83! That is a pretty weird number. The role of the mean in inferential statistics, the sum of the deviations from the mean, and the meaning of the word reliable will be discussed later.

How about the median? The median has two advantages: It is sometimes computable when the mean is not and it is less affected by outliers. The disadvantage of the median is that it does not play a large role in inferential statistics. Rats and college sophomores may misbehave in research. Sometimes rats sit interminably when they are expected to run. Suppose a rat ran on five trials and sat on the sixth trial. What is the rat's mean running time? As a result of the indeterminate number for the sixth trial, the mean cannot be calculated. While we do not have a time for the sixth trial, we do know that it is the largest number. Thus, the median can be found. It is the number halfway between the third and fourth numbers when the six times have been ordered with the "sat" time as the largest number.

In the previous set of numbers showing the effect of an outlier, the inclusion of 57 produced a badly distorted mean. Notice that the median is unaffected by the extreme score. The median for the numbers 6, 7, 8, 8, 9, and 57 is 8. If the extreme number were 570 or 5700, the median would still be 8.

Should you use the median or the mean? Sometimes you can resolve an either/or conflict by a "both" solution. Should we use the mean or median? Why not both? If we compare the mean and median for a batch of numbers, the comparison will be informative about the *shape* of the batch. Earlier we said that the mean and median were the same in a symmetrical distribution and different in a skewed distribution. Therefore, it is advantageous to compare the mean and median. If the two numbers are about equal, then the distribution is nearly symmetrical. If the two numbers differ considerably, then you are probably dealing with a skewed distribution. But we can squeeze even more information from the mean/median comparison. Numbers in the tail of a distribution "pull" the mean toward the tail. Accordingly, if the $\overline{X} > Mdn$, the distribution is *positively* skewed. If the $\overline{X} < Mdn$, the distribution is *negatively* skewed. In summary

1. The reliable mean is the most widely applied measure of the center
2. The median is handy for dealing with skewed distributions
3. Comparing the mean and median can provide information about symmetry.

NEW TYPES OF CENTERS

3.7

Statistics is not static. Statisticians are continuously engaged in research. Today a topic of considerable concern is "robust statistics," or statistics that behave properly with data that are not normal. We have just seen that the mean is badly distorted by outliers. Statisticians have sought better centers and these new centers are being incorporated into computer packages like the BMDP statistics package (Dixon, Brown, Engelman, Frane, Hill, Jennrich, & Toporek, 1983). Consider some of the ideas here. When a mean is calculated, each number in the batch is treated democratically—it is accorded a weight of one. New centers can be devised by tinkering with the weights (Andrews, Bickel, Hampel, Huber, Rogers, & Tukey, 1972), assigning extreme numbers less weight than those in the middle. Centers can be based on other numbers or on parts of the batch. As an example of the former strategy, Tukey (1977) proposed a *trimean*, a center based on three percentiles:

$$\text{trimean} = \frac{\text{25th percentile}}{4} + \frac{Mdn}{2} + \frac{\text{75th percentile}}{4}$$

An example of center based on part of the data is a *trimmed mean*. A trimmed mean is obtained by ordering a batch of numbers, deleting a percentage of the numbers (e.g., 15%) from each end of the distribution, and calculating a mean of the remaining numbers. What will be the outcome of these attempts to find better centers? We will have to see if they are applied by researchers in the next decade and with what results.

Another very active area of statistical research is on the troublesome problem of outliers (see Barnett & Lewis, 1978; Beckman & Cook, 1983). We hope this work will produce some solutions because outliers are a formidable problem. And since outliers influence centers, the current research emphasis on outliers is related to the research upon robust centers.

SUMMARY

A center reduces and summarizes a batch of numbers. Exact measures of centers provide a more accurate portrait of reality and help us to compare and relate things. The three most common centers are the arithmetic mean, the median, and the mode. The mean is the balance point for a frequency distribution. The median is a point that divides N ordered numbers into two piles of $N/2$ numbers. The mode is the most frequently occurring number in a batch. The mean can be calculated from raw scores or from grouped data (see the appendix). The latter procedure assumes that the numbers are evenly distributed within the classes. The two procedures generally produce different means because different batches of numbers are being analyzed. Calculators with an algebraic and a reverse or RPN logic can compute means. The mean of a population, μ, is an example of a parameter. The mean of a sample, $\bar{X}$, is an example of a statistic. The sample mean provides an estimate of the population mean.

There are three ways to determine a median. In the sorting-and-counting method the median is the $(N+1)/2$th number in an ordered batch. A median of a frequency distribution is the number corresponding to the 50th percentile. A median can also be found graphically from a cumulative percentage polygon in which the cumulative percentages for the classes are plotted against the upper real limits of the classes.

The mode is a center with limited utility because it is restricted to descriptive statistics. The mean is the most widely used center, it is the most reliable center, it is an integral part of inferential statistics, and the sum of the deviations from it is zero.

The disadvantage of the mean is that extreme numbers distort it badly. The median can be calculated on occasions (extreme unknown score) when the mean cannot, and it is less affected by extreme numbers. But the median is not widely incorporated into inferential statistics. Comparing the mean and median supplies information about the symmetry of a distribution of numbers. If the mean and median are nearly the same, the distribution is close to symmetrical. If the $\overline{X} >$ Mdn, there is positive skewing. If the $\overline{X} <$ Mdn, there is negative skewing. While the mean is used most often, the median is useful with skewed distributions.

Much research is currently in progress on robust statistics and on outliers. Some new centers weight numbers differentially in contrast to the mean, which gives each number a weight of one. Other new centers are based upon percentiles (trimeans) or a reduced batch of numbers (trimmed means). The present focus on outlier problems is important because outliers are related to centers.

PROBLEMS

1. A consumer-research organization tested 19 inexpensive loudspeakers and measured their "efficiencies" as follows: 73, 77, 79, 80, 82, 82, 82, 83, 85, 86, 87, 87, 88, 89, 90, 90, 91, 91, 92. Using your calculator, compute the mean. Does your answer seem reasonable? Why?

2. (a) Find the mode of the efficiency scores. (b) Using the sorting-and-counting method, determine the median. (c) Compare the median and mean—what does this comparison tell you?

3. In a learning experiment 35 subjects committed the following numbers of errors:

Stem	Leaves	f	cf
50–54	51, 53	2	35
45–49	47	1	33
40–44	42, 43, 40	3	32
35–39	36, 36, 38, 35, 36	5	29
30–34	31, 34, 30, 32	4	24
25–29	25, 27, 29, 25, 26, 29, 26, 26	8	20
20–24	20, 21, 24, 20, 23	5	12
15–19	19, 19, 16, 15	4	7
10–14	11, 14, 10	3	3

With your calculator, find the mean by the raw-score method.

4. By sorting within the critical class, calculate the median by the sorting-and-counting method.

5. Find the 50th percentile of the error distribution in Problem 3.
6. Compare the mean errors with the median errors. Compare the mean with the 50th percentile. What do these two comparisons tell you about the error distribution? Is this answer reasonable from a visual inspection of the error distribution?
7. Construct a cumulative percentage polygon of the error data on a sheet of graph paper. Find the 50th percentile graphically.
8. Show that $\Sigma(X - \overline{X}) \cong 0$ for the speaker efficiencies in Problem 1. Demonstrate that $\Sigma(X - Mdn) \neq \Sigma(X - \overline{X})$.
9. Add X' and fX' columns to the error distribution in Problem 3. Compute the mean using the grouped-data method. When should the grouped-data method be applied?

APPENDIX: FINDING THE MEAN FROM A FREQUENCY DISTRIBUTION

In Section 3.3, the most frequently employed center—the mean—was calculated from "raw scores." The formula, $\overline{X} = \Sigma X/N$, can be applied to a batch of numbers whether they are ordered or are in their common state of disarray. Suppose, on the other hand, that we wish to calculate the mean of a batch of numbers arranged in a frequency distribution. This requires a new formula:

$$\overline{X} = \frac{\Sigma fX'}{N}$$

Here Σ is a summation *over the classes*, f is the *frequency* in each class, and X' is the *midpoint* of each class. Let us calculate the mean from the frequency distribution for the men's time estimates from Chapter 2. Columns for X' and fX' have been added to the previous frequency distribution. Recall from Chapter 2 that a class midpoint is

$$X' = LRL + \frac{i}{2}$$

where LRL is the lower real limit of a class and i is the class size. The LRL for the lowest class is LL (the stated lower limit) $- 0.5$ or $40 - 0.5 = 39.5$. Since $i = 20$, $i/2 = 10$. Thus, for the lowest midpoint:

$$X' = 39.5 + 10 = 49.5.$$

It is not difficult to get the remaining midpoints because the midpoints are simply i units apart. In other words, the second midpoint is the first $X' + i = 49.5 + 20 = 69.5$. The next is $69.5 + 20 = 89.5$, and so on. Multiplying the frequency for each class by the midpoint produces the entries in the fX' column (see Table 3.2).

CH. 3 □ BEING MORE EXACT ABOUT CENTERS

TABLE 3.2
Frequency distribution for computing the mean

Class	f	X'	fX'
180–199	1	189.5	189.5
160–179	0	169.5	0.
140–159	1	149.5	149.5
120–139	4	129.5	518.0
100–119	7	109.5	766.5
80–99	16	89.5	1432.0
60–79	8	69.5	556.0
40–59	3	49.5	148.5

Summing the numbers in column four, $\Sigma fX' = 3760$. Then,

$$\overline{X} = \frac{\Sigma fX'}{N}$$

$$= \frac{3760}{40} = 94$$

Alternatively, the mean can be found by a calculator with algebraic logic without recording the fX' values in the table.

1. List the formula: $\overline{X} = \frac{\Sigma fX'}{N}$

2. Calculate fX', enter in memory, and sum:
    3 [X] 49.5 [=] 148.5 [M+]
    8 [X] 69.5 [=] 556.0 [M+]
    16 [X] 89.5 [=] 1432.0 [M+]
    .
    .
    1 [X] 189.5 [=] 189.5 [M+]

3. Recall $\Sigma fX'$:   [ΣX] 3760

4. List formula 1: $\overline{X} = \frac{\Sigma fX'}{N}$

5. Substitute and calculate: 3760 [÷] 40 [=] 94.0

There is a potential disaster here. Suppose you obtained [N] after step 3 and divided $\Sigma fX'$ by that N. The resulting mean, $\overline{X} = 470$, is truly a large number! Or suppose that in step 3 you got the mean directly by keying [X̄]. Again it is a large number, 470. Why are we finding this bizarre answer? Because we are actually computing the average of the midpoint frequency products when we divide their sum by 8, the number of classes. However, pun-

APPENDIX: FINDING THE MEAN FROM A FREQUENCY DISTRIBUTION

ching the $\boxed{N}$ key can be of value—it tells us how many midpoint frequency products have been stored in memory. Just do not use the number of classes as a divisor to try to get the mean.

Previously, the raw-score method produced an $\overline{X}$ of 91.88. The formula for a frequency distribution produced an $\overline{X}$ of 94.0. Why do these two values for the center differ? They differ because they are the means of *different* batches of numbers. When the grouped-data formula, $\overline{X} = \Sigma fX'/N$, is applied, we assume that the scores are evenly distributed within each class. It follows then that the midpoint, X', of each class can represent every number in the class. In the raw-score batch the numbers being averaged were: 40, 45, 50, 60, 60, . . . , 180. In the frequency distribution the corresponding numbers being averaged were: 49.5, 49.5, 49.5, 69.5, 69.5, . . . , 189.5. The grouped-data method will generate the same mean as the raw-score method only if the numbers are evenly distributed in the classes or are unequally distributed in the classes but balance out across all classes. It should also be obvious that the mean can vary in the grouped-data method as a result of how many classes are used and what their limits are. The grouped-data method is an anachronism. It is a relic from precalculator, precomputer days. The only time that you should apply this relic is when you are confronted by a frequency distribution and do not know the actual numbers in the batch. Note that a stem-and-leaf display permits you to use the raw-score procedure for $\overline{X}$.

CHAPTER OUTLINE

4.1 WHY SPREAD IS THE FUNDAMENTAL CONCEPT IN STATISTICS

4.2 SOME SPREADS—STANDARD DEVIATION, SEMI-INTERQUARTILE RANGE, AND RANGE

4.3 KEY CONCEPTS—SUM OF SQUARES, STANDARD DEVIATION, AND VARIANCE

4.4 GETTING STANDARD DEVIATIONS AND VARIANCES

4.5 GETTING THE SEMI-INTERQUARTILE RANGE

4.6 A QUICK-AND-DIRTY SPREAD—THE RANGE

4.7 WHENS AND WHYS FOR THE THREE SPREADS

4.8 SUMMARIZING BATCHES OF NUMBERS

4.9 THE COMPARED-TO-WHAT PROBLEM AGAIN—THE STANDARD-SCORE APPROACH

4.10 THE NORMAL DISTRIBUTION AND z SCORES

CHAPTER FOUR
NAILING DOWN SPREADS

WHY SPREAD IS THE FUNDAMENTAL CONCEPT IN STATISTICS

4.1

After attempting to "mold young minds" in statistics for many years, it slowly became clear to me that there is a fundamental concept—the concept of *spread*. Spread is a measure of how much numbers scatter or vary. For example, consider again Figure 2.9 and Figure 2.10 in which we have two batches of numbers with the same center. In one batch (Figure 2.9) the numbers are clustered closely about the center, while in the other batch (Figure 2.10) the numbers extend further in both directions from the center. This simple example demonstrates the need for an index of spread to more fully describe a batch of numbers. It is not sufficient to describe a distribution by presenting only its center. And, as will become apparent later, spread is also the fundamental concept in inferential statistics. Everyone understands the notion of a center or an average. Spread, however, is a less intuitive concept than center, and since it is such a critical idea, you should pay close attention to it.

SOME SPREADS—STANDARD DEVIATION, SEMI-INTERQUARTILE RANGE, AND RANGE

4.2

There are three commonly used spreads: the standard deviation, the semi-interquartile range, and the range. The most complicated and most frequently employed spread is the *standard deviation*. The standard deviation is the positive square root of the average of the squared deviations of the numbers in a batch from the batch's mean. How do the squared deviations from the mean tell us about spread? Here are five numbers: 11, 13, 15, 17, 19 with a mean of 15. The respective *deviations from the mean, $(X - \overline{X})$*, are $11 - 15, 13 - 15, 15 - 15, 17 - 15, 19 - 15$ or $-4, -2, 0, 2, 4$. The sum of the deviations is $\Sigma(X - \overline{X}) = (-4) + (-2) + 0 + 2 + 4 = 0$. Remember we said in Chapter 3 that the deviations from the mean sum to 0. Here are five numbers that scatter more: 5, 10, 15, 20, 25. Their mean is also 15 but the deviations are larger: $-10, -5, 0, 5, 10$. Again, the $\Sigma(X - \overline{X}) = 0$. These two sets of numbers show how $\Sigma(X - \overline{X})$ fails to "capture" the difference between the spreads as $\Sigma(X - \overline{X}) = 0$ in both cases. If we square the deviations, however, then the story is different. Now we have $\Sigma(X - \overline{X})^2 = (-4)^2 + (-2)^2 + (0)^2 + (2)^2 + (4)^2 = 40$ for the first batch, and $\Sigma(X - \overline{X})^2 = (-10)^2 + (-5)^2 + (0)^2 + (5)^2 + (10)^2 = 250$ for the second batch. When we square the deviations from the mean, we observe that the answers clearly reflect the difference in the spreads of the numbers in the two batches. For the moment we will ignore the part of the definition of the standard deviation concerning the positive square root of the average of the squared deviations.

The *semi-interquartile range* has a formidable name, but the idea is simple. The interquartile range is the difference between the number corresponding to the 75th percentile and the number corresponding to the 25th percentile. The interquartile range is the amount of space occupied by the middle 50% of the numbers. If a distribution varies widely, the interquartile range will be a large number; if it is a tightly packed distribution, the interquartile range will be a smaller number. The prefix *semi-* means half. Thus, we simply divide the interquartile range by 2 to find the semi-interquartile range. As we shall see later in Chapter 14, some statisticians like Tukey and Mosteller employ the interquartile range as a measure of spread.

Just as the mode was the poor relation of the center set, the *range* is the poor relation of the spread set. It is defined, as stated before, as the difference between the largest and smallest numbers in a batch.

KEY CONCEPTS—SUM OF SQUARES, STANDARD DEVIATION, AND VARIANCE

4.3

In the example of the squared deviations from the mean, $\Sigma(X - \overline{X})^2$, we showed that the distribution with the greater variability had a larger $\Sigma(X - \overline{X})^2$. Actually, the quantity $\Sigma(X - \overline{X})^2$ has a name, *SS*, which stands for the *sum of squares* or the sum of the squared deviations from the mean:

$$SS = \Sigma(X - \overline{X})^2$$

The sum of squares, *SS*, is a measure of spread that will appear frequently in later chapters of the book.

If we "average" the squared deviations from the mean and take the positive square root of the result, we find a spread called the *standard deviation*. Average was put in quotes because in the case of a *sample* the divisor is $N - 1$ and the standard deviation is labeled s. That is

$$s = \sqrt{\frac{\Sigma(X - \overline{X})^2}{N - 1}}$$

When dealing with a population, however, the divisor is N_p and the standard deviation is termed σ (a small sigma):

$$\sigma = \sqrt{\frac{\Sigma(X - \mu)^2}{N_p}}$$

Here N_p is the count for the whole population and μ is the population mean. Why, you might ask, is the divisor $N - 1$ for s rather than N? The $N - 1$ divisor produces an index (s) that is a better estimate of the parameter (σ).

As is frequently the case, things are more complicated than they seem at first. σ is a population *parameter*; s is an *estimate* of the population parameter, σ, based on a sample. We need a statistic reflecting the *actual variability of a sample*. Let us term it sd. It is the actual standard deviation of a sample. The formula for this measure of spread is

$$sd = \sqrt{\frac{\Sigma(X - \overline{X})^2}{N}}$$

where N is the count for the sample. To recapitulate: σ, a parameter, is the standard deviation for a population (the divisor is N_p); s is an estimate of σ (the divisor is $N - 1$); and sd, a statistic, is the standard deviation for a sample (the divisor is N). While sd is appropriate for describing a sample, s is generally reported as if it were a statistic even though it is an estimator. It is also widely used in inferential statistics as will become apparent. Accordingly, we will employ s almost exclusively throughout the book. Parenthetically, it should be noted that some calculators have keys for the two standard deviations. My calculator with an algebraic logic has these keys: $\boxed{\sigma_n}$ $\boxed{\sigma_{n-1}}$. The first key produces sd (or σ if the numbers entered constitute a population). The second key outputs s. The other calculator with reverse logic only has an s key.

Lastly, we can deal with the *variance* quickly. The *variance is the standard deviation squared*. For a population, the variance is

$$\sigma^2 = \frac{\Sigma(X - \mu)^2}{N_p}$$

The estimate of the population variance based upon a sample is

$$s^2 = \frac{\Sigma(X - \overline{X})^2}{N - 1}$$

The actual variance of a sample is

$$(sd)^2 = \frac{\Sigma(X - \overline{X})^2}{N}$$

Once more, because s^2 is widely employed as if it were a statistic rather than an estimator and because of its role in inferential statistics, our focus will be upon s^2. Just as s provides a better estimate of σ than sd, s^2 is a better estimator of σ^2. In fact, s^2 is termed an *unbiased estimator* of σ^2 because it results in estimates of σ^2 that are systematically neither too large nor too small (Ferguson, 1981).

GETTING STANDARD DEVIATIONS AND VARIANCES

4.4

The formulas above for the population standard deviation and variance are not ordinarily applicable in research where we usually work with samples. In addition, the formulas for both populations and samples are *rational formulas*. They should never be applied. We need *computational formulas* for the population estimates, s and s^2. It can be shown that

$$\Sigma(X - \overline{X})^2 = \Sigma X^2 - \frac{(\Sigma X)^2}{N}$$

Note that it is unnecessary in this formula to get the actual deviations from the mean. Substituting in the rational formula produces a useful pair of computational formulas:

$$s = \sqrt{\frac{\Sigma X^2 - \frac{(\Sigma X)^2}{N}}{N - 1}}$$

$$s^2 = \frac{\Sigma X^2 - \frac{(\Sigma X)^2}{N}}{N - 1}$$

Two words of warning here. First, observe carefully that $\Sigma X^2 \neq (\Sigma X)^2$. ΣX^2 is the sum of the squared numbers; $(\Sigma X)^2$ is the sum of the numbers squared. Confuse these two terms and you will get wrong answers. Second, do not apply the rational formulas on the grounds that they look simpler. The longer computational formulas will decrease the chances of making errors. Trust me.

It is time to take out your handy algebraic calculator again and find s and s^2 for the men's time estimates.[1]

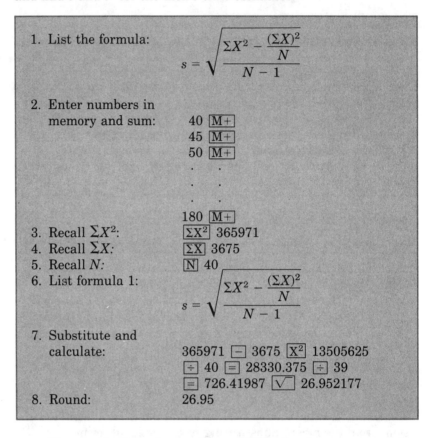

If your calculator has an $\boxed{s}$ or a $\boxed{\sigma_{n-1}}$ key, the answer can be found directly in step 3.
3. Read s: $\boxed{\sigma_{n-1}}$ 26.95
If $\boxed{N}$ is keyed, a count of the numbers in memory is output as a check. Note that if you wished to compute the actual standard deviation of the numbers (sd) the formula would contain a final divisor of N rather than $N - 1$ and in step 3 the $\boxed{\sigma_n}$ key would be pressed.

The other calculator produces s differently.

1. In doing calculations it's a sound tactic to carry numbers out to many decimal places and then round at the end.

SEC. 4.4 □ GETTING STANDARD DEVIATIONS AND VARIANCES

1. List the formula:
$$s = \sqrt{\frac{\Sigma X^2 - \frac{(\Sigma X)^2}{N}}{N-1}}$$

2. Enter numbers in memory and sum:
   40 $\boxed{\Sigma +}$
   45 $\boxed{\Sigma +}$
   50 $\boxed{\Sigma +}$
   . .
   . .
   . .
   180 $\boxed{\Sigma +}$

3. Recall ΣX^2: $\boxed{RCL}$ $\boxed{4}$ 365971
4. Recall ΣX: $\boxed{RCL}$ $\boxed{3}$ 3675
5. Recall N: $\boxed{RCL}$ $\boxed{2}$ 40
6. List formula 1:
$$s = \sqrt{\frac{\Sigma X^2 - \frac{(\Sigma X)^2}{N}}{N-1}}$$

7. Substitute and calculate:
   365971 $\boxed{ENTER}$ 3675 $\boxed{x^2}$ 13505625
   40 $\boxed{\div}$ 337640.625 $\boxed{-}$ 28330.375 39
   $\boxed{\div}$ 726.4199 $\boxed{\sqrt{x}}$ 26.9522

8. Round: 26.95

If the calculator has an $\boxed{s}$ key, step 3 becomes
3. Read s: $\boxed{s}$ 26.95
Checking on the count in memory is again advisable by $\boxed{RCL}$ $\boxed{2}$.

Our job in this section is to calculate s and s^2. En route to s we calculated an s^2 of 726.4199. It was the number we obtained just before we calculated the positive square root. Any time you have s^2 you can find s by taking the positive square root: To get s^2, simply square s. It will also be necessary later to know how to calculate the $SS = \Sigma(X - \overline{X})^2$. It was done above in the calculator runs. Can you find $\Sigma(X - \overline{X})^2$ in step 7? Hint: $\Sigma(X - \overline{X})^2 = \Sigma X^2 - (\Sigma X)^2/N$.

As always, we must ask: Is our standard deviation a reasonable number? By comparing an obtained mean with a sorted batch of numbers, it is possible to decide that the mean is not a strange number. Checking a standard deviation is not as easy.

Here is one approach to the problem. In a large sample that is normally distributed about two-thirds of the numbers should fall within the bounds of $\overline{X} \pm s$. Although our sample ($N = 40$) is not very large and we are uncertain about normality, we can still apply this procedure as a rough test of s. For the actual numbers, $\overline{X} = 91.88$ and $s = 26.95$, therefore

$$\overline{X} - s = 91.88 - 26.95 = 64.93$$
$$\overline{X} + s = 91.88 + 26.95 = 118.83$$

Thus, we would expect that approximately two-thirds of our 40 numbers would fall between the bounds of about 65 and 119. Turning to the sorted time estimates in Chapter 2, we count 29 numbers that fell between 65 and 119. This result, $29/40 = .725$, is close enough to .67 (two-thirds converted to a proportion) to suggest that the obtained s is not an impossible number.

In Chapter 1 we stressed the importance of checking answers and looking for errors. When calculating standard deviations or variances you may receive a lucid message that your ship is sinking. If $\Sigma(X - \overline{X})^2$, $\Sigma X^2 - (\Sigma X)^2/N$, SS, s, or s^2 are *negative*, all is not well. None of these quantities can be negative. Go back immediately and start checking your numbers. Important items to check are the divisors. Don't try to ignore the negative sign. You can't sweep it under the rug—you have to find the error(s).

Finally, as in the case of the mean, it is possible to calculate s and s^2 from a frequency distribution. Since this is an outmoded procedure, we have relegated it to the appendix for this chapter. The procedure is of value when you only have a frequency distribution. But if you know the numbers in a batch, apply the raw-score method described above.

GETTING THE SEMI-INTERQUARTILE RANGE

4.5 Now we come to a spread, the *semi-interquartile range, SIR*, that has been making a comeback recently as a result of Tukey's book, *Exploratory data analysis* (1977). The *SIR* is half of the difference between the number corresponding to the 75th percentile and the number corresponding to the 25th percentile. The 75th percentile is the point at or below which 75% of the numbers fall when numbers have been ordered; the 25th percentile is the point at or below which 25% of the numbers fall. We will denote these points (numbers) as Q_3 and Q_1, respectively. Then the median or the

SEC. 4.5 □ GETTING THE SEMI-INTERQUARTILE RANGE

number corresponding to the 50th percentile will be labeled Q_2. The semi-interquartile range is

$$SIR = \frac{Q_3 - Q_1}{2}$$

When we considered the median, we showed how to find it in three ways: (a) from a sorted batch of numbers, (b) from a frequency distribution, and (c) from a cumulative percentage polygon. These same three techniques enable us to find Q_3 and Q_1. From these Qs we can then calculate the semi-interquartile range.

First, we will apply the sorting-and-counting method. (The sorting-and-counting method should be used when the numbers are in the form of a sorted batch or a stem-and-leaf display.) When we found the median from a sorted batch we located the number corresponding to the $(N + 1)/2$th rank. In the case of Q_1 we want the number corresponding to the $(N + 1)/4$th rank. For men's time estimates $N = 40$ and $(40 + 1)/4 = 10.25$. In the sorted distribution the 10th number is 75. The 11th number is 77. Q_1 is 75 plus one-fourth of the difference between 77 and 75. What we are doing is a linear interpolation to find Q_1. $Q_1 = 75 + (77 - 75)/4 = 75.5$. For Q_3 we need the number corresponding to the $3(N + 1)/4$th rank or the $3(40 + 1)/4 = 30.75$th number. The 30th number is 105 and the 31st number is 105. Therefore, no interpolation is necessary, $Q_3 = 105$, and SIR is

$$SIR = \frac{Q_3 - Q_1}{2}$$
$$= \frac{105 - 75.5}{2}$$
$$= 14.75$$

What does the SIR mean? It is a measure of the spread of the batch of numbers in question. If we have more than one batch, we can compare their SIRs to evaluate their spreads. For example, we could calculate the SIR for the women's time estimates and see how that number compares with the men's SIR of 14.75.

Second, we can find Q_1 and Q_3 from a frequency distribution with the percentile-to-score method described in Chapter 2. There we discovered that the number (X) corresponding to the 75th percentile (Q_3) was 108.07. As a review, we will compute Q_1, the X

TABLE 4.1
Distribution of the men's time estimates (critical class boxed)

Class	f	cf
180–199	1	40
160–179	0	39
140–159	1	39
120–139	4	38
100–119	7	34
80–99	16	27
60–79	8	11
40–59	3	3

corresponding to the 25th percentile. The men's time estimates are displayed in Table 4.1. The critical class is 60–79. Why? The formula is

$$X = Q_1 = LRL + \frac{i(\text{percentile} - B)}{K}$$

$$Q_1 = 59.5 + \frac{20(25 - 7.5)}{20}$$

$$= 59.5 + 17.5 = 77$$

Where did the numbers for B and K come from (see section 2.6)? Does this answer seem reasonable to you? Why? Thus, the semi-interquartile range from the frequency-distribution method is

$$SIR = \frac{Q_3 - Q_1}{2}$$

$$= \frac{108.07 - 77}{2}$$

$$= 15.54$$

Again you will notice that the SIRs from the two methods are not identical. The sorting-and-counting method involves the actual numbers. The grouped-data method is based on the frequencies in the classes and on the assumption that the numbers are evenly spread in each class.

The third method of finding the SIR consists of reading Q_3 and Q_1 from the baseline of a cumulative percentage polygon (see Figure 3.3). When horizontal lines are drawn from the 75% and 25% points on the vertical scale to the curve and vertical lines are dropped from the intersections to the baseline, $Q_3 = 109$ and $Q_1 = 77$. The resulting *SIR* is

$$SIR = \frac{Q_3 - Q_1}{2}$$

$$= \frac{109 - 77}{2} = 16$$

The graphical method is fast when you have an ogive, but it is a bit rough.

A QUICK-AND-DIRTY SPREAD—THE RANGE

4.6

The last spread is truly a quick-and-dirty statistic—the *range*. It is readily obtained from a sorted batch of numbers and it is simply the difference between the largest and smallest numbers in a batch. In the case of the men's estimates the largest score is 180, the smallest is 40, and the range is $180 - 40$, or 140. The range has been the scapegoat of the authors of statistics books. Despite its definite weaknesses, the range should not be ignored.

WHENS AND WHYS FOR THE THREE SPREADS

4.7

The estimated standard deviation (and its offshoots, the variance and sum of squares) occupies a major position in inferential statistics. With samples from a population of normally distributed numbers the standard deviation is highly reliable. "Highly reliable" means that if we draw, for example, 1000 random samples of size N and calculate an s for each sample, the spread of the frequency distribution of the resulting 1000 s values will be smaller than for some other spread like the 1000 ranges for the same samples. Embedded in this sentence is a fundamental idea in statistics: The distribution of 1000 s values is termed a *sampling distribution*. More particularly, it is a *sampling distribution of the standard deviation*.

The advantages of the standard deviation are (a) it is central in inferential statistics; (b) it is a highly reliable spread; and (c) as noted earlier, s^2 is an unbiased estimator of the population parameter, σ^2. The disadvantage is that s may be distorted by outliers. In this context Mosteller and Tukey assert that ". . . a few straggling values scattered far from the bulk of the measurement can, for example, alter a sample mean drastically, and a sample

s^2 catastrophically" (1968, p. 93). Why do extreme numbers have such tremendous effects upon s^2? The reason is that in calculating s and s^2 the deviations from the mean are squared. Once more, we can observe the impact of outliers.

In Chapter 3 we asserted that the sample mean is highly reliable. As in the case of the standard deviation, 1000 random samples could be drawn, the means of each sample computed, and a distribution of these 1000 means constructed. This distribution would be called the *sampling distribution of the mean*. It would have a smaller spread than a sampling distribution of the median. It should also be stressed that $\overline{X}$ is an unbiased estimator of μ. The $\overline{X}$ yields estimates of μ that are systematically neither too large nor too small.

The semi-interquartile range, in contrast to the standard deviation, is unaffected by extreme scores. Recall again that we are quartering a batch of numbers and the *SIR* is half of the spread of the middle 50% of the numbers. Accordingly, the semi-interquartile range is a better indicator of spread when the batches are from a population of nonnormal numbers. Distributions of such numbers are typically pushed down in the middle, straggle, and are full of holes (Tukey, 1962; Mosteller & Tukey, 1968). I know of no direct evidence, but one would suspect that the semi-interquartile range might be more reliable than the standard deviation under these population conditions. In Chapter 3 the median was advocated as an appropriate center for skewed distributions. The semi-interquartile range and the median are a useful pair of statistics for coping with what Tukey (1962) labels the "spotty" data that often result from research.

While the semi-interquartile range, like the median, helps describe spotty numbers, it too suffers from a fatal flaw. The fatal flaw is that the semi-interquartile range, like the median, is an outsider as far as inferential statistics is concerned. Does that mean that we should forget the semi-interquartile range? Definitely not! Tukey and others have urged us to explore our numbers thoroughly *before* rushing into inference. The semi-interquartile range is a valuable tool in these exploratory endeavors.

Now we will turn to the much-maligned range. The range is a highly unreliable statistic (as defined in reference to a sampling distribution). It is most unreliable for large batches of numbers. The unreliability of the range is not surprising: It uses the two extreme numbers, while the standard deviation is derived from all numbers in a batch. Another problem with the range is that its formula contains no reference to the size of the sample, N. Because the range may be influenced by sample size, it is probably most suitable for roughly comparing the spreads of batches whose sizes are equal or nearly equal. Finally, although there have been some limited efforts to incorporate the range into inferential statistics (see Dixon & Massey, 1951), it is clearly another outsider. In view of these serious maladies, should the range be permitted to die quietly? I think not. Why? It is valuable in exploratory data analysis since it forces us to examine the two most extreme numbers in a batch. If we also poke around at the neighbors of these extremes, we may pick up hints that we have an outlier problem with its attendant severe consequences. If we rush into calculating means and variances without an examination of the extreme numbers, our means and variances could be badly distorted numbers.

The bottom lines are these: most of the time we are engaged in inferential statistics. Here means and standard deviations (or variances) are the statistics of choice. Before computing these statistics, however, it is wise to do a sort or stem-and-leaf display of your numbers and study a picture of them. From the sort you can quickly find the two extremes, Q_1, Q_2 (median), and Q_3. From these numbers you can calculate the range and semi-interquartile range. These various statistics, despite their ills, can provide information about the nature and degree of spottiness of your numbers. Sometimes, as will be evident, your numbers can be made more normal. You have to spend a lot of time and money to get numbers. It is worthwhile to spend some more time exploring your numbers. As any horse player will tell you, nothing is sadder than watching your horse fade in the stretch.

SUMMARIZING BATCHES OF NUMBERS

4.8

We must not forget what we are trying to do. We are concerned with what our numbers are like. Our goal is to make sense of our numbers. As indicated earlier, one step in attaining this goal is to reduce a batch of numbers to a few numbers—centers and spreads—that summarize the batch. In this final section we will talk about two ways of summarizing numbers: a five-number plan proposed by Tukey (1977) and a more traditional two-number summary.

Throughout the last two chapters we have advocated the use of some statistics that suffer from various ills. These statistics, such as the median, the semi-interquartile range, and the range, are often not reliable and they are not incorporated into inferential statistics. But the median and semi-interquartile range are not distorted by extreme scores like the mean and the standard deviation. Accordingly, such statistics are valuable for summarizing spotty numbers, skewed distributions, and the like. They also inform us about the state of our numbers. For example, a consideration of the extreme numbers and their neighbors can illuminate the search for outliers. In short, despite their shortcomings these statistics are helpful in exploratory data analysis. Tukey (1977) devised a new picture, a *box-and-whisker plot,* that includes five numbers: the smallest number, Q_1 (25th percentile), Q_2 (*Mdn*), Q_3 (75th percentile), and the largest number. When these numbers are plotted on either a horizontal or vertical scale, you have a concise and sometimes revealing picture of your numbers. The "box" displays the three quartiles, and the "whiskers" depict the distances from the two outer quartiles to the extremes. Here is a box-and-whisker plot of the men's time estimates:

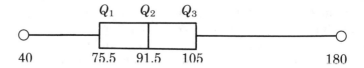

This plot suggests that the time estimates are extremely variable and positively skewed. Numbers spread out like this are troublesome in research as we will show in Chapter 14. Regarding skewness, notice that the right whisker (Q_3 to the largest number) is much longer than the left whisker (Q_1 to the smallest number). Note also that while 50% of the numbers are somewhat narrowly confined in the box, the whiskers, each containing 25% of the numbers, extend out. With the right whisker, we know that 25% of the numbers are out there, but we do not know where. For example, all the numbers, except for the largest number, could be located close to Q_3. Inspecting the sorted numbers would clarify this problem. Some statistical packages, such as MINITAB (Ryan et al., 1981) produce box-and-whisker plots. Naturally, when one wants to compare several batches of numbers their box plots can be contrasted.

The traditional way to summarize batches is in terms of two numbers: the mean and standard deviation (or variance). This is entirely appropriate when the numbers are reasonably normal. How do we know this? Explore them with sorts, histograms or frequency polygons, box-and-whisker plots, and so on.[2] If we have very large samples like several hundred numbers, it is possible to augment the two-number summary with quantitative indices of kurtosis and skewness. These indices are based on moments. A *moment*, *M*, is the average of the deviations from the mean raised to a certain power. The first moment about the mean is 0:

$$M_1 = \frac{\Sigma X - \overline{X}}{N}$$

The variance is the second moment:

$$M_2 = \frac{\Sigma (X - \overline{X})^2}{N}$$

The third moment is

$$M_3 = \frac{\Sigma (X - \overline{X})^3}{N}$$

and the fourth moment is

$$M_4 = \frac{\Sigma (X - \overline{X})^4}{N}$$

Proper combinations of these moments describe precisely the degree of skewness and kurtosis. Skewness, β_1, is

$$\beta_1 = \frac{M_3}{M_2 \sqrt{M_2}}$$

Kurtosis, β_2, is

$$\beta_2 = \frac{M_4}{(M_2)^2} - 3$$

In a normal distribution β_1 and β_2 should be close to 0. A $\beta_1 > 0$ indicates positive skewness; a $\beta_1 < 0$ indicates negative skewness. A $\beta_2 > 0$ indicates a leptokurtic distribution; a $\beta_2 < 0$ indicates a platykurtic distribution. The best way to obtain β_1 and β_2 is to use a computer package such as SPSS[x] or BMDP.

Batches of numbers may be summarized by either five numbers (extremes and quartiles) or by two numbers (the mean and standard deviation). Which should be done? The answer is simple—do both! It does not cost much to explore your numbers before playing the inference game. As an old psychiatrist friend always said when he handed me the lunch check: *"Remember,* it only costs a little more to go first class!"

2. Chapter 9, Section 9.2, presents some specific techniques for assessing normality.

THE COMPARED-TO-WHAT PROBLEM AGAIN— THE STANDARD-SCORE APPROACH

4.9

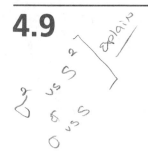

In Chapter 2 we encountered the compared-to-what problem. How can the meaning of a number like a test score be specified more precisely? How can numbers from two different tests be compared? The solution proposed in Chapter 2 was to convert the numbers into *percentiles*. A percentile score indicates what percentage of the class falls at or below the test score. Comparing the percentiles for two scores from different tests reveals the test in which performance was better.

Measures for centers were described in Chapter 3. This chapter discussed spreads. By relating numbers to the mean and standard deviation of a batch, it is possible to convert numbers to *standard scores* (Zs). Transforming numbers to Z scores, then, constitutes a second solution to the compared-to-what problem.

A standard, or Z score is a *deviation score*, $X - \overline{X}$, divided by the standard deviation:

$$Z = \frac{X - \overline{X}}{s}$$

where s is an estimate of the population standard deviation, σ, based on the batch. If we know the population parameters, which is a rare occurrence, then

$$Z = \frac{X - \mu}{\sigma}$$

Suppose a student had a test score of 18 in history and 47 in sociology. If the history class had a $\overline{X} = 19$ and an $s = 4$, and the sociology class had a $\overline{X} = 31$ and $s = 8$, then the standard scores in history and sociology, respectively, would be

$$Z = \frac{X - \overline{X}}{s}$$

$$= \frac{18 - 19}{4} = -0.25$$

$$= \frac{47 - 31}{8} = 2.0$$

How are these Z scores interpreted? Answering this question requires more information about Z scores. All Z scores have two important characteristics: (a) the mean of a set of Z scores is equal to zero, or $\overline{Z} = 0$; and (b) the standard deviation of a set of Z

scores is equal to one, or $s_z = 1$. We can have variables based on different scales such as pupil diameter in millimeters, age in years, salary in dollars, and so on. If we convert them to standard scores, the resulting set of Z scores for each variable has a $\overline{Z}$ of 0 and s_z of 1. The student is a little below (0.25 of a standard deviation) the mean in history and well above (two standard deviations) the mean in sociology. The student was more successful in sociology. When transforming numbers to Zs, the distance of the number from the mean is being scaled in standard deviation units.

What underlies the transformation of numbers to standard scores? It is related to the effects on numbers of adding, subtracting, multiplying, and dividing by a constant, c. If we add c to every number in a batch, the new mean would be $\overline{X}_{new} = \overline{X}_{old} + c$. We simply move the whole distribution up the scale by c units and do not change s: $s_{new} = s_{old}$. If we subtract c, $\overline{X}_{new} = \overline{X}_{old} - c$, again with no effect upon s. If we multiply by c, $\overline{X}_{new} = c\overline{X}_{old}$ and $s_{new} = cs_{old}$, so that both the new mean and new standard deviation are c times larger. If we divide by c, $\overline{X}_{new} = \overline{X}_{old}/c$ and $s_{new} = s_{old}/c$. Welkowitz, Ewen, and Cohen (1982, p. 73) showed how the rules regarding constants produce the values of $\overline{Z} = 0$ and $s_z = 1$ in a set of Z scores. In calculating Z scores a constant ($c = \overline{X}_{old}$) is subtracted from every score. The mean of the deviation scores ($\overline{X}_{new}$) is $\overline{X}_{new} = \overline{X}_{old} - c = \overline{X}_{old} - \overline{X}_{old} = 0$. Each deviation score is then divided by a constant ($c = s_{old}$) to compute the Z scores whose mean is

$$\overline{Z} = \frac{0}{c} = \frac{0}{s_{old}} = 0$$

The standard deviation of a set of Z scores, s_{new} is

$$s_{new} = \frac{s_{old}}{c} = \frac{s_{old}}{s_{old}} = 1$$

In the real world, Z scores are not readily understood. Why? They are small numbers, and half of them are negative. Applying the previously mentioned rules regarding constants, Z scores may be transformed into different kinds of standard scores without the undesirable qualities of Z scores. The general formula is

_____ score = (standard deviation)(Z) + mean

Here _____ is the name of the new standard score, or your own name, if you are a trifle narcissistic. The standard deviation is the *desired* standard deviation for the scale and the mean is the *desired* mean. One such scale is the *T scale* with a mean of 50 and a standard deviation of 10. The student's *T* score in history is

$$T \text{ score} = 10Z + 50$$
$$T \text{ score} = 10(-0.25) + 50 = 47.5$$

The student is still below the mean in history.

Another type of standard score is an *SAT score,* which has a mean of 500 and a standard deviation of 100. In this case the student's history score converted to a *SAT* score is

$$SAT \text{ score} = 100Z + 500$$
$$SAT \text{ score} = 100(-0.25) + 500 = 475$$

While the *SAT* score certainly looks more impressive than the original score of 18, the student is still below the mean in history.

Should the compared-to-what problem be solved by percentiles or some variety of a standard score? For tasks like evaluating the test scores of applicants for graduate school, I find that both kinds of information are helpful. For *GRE*s (which are *SAT* scaled) I like to see both the percentiles and the *GRE* scores. Hinkle, Wiersma, and Jurs (1979, p. 28) cautioned that differences in percentiles are unequal in different portions of a percentile scale. But they accept percentiles as useful for locating numbers in a batch. Percentiles are readily understood by real people like bartenders, undertakers, used car salespersons, and bankers. That is an advantage. On the other hand, *Z* scores can solve some other problems as we shall observe in Section 4.10.

Here is an example to demonstrate a third important characteristic of *Z* scores. In the frequency distribution (Table 4.2) the 34 numbers ranging from 2 to 24 are positively skewed and have a $\overline{X} = 8.76$ and an $s = 4.72$. In the last column the numbers have been converted to standard scores:

$$Z = \frac{X - \overline{X}}{s}$$

$$Z = \frac{2 - 8.76}{4.72} = -1.43$$

$$\vdots$$

$$Z = \frac{24 - 8.76}{4.72} = 3.23$$

TABLE 4.2
Frequency distribution of numbers and standard scores

X	f	Z
2	2	−1.43
4	4	−1.01
6	6	−0.58
8	10	−0.16
10	5	0.26
12	3	0.69
16	2	1.53
20	1	2.38
24	1	3.23

If you study the Z scores, you will conclude that they are positively skewed, too. Transforming to Z scores does *not* change the shape of the distribution. If the distribution of a batch of numbers is approximately normal, then the distribution of its Z scores will also be approximately normal. This characteristic of Z scores has important implications for solving some problems with Z scores.

THE NORMAL DISTRIBUTION AND Z SCORES

4.10

Many variables—heights, weights, intelligence, and so forth—are approximately normally distributed. What is a *normal distribution?* It is a family of symmetrical, bell-shaped curves whose equation is

$$Y = \frac{1}{\sigma\sqrt{(2\pi)}} e^{-(X - \mu)^2/2\sigma^2}$$

where Y is the height of the curve for X, σ is the population standard deviation, μ is the population mean, and π (3.1416) and e (2.7183) are constants.

Since μ and σ vary, there is a family of normal distributions. If, however, the numbers are converted to Z scores, then a single normal distribution will produce a $\mu = \overline{Z} = 0$, $\sigma = s_z = 1$, and a simpler equation:

$$Y = \frac{1}{\sqrt{(2\pi)}} e^{-Z^2/2}$$

Proportions of the area under the *standard normal distribution* are presented in Table D. Column A consists of Z values. In column B the proportions of the area from the mean (Z = 0) to Z are shown. In column C the proportions of the area beyond Z (i.e., the tail areas) are displayed. Remember that the normal distribution is symmetrical so that negative Z values will cut off the same proportions in columns B and C as the corresponding positive Z values. If the normal distribution has an area of 1, then the areas in Table D denote proportions of the whole curve.

Figure 4.1 is a *unit normal distribution*. What proportion of the area falls between the mean, $Z = 0$, and $Z = 1.00$? Entering Table D for $Z = 1.00$, the proportion in column B is .3413.

What proportion of the area falls beyond $Z = 1.00$? From column C it is .1587, or this answer may be found by subtraction. Since the whole area is 1, the area above $Z = 0$ is .5. From the mean to $Z = 1.00$ the proportion is .3413. Therefore, the tail proportion beyond $Z = 1.00$ is $.5 - .3413 = .1587$.

How much of the area falls between $Z = 1.00$ and $Z = -1.00$? The proportion from the mean to $Z = 1.00$ is .3413. The proportion from the mean to $Z = -1.00$ is .3413. Accordingly, the proportion between $Z = 1.00$ and $Z = -1.00$ is $.3413 + .3413 = .6826$, or it is $2(.3413) = .6826$.

Yesterday's paper contained an article on the Mensa Society. It stated that to become a member a person had to score at least 132 on a standard intelligence test and that members were in about the top 2% of the population. If we assume that intelligence is approximately normally distributed with a population mean (μ) of 100 and a population standard deviation (σ) of 16, we can check

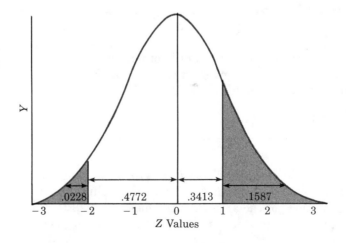

FIGURE 4.1
Normal curve showing proportions corresponding to different Z values

the 2% claim. We need to convert 132 to a Z score and find the proportion falling beyond Z:

$$Z = \frac{X - \mu}{\sigma}$$

$$= \frac{132 - 100}{16} = 2.00$$

From Table D, the tail area (column C) for $Z = 2.00$ is .0228. To convert this proportion to a percentage we multiply the proportion by 100: $100(.0228) = 2.28$. The claim that Mensa members are in about the top 2% of the population is thus verified.

"Normal" intelligence is said to range from 90 to 110. What proportion of the population falls between these limits? What percentage? To answer these questions the scores must be converted to Z scores:

$$Z = \frac{90 - 100}{16} = -0.625$$

$$= \frac{110 - 100}{16} = 0.625$$

The proportion of the area from the mean to $Z = -0.62$ is .2324 and from the mean to $Z = -0.63$ is .2357. Thus, for $Z = -0.625$ the proportion is $(.2324 + .2357)/2 = .23405$ or halfway between $Z = -0.62$ and $Z = -0.63$. Because the curve is symmetrical the proportion for $Z = 0.625$ will also be .23405 and the proportion between $Z = -0.625$ and $Z = 0.625$ will be $.23405 + .23405 = 0.4681$. In percentage terms it is $100(0.4681) = 46.81$. Thus, almost half (47%) of the population falls in the range from 90 to 110.

An investigator studying creativity wishes to limit a sample to subjects who fall in the upper 20% of the population in intelligence. What cutting score should be enforced to select the sample? This problem must be done in a reverse manner to the previous problems: (a) convert 20% to a tail *proportion,* (b) find the Z corresponding to the tail proportion, and (c) find the X for the Z. In proportion terms, 20% is: $20/100 = .20$. The Z in column A corresponding to .20 in column C is 0.84. By substituting this Z value in the standard-score formula, X can be determined:

$$0.84 = \frac{X - 100}{16}$$
$$X = 100 + 13.44 = 113.44$$

So selecting subjects with an intelligence score of 113 or 114 would achieve the investigator's goal.

These problems demonstrate the utility of the standard normal distribution with numbers that have been converted to standard scores. But the value of the normal distribution extends to other problems. Often, test statistics become normally distributed as N increases. When this happens, Table D can be employed to find the *probability* of an observed test statistic. How can probabilities be obtained from Table D? If a test statistic is normally distributed and is standardized (i.e., converted to a Z score), then the tail area in Table D is the probability of the test statistic's occurring by chance. When the probability is sufficiently small, then the investigator asserts that something other than chance is at work. If the probability is large, then the investigator does not reject the chance hypothesis. This is a brief, oversimplified description of statistical inference, which will be our concern beginning in Chapter 5.

SUMMARY

Spread is a measure of the variability of a batch of numbers. It is a basic concept in descriptive statistics and is fundamental also in inference. The three most common spreads are: the standard deviation, the semi-interquartile range, and the range. The standard deviation is the positive square root of the average of the squared deviations of the numbers from their mean. The semi-interquartile range is a half of the difference between a number corresponding to the 75th percentile (Q_3) and a number corresponding to the 25th percentile (Q_1). The range is the difference between the largest and smallest numbers in a batch. The standard deviation based on a sample (s) is an estimate of σ and its divisor is $N - 1$. The sd is that actual standard deviation of a batch of numbers and its divisor is N. The standard deviation of a population (σ) is a parameter and its divisor is N_p. The variance is a squared standard deviation, s^2, $(sd)^2$, or σ^2. The sum of the squared deviations is also known as the sum of squares, SS. The widely applicable spreads, s or s^2, should be calculated with computational formulas that bypass obtaining the actual deviations from the mean. There are different computational formulas for raw scores and for grouped data (see the appendix). The latter method should be used when only a frequency distribution is available. A rough check on s can be done by determining whether approximately two-thirds of the numbers fall within the bounds of $\overline{X} \pm s$.

The semi-interquartile range can be found from a sorted batch of numbers, from a frequency distribution, and from a cumulative percentage polygon. The range is readily obtained from a sorted batch of numbers. The estimated standard deviation, s (or variance), is highly reliable, is an estimator of σ, and plays a major role in inferential statistics. Its weakness is that it is strongly affected by extreme numbers. High reliability was defined by comparing the smaller spread in the sampling distribution of the standard deviation to the sampling distribution of the range. The semi-interquartile range is unaffected by extreme numbers, best suited for spotty numbers, and is not part of inferential statistics. The range is highly unreliable in large samples, affected by sample size, and not integrated into inferential statistics. Its consideration is worthwhile, however, because it may lead to the detection of outliers.

Two ways of summarizing a batch of numbers were described: a five-number summary, and a two-number summary. In the first summary the numbers are the smallest number, Q_1, Q_2 (Mdn), Q_3, and the largest number. Tukey has combined these numbers into a new picture—a box-and-whisker plot. The traditional two-number summary consists of the mean and standard deviation (or variance). With large samples, proper combinations of moments serve as quantitative measures of kurtosis and skewness. A moment is an average of the deviations from the mean raised to a certain power: The third moment is $\Sigma(X - \overline{X})^3/N$.

Formulas for kurtosis and skewness were presented. Both types of summaries should be utilized, and numbers should be explored thoroughly with sorts, pictures, and by less-than-perfect statistics prior to doing inferential statistics.

A second solution to the compared-to-what problem—converting numbers to standard scores (Zs)—was described. A Z score is the distance a number is from the mean scaled in standard deviation units. All Z scores have a mean of 0 and a standard deviation of 1. The sign of a Z score shows whether a test score is above or below the mean and the magnitude of Z signifies how far the number is from the mean in s or σ units. Numbers from different tests or variables can be evaluated by comparing their Z scores. Other types of standard scores are T scores, which have a mean of 50 and a standard deviation of 10, and SAT scores, which have a mean of 500 and a standard deviation of 100.

Converting numbers to Z scores does not change the shape of the distribution: A normally distributed batch of numbers will yield a normally distributed set of Z scores. A normal distribution is a bell-shaped, symmetrical distribution having a known function. Since the normal distribution depends upon μ and σ, there is a family of normal distributions. If numbers are converted to Z scores, a standard normal distribution with a mean of 0 and a standard deviation of 1 results. When the distribution of a variable is approximately normal, the standard normal distribution with an area of 1 (the unit normal distribution) can be applied to solve a variety of problems like: What proportion of the area is included between the mean and Z? Or, what cutting score is needed to select subjects in the upper 10% of the population on some trait? Because some test statistics become normally distributed as N increases, the normal distribution plays an important part in statistical inference.

PROBLEMS

1. The times in seconds for 20 subjects to solve a problem on a creativity test are: 11, 13, 17, 19, 20, 27, 35, 36, 38, 42, 44, 45, 51, 53, 55, 56, 59, 61, 63, 64. Find (a) the sum of squares (SS), (b) standard deviation (s), and (c) variance (s^2) of the times using the computational formulas and your calculator.
2. Demonstrate that your answer for s is reasonable.
3. Find the actual standard deviation (sd) for the time scores.
4. What is the range of the time scores?
5. Using the sorting-and-counting method, obtain Q_1 and Q_3. What is the semi-interquartile range (SIR) for the time scores?
6. Construct a box-and-whisker plot of the time scores.
7. Below is stem-and-leaf display for 40 test scores. Record the amount of time it takes you to calculate s using the leaves and the computational formula.

Stem	Leaves	f	cf
80–89	85, 88	2	40
70–79	75, 76, 78, 79	4	38
60–69	62, 63, 68, 69	4	34
50–59	50, 51, 53, 54, 56, 59	6	30
40–49	40, 41, 42, 43, 47, 48, 48, 49, 49	9	24
30–39	31, 32, 38, 39, 39	5	15
20–29	23, 27, 29, 29	4	10
10–19	10, 12, 17	3	6
0–9	5, 7, 9	3	3

8. After adding the necessary columns to the table of 40 test scores, see how long it takes you to calculate s by the grouped-data method. Which method takes longer and which is likely to lead to more errors?
9. Assume the test scores to be normally distributed. If a student had a score of 78 on the test, what is the corresponding Z score? T score? SAT score?
10. If the normality assumption is true, what proportion of scores should fall between $Z = -1.0$ and $Z = 1.0$? Beyond $Z = 1.645$? Beyond $Z = -1.96$ and $Z = 1.96$?
11. To enroll in the follow-up course a student must be in the top third of the class. Using the normal distribution find the cutting score required for permission to take the follow-up course.
12. For the test score distribution determine the actual percentage of students whose scores fell between $Z = -1.0$ and $Z = 1.0$.

APPENDIX: CALCULATING s FROM A FREQUENCY DISTRIBUTION

It is possible to calculate s either from the raw scores or grouped data. The formula for calculating the standard deviation from a frequency distribution is

$$s = \sqrt{\frac{\Sigma fX'^2 - \frac{(\Sigma fX')^2}{N}}{N-1}}$$

where X' is the midpoint of the classes. To the men's time estimates we have added columns for X'^2 and fX'^2 in Table 4.2.

TABLE 4.2
Frequency distribution for computing the standard deviation

Class	f	X'	X'²	fX'	fX'²
180–199	1	189.5	35910.25	189.5	35910.25
160–179	0	169.5	28730.25	0.0	0.00
140–159	1	149.5	22350.25	149.5	22350.25
120–139	4	129.5	16770.25	518.0	67081.00
100–119	7	109.5	11990.25	766.5	83931.75
80–99	16	89.5	8010.25	1432.0	128164.00
60–79	8	69.5	4830.25	556.0	38642.00
40–59	3	49.5	2450.25	148.5	7350.75

The sum for the fifth column is $\Sigma fX' = 3760$. The sum for the sixth column is $\Sigma fX'^2 = 383430$. Solving for s,

$$s = \sqrt{\frac{\Sigma fX'^2 - \frac{(\Sigma fX')^2}{N}}{N-1}}$$

$$= \sqrt{\frac{383430 - \frac{(3760)^2}{40}}{39}}$$

$$= \sqrt{\frac{29990}{39}}$$

$$= \sqrt{768.97436} = 27.73$$

As shown in Chapter 3, it is possible to obtain the fX' values on your calculator, put them in memory, and find $\Sigma fX'$ with the $\boxed{\Sigma X}$ key. Likewise, the fX'^2 values may be obtained, put in memory, and $\Sigma fX'^2$ may be found with the $\boxed{\Sigma X}$ key. For example, for the 180–199 class the fX'^2 value put in memory would be
1 $\boxed{X}$ 189.5 $\boxed{X^2}$ 35910.25 $\boxed{=}$ 35910.25 $\boxed{M+}$

Once more, we must emphasize that the grouped-data method is a relic from the hand-cranked Monroe calculator days. It should be applied only when the actual numbers are unavailable. Why does the raw-score method yield an $s = 26.95$ in contrast to an $s = 27.73$ with the grouped-data method? Because two different batches of data are being analyzed as was demonstrated for the mean in Chapter 3.

ADDENDUM: USING COMPUTERS IN STATISTICS

MINITAB is employed here to produce some statistics on the previously input time estimates for the men. The $\overline{X} = 91.875$ (see 1) is output immediately upon the command, "AVERAGE C1." C1 is the column in the "worksheet" where the time estimates are stored. The box-and-whisker plot (2) is output on the command, "BOXPLOT C1." The asterisk at the end of the right whisker suggests to the analyst that the largest number, 180, may be an outlier. The only problem with this picture is reading the scale at the bottom.

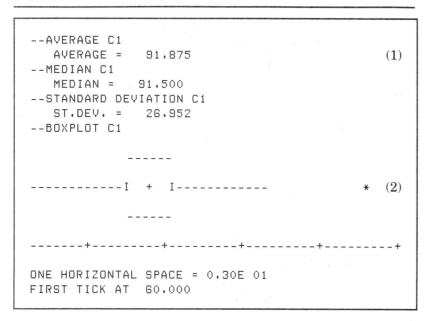

```
--AVERAGE C1
   AVERAGE =   91.875                              (1)
--MEDIAN C1
   MEDIAN =    91.500
--STANDARD DEVIATION C1
   ST.DEV. =   26.952
--BOXPLOT C1


                  ------
 ------------I  +  I------------          *       (2)
                  ------
 -------+---------+---------+---------+---------+
ONE HORIZONTAL SPACE = 0.30E 01
FIRST TICK AT   60.000
```

CHAPTER OUTLINE

5.1 THE SUDS SCENE

5.2 PUTTING THE SCENE INTO A STATISTICAL CONTEXT

5.3 TESTING HYPOTHESES WITH THE BINOMIAL DISTRIBUTION

5.4 TESTING HYPOTHESES WITH THE NORMAL DISTRIBUTION

5.5 TESTING HYPOTHESES ABOUT CENTERS, SPREADS, AND OTHER STATISTICS

CHAPTER FIVE
FLIPPIN' COINS AND BUYIN' BEERS

THE SUDS SCENE

5.1
One afternoon I was sitting in a dimly lighted cantina, having a cool pop, and listening to the juke box wail. Through the swinging door came a tall, skinny stranger. He wore a Penney's shirt with a pocketful of writing utensils and he had that lost look. I knew immediately that he was probably a college professor. Another dude slithered over to join "the professor" at the bar. This slick dude was a bit shifty-eyed and was coated with a thin layer of blue pool chalkdust. I pegged him right away as a "hustler." Soon the professor and the hustler were chatting in a friendly manner about this and that. It wasn't long before the hustler proudly displayed his "lucky" silver dollar. In the next couple of hours he flipped it four times and called "heads" each time. And the professor was down four beers to zip as dusk fell on the friendly cantina.

PUTTING THE SCENE INTO A STATISTICAL CONTEXT

5.2
What can we do with this scene statistically? We can treat the coin-flipping episode as a prototype of *statistical inference*. We start with a *research hypothesis:* The lucky silver dollar is loaded (the hustler is indeed a hustler!). Another hypothesis is that the dollar is not loaded. The hypothesis of an unbiased coin can be brought into the realm of statistics and labeled the *null hypothesis* or H_0. Another statistical hypothesis, called the *alternative hypothesis*, H_1, coincides with the research hypothesis that the coin is loaded. We want to decide which statistical hypothesis is true and which is false and relate this decision to our research hypothesis. In addition, we are concerned with generalizing to a *population* of events—a very large number of flips of the dollar—from the *sample* of four coin flips.

Consider the null hypothesis, H_0. If chance alone were operating (the coin were unbiased), we would expect about as many heads as tails to turn up. In statistical parlance, the *probability* of a head should be equal to the probability of a tail. If the number of heads is equal or close to the number of tails, we will decide *not to reject* H_0 and to *reject* H_1. We will regard H_0 as *true* and H_1 as *false*. On the other hand, what if the number of heads was much larger than the number of tails? When this happens with an unbiased coin, we are witnessing an unusual event. If the event is sufficiently unusual, then we will reach the opposite decision; namely, we will *reject* H_0 and *accept* H_1. Then, we are asserting that H_0 is *false* and H_1 is *true*.

SEC. 5.2 □ PUTTING THE SCENE INTO A STATISTICAL CONTEXT

Obviously, we require a method for determining the probabilities of events and a criterion of unusualness. One way to find the *probability of an event,* such as four heads in a row, is to divide the number of the ways *(W)* this event can occur by the total number of possible mutually exclusive and equally likely outcomes *(T)*. Probabilities range from zero to unity.[1] The criterion of unusualness will be termed the *criterion of significance,* α (a small alpha). Almost always in the behavioral sciences α is set at .05. The *probability of the test statistic* (e.g., probability of getting four heads) is then compared with the criterion of significance, α, to reach decisions about H_0 and H_1. If the test statistic's probability is greater than α, then H_0 is not rejected and H_1 rejected. If the test statistic's probability is less than or equal to α, then H_0 is rejected and H_1 is accepted. Suppose the outcome of a series of coin flips had a probability of .03 and that we had set α at .05. Since $.03 < .05$, we would reject H_0 and accept H_1. We would also relate these statistical decisions to our research hypothesis by assuming that the lucky silver dollar was loaded. And, generalizing to the population, we would expect that the professor would be doing a lot of buying if the game went on.

Unfortunately, there is more to statistical inference, and it does not simplify the situation. We can fail to reject H_0, that is, decide that H_0 is true and H_1 is false; or we can reject H_0, that is, decide that H_0 is false and H_1 is true. These possibilities represent the statistical decisions. How about reality? Well, H_0 can be true and H_1 false, or H_0 can be false and H_1 true. If we put these statistical decisions and states of reality into a 2×2 matrix, we have

		States of Reality	
		H_0 true	H_0 false
Statistical Decisions	H_0 not rejected	correct decision $1 - \alpha$ (1)	Type 2 error β (3)
	H_0 rejected	Type 1 error α (2)	correct decision $1 - \beta$ (4)

1. If the probability of an event is 1, the event is certain to occur. If the probability of an event is 0, the nonoccurrence of the event is certain. If the probability of an event is .5, then the event is as likely to occur as it is not to occur.

Whenever a statistical decision is made, we are caught in this matrix. We can be wrong in two ways (cells 2 and 3) and be right in two ways (cells 1 and 4).

We will consider the correct decisions and errors in order. In cell 1 we do not reject H_0, deciding the coin is unbiased, and in truth it is unbiased. We are right on target—we have made a correct decision.

In cell 2 we erred—we rejected H_0, deciding the coin was loaded, but it really was not. This is called a *Type 1 error of inference*—finding something that is not there. The probability of committing a Type 1 error is equal to α, the criterion of significance. Then, the probability of attaining a correct decision in cell 1 is $1 - \alpha$. If we set $\alpha = .05$, then 5 times out of 100 experiments we will find something that isn't there, and 95 times out of 100 we will correctly fail to reject H_0 when H_0 is true.

In cell 3 a different kind of an error is committed, a *Type 2 error of inference*. Here H_1 is true: The coin is loaded, but we decide that it is not. The probability of a Type 2 error is β (a small beta).

Cell 4 is the second type of correct decision: H_1 is true, the coin is indeed loaded, and we decide that it is. The probability of this correct decision is $1 - \beta$. This probability, $1 - \beta$, is also termed *statistical power*. It is the probability of finding something when something is there to be found.

We wish to arrive at correct decisions and avoid committing errors. Why not try to reduce the Type 1 errors and thereby increase the number of correct decisions? We stated that it is customary to set α at .05. The researcher, however, has the power to set α. If we set $\alpha = .10$, then $1 - \alpha = .90$ and we would be making *fewer* correct decisions and committing *more* Type 1 errors when H_0 is true in comparison to applying the conventional $\alpha = .05$. But if we set $\alpha = .001$, then $1 - \alpha = .999$ and we have *increased* the number of correct decisions and *reduced* the number of Type 1 errors in comparison to $\alpha = .05$. It looks like we are on the right track by setting $\alpha = .001$, but the success is illusory.

The source of difficulty is this: α and β are *inversely* related. If α is set more conservatively, such as at .001, to *lower* Type 1 errors, then Type 2 errors will be *increased*. If α is relaxed, for example, to .10, then Type 1 errors will be *increased* and Type 2 errors *decreased*. What are the consequences of Type 1 and Type 2 errors? The consequence of a Type 1 error is that the investigator finds something that is not there. He or she will be embarrassed when other researchers fail to corroborate the finding. The consequence of a Type 2 error is that the investigator will overlook something that is there. Doing research is expensive and failing to find something does not motivate researchers to go down the same road again. Which type of error is more serious? It depends upon the something that is found or is overlooked and what might be done with that something. If an investigator is exploring a new research area and looking for ideas, then α might be relaxed to prevent overlooking effects that could be investigated further. If an investigator is testing a critical theory or an important product (e.g., a medicinal drug for human use), then finding something that is not there (a Type 1 error) might not be desirable. In these latter two situations α might better be set more conservatively (e.g., at .001). It should be stressed again that the setting of the α level is an *arbitrary* decision—employing α = .05 is merely a convention.

Social scientists have paid attention to the left half of the 2 × 2 decision-reality matrix and this half will be our concern in much of the book. However, we cannot ignore the consequences of setting α for the right half of the matrix. We must recognize too that a very common and deadly research error is doing research with inadequate statistical power. That is, the investigator tries to find something when the chances of detection are slim even when that something is there. We will return to the critical topic of power in Chapter 14.

Finally, it should be emphasized again that the 2 × 2 decision-reality matrix is involved every time a statistical decision is made. You do not know where you are in the matrix. Earlier it was argued that the outcome of a single experiment is never conclusive. Thinking about the 2 × 2 matrix should inspire caution and should convince you that replications are essential for achieving conclusions. Do not forget that when α = .05, on 5 occasions out of 100 you can be rejecting H_0 when H_0 is true and finding something that is not there. Statistics isn't rigorous like plumbing.

TESTING HYPOTHESES WITH THE BINOMIAL DISTRIBUTION

5.3

With this background we will return to the cantina. The outcome of "the experiment" was that the lucky silver dollar turned up heads on four flips in succession. What is the probability *(P)* of this event happening with an unbiased coin? To test hypotheses regarding this event we need to compare P with α, the criterion of significance. We said that

$$P = \frac{W}{T}$$

where W is the number of ways the event can occur and T is the total number of possible mutually exclusive and equally likely outcomes. If we flip an unbiased coin four times, there are 16 possible outcomes shown in Table 5.1.

Note that outcome 1 (4H) can occur in one and only one way; likewise for outcome 16 (4T). Outcomes 2–5 (3H, 1T) can occur four ways as can outcomes 12–15 (1H, 3T). Lastly, outcomes 6–11 (2H, 2T) can occur six ways. The probabilities of the five events or possible *test statistics* are as follows:

$$4H \qquad P = \frac{W}{T} = \frac{1}{16} = .0625$$

$$3H, 1T \qquad = \frac{W}{T} = \frac{4}{16} = .25$$

$$2H, 2T \qquad = \frac{W}{T} = \frac{6}{16} = .375$$

$$1H, 3T \qquad = \frac{W}{T} = \frac{4}{16} = .25$$

$$4T \qquad = \frac{W}{T} = \frac{1}{16} = .0625$$

Note that the sum of these probabilities is $.0625 + .25 + \ldots + .0625 = 1.0$. For every binomial distribution, $\Sigma P = 1.0$.

Computing the probabilities of the test statistics was accomplished by enumerating the 16 possible outcomes for four coin

TABLE 5.1
Outcomes from flipping a fair coin four times

Flips	Outcomes															
	1	2	3	4	5	6	7	8	9	10	11	12	13	14	15	16
1	H	T	H	H	H	T	H	H	H	T	T	H	T	T	T	T
2	H	H	T	H	H	T	T	H	T	H	H	T	H	T	T	T
3	H	H	H	T	H	H	T	T	H	T	H	T	T	T	H	T
4	H	H	H	H	T	H	H	T	T	H	T	T	T	T	H	T

flips. More generally, the probability of a test statistic is obtained by referring it to a *theoretical distribution*. The appropriate theoretical distribution for a problem like coin tossing which has only two possible outcomes, heads or tails, on each trial (flip) is called the *binomial distribution*. Actually, the binomial distribution is a *family of distributions.* The binomial distributions depend upon (a) N_T, the number of trials or flips, and (b) the probabilities of the two outcomes, such as getting a head versus getting a tail. What do these distributions look like and how can they be found? We have just developed one binomial distribution—the one for $N_T = 4$ when the two outcomes are equally likely. We can picture this binomial distribution by constructing a histogram with the five outcomes (0T, ..., 4T) on the horizontal axis and the probabilities with which these events can occur by chance on the vertical axis. Figure 5.1 shows the binomial distribution for $N_T = 4$.

Three properties of binomial distributions are worthy of mention.

1. The distributions are symmetrical when the two outcomes are equally likely.
2. The distributions are discrete (it is impossible to observe 3.25 tails, for example).
3. As N_T increases, the binomial distribution becomes increasingly like the normal distribution described in Chapter 4.

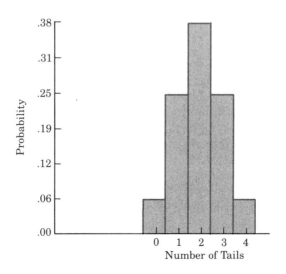

FIGURE 5.1
The binomial distribution for $N_T = 4$

An implication of 3 is that when N_T is reasonably large (e.g., $N_T \geq 10$), the well-known and widely available *normal distribution,* another theoretical distribution, can be used to approximate binomial probabilities. The probabilities from the binomial distribution are exact in contrast to the approximate probabilities from the normal distribution. But when N_T is reasonably large, the discrepancies between the exact probabilities and the approximate probabilities are trivial in magnitude.

How can the binomial distributions and the necessary probabilities be obtained? There are three approaches to this problem. First, we can calculate the probabilities by "expanding the binomial." Second, we can look up the probabilities in tables for the binomial distribution. Third, we can calculate approximate probabilities by utilizing the normal distribution. The third approach will be presented in Section 5.4.

The binomial expansion is $(p + q)^{N_T}$, where p is the probability of one event (e.g., a head) and $q = 1 - p =$ the probability of the other event (e.g., a tail). Suppose we consider an easy problem where $N_T = 2$. We want $(p + q)^2$, so we multiply $(p + q)$ by $(p + q)$ and get $p^2 + 2pq + q^2$. The number of ways p^2 (two heads) can occur is that term's implicit coefficient of 1, so that p^2 means $1p^2$. Similarly, the coefficient for q^2 (two tails) is 1. The set of coefficients (Ws) for the terms p^2, pq, and q^2 is 1, 2, 1. The sum of the set of coefficients (T) is 4. When $p = q = .5$, the probability of getting 2H (i.e., p^2) = W/T = 1/4 = .25. The probability of getting 1H,1T (i.e., $2pq$) = W/T = 2/4 = .5. The probability of getting 2T (i.e., q^2) = W/T = 1/4 = .25. If we plotted these probabilities in the same manner as in Figure 5.1, the result would be the binomial distribution for $N_T = 2$ and $p = q = .5$.

There is a second method of calculating these probabilities: Substitute the values for p and q into the expansion. The probability of obtaining 2H (i.e., p^2) = $(.5)^2$ = .25; 1H,1T (i.e., $2pq$) = $2(.5)(.5)$ = .5; and 2T (i.e., q^2) = $(.5)^2$ = .25. When $p = q = .5$, either technique will work. But when $p \neq q$, the second method must be applied. Suppose $p = .67$. Then $q = 1 - .67 = .33$. The probability of getting 2H is $p^2 = (.67)^2 = .4489$. The probability of getting 1H,1T is .4422, and of 2T is .1089. As always, the probabilities sum to 1, but when $p \neq q$ the binomial distribution is *not symmetrical.*

What if we wanted the binomial distribution for $N_T = 6$? We could raise $(p + q)$ to the sixth power, but that would be laborious. So the question becomes: How can we expand the binomial without doing it the hard way? Looking again at $p^2 + 2pq + q^2$, we observe two things: There are three terms in the expansion when $N_T = 2$, and the powers of p and q change in an orderly manner. For the first term, p's power $= N_T$ and q's is 0. (Remember that a number raised to the 0 power is 1.) In the second term, the power of p is decreased by 1 and q is increased by 1, yielding pq. In the third term, these orderly changes in powers continue, yielding q^2. Generalizing these rules to $N_T = 6$, we can list the $N_T + 1$ or 7 terms and their powers:

$$p^6 + p^5q + p^4q^2 + p^3q^3 + p^2q^4 + pq^5 + q^6$$

Now how can we find the coefficients (Ws) for the terms? The mathematician, Pascal, developed a useful device, *Pascal's triangle,* for obtaining the coefficients for the binomial expansion. Start your triangle with: 1

Add two 1s: 1   1

Add two more 1s: 1   2   1

and sum the previous 1s. Note that the last three numbers (1, 2, 1) are the coefficients for $(p + q)^2 = p^2 + 2pq + q^2$. Keep adding 1s on the outside and summing the numbers from the previous row:

                      coefficients for $N_T = 3$

Continuing this procedure results in a triangle (Table 5.2) that enables us to find the coefficients for $N_T = 6$.

TABLE 5.2
Pascal's triangle for $N_T = 6$

N_T				1				$\Sigma = T$
1				1 1				2
2				1 2 1				4
3				1 3 3 1				8
4				1 4 6 4 1				16
5				1 5 10 10 5 1				32
6				1 6 15 20 15 6 1				64

Notice that in Pascal's triangle: (a) the second diagonal on the left and right is 1, 2, 3, 4, 5, 6; (b) the second coefficient in each row always equals N_T; and (c) the sum of the coefficients across the rows is equal to T, the total number of possible outcomes. It should also be pointed out that $T = (2)^{N_T}$. For example, $T = (2)^6 = 64$.

Knowing the coefficients, we can complete the binomial expansion for $N_T = 6$. Beneath the terms are the outcomes.

$$p^6 + 6p^5q + 15p^4q^2 + 20p^3q^3 + 15p^2q^4 + 6pq^5 + q^6$$
$$\text{(6H)} \quad \text{(5H, 1T)} \quad \text{(4H, 2T)} \quad \text{(3H, 3T)} \quad \text{(2H, 4T)} \quad \text{(1H, 5T)} \quad \text{(6T)}$$

Figure 5.2 displays the binomial distribution for $N_T = 6$.

The heights of the bars in Figure 5.2 are the *probabilities* for various outcomes. Again, these probabilities can be calculated by either of the two methods described above for the binomial distribution for $N_T = 2$. The $P(6H) = W/T = 1/64 = .0156$. By the other method, the $P(6H) = p^6 = (.5)^6 = .0156$. What is the probability of getting five or more heads in six flips? We must find $P(5H,1T \text{ or } 6H)$. The P is $P(5H,1T) + P(6H) = 6/64 + 1/64 = 7/64 = .1094$. What is the probability of getting six heads or six tails in six flips? $P(6H \text{ or } 6T) = P(6H) + P(6T) = 1/64 + 1/64 = 2/64 = .0312$. Doing the last two problems by the second method, $P(5H,1T \text{ or } 6H) = 6p^5q + p^6 = 6(.5)^5(.5) + (.5)^6 = 6(.5)^6 + (.5)^6 = .0938 + .0156 = .1094$, as before. The $P(6H \text{ or } 6T) = p^6 + q^6 = (.5)^6 + (.5)^6 = .0156 + .0156 = .0312$, as before.

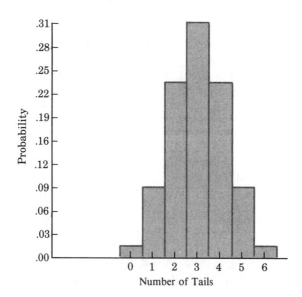

FIGURE 5.2
The binomial distribution for $N_T = 6$.

SEC. 5.3 □ TESTING HYPOTHESES WITH THE BINOMIAL DISTRIBUTION

The latter two "or" problems demonstrate the *additive rule of probabilities*:

$$P(A \text{ or } B) = P(A) + P(B) - P(AB)$$

where $P(AB)$ is the probability of both A and B occurring. Since a coin cannot be both a head and a tail on one trial, $P(AB) = 0$. Notice that *or* and the *addition* of probabilities go together. The $P(6H)$ is an example of what is called the *multiplicative rule of probabilities*:

$$P(A \text{ and then } B) = P(A) \times P(B)$$

In the case of $P(6H)$, $(.5)(.5)(.5)(.5)(.5)(.5) = (.5)^6 = .0156$. Notice that *and* and the *multiplication* of probabilities go together. Computing probabilities can quickly become mind-boggling. We will, however, be testing hypotheses involving a variety of statistics in the remainder of the book. These hypothesis tests, of necessity, require knowing probabilities, but fortunately the probabilities will generally be read from the tables for various theoretical distributions.

Now that we know how to determine binomial probabilities we can apply them to test statistical hypotheses. Imagine that the hustler's coin turned up heads six times in succession. Suppose further that the professor had watched the hustler at work previously and suspected that the lucky coin was loaded to turn up heads. This suspicion (a loaded coin) is the professor's research hypothesis and generates the alternative hypothesis, H_1. The alternative hypothesis, H_1, is stated so it coincides with the research hypothesis, and it is convenient to formulate H_1 first.[2] In this case, H_1 is $P(H) > .5$. The null hypothesis, H_0, is that the coin is unbiased and is often stated as $P(H) = .5$. However, we prefer to formulate H_0 as $P(H) \leq .5$. If fewer heads than tails are observed, we do not need to do a statistical test. This outcome makes H_1 untenable. In formulating H_0 note that the greater than or less than sign is reversed from the sign in H_1. We will set $\alpha = .05$. The test statistic is 6H and the $P(6H)$ with an unbiased coin ($p = q = .5$) is .0156. Since the $P(6H) < \alpha$, the decisions are to reject H_0 and accept H_1. Observing six heads in a row is too unusual an event for us to believe that the coin is unbiased ($p = $

2. While it is convenient to state H_1 first, H_0 is evaluated first. The decision regarding H_0 determines the decision regarding H_1.

$q = .5$). To repeat: The criterion for judging an event to be unusual is α, the criterion of significance. Instead of believing that the coin is unbiased (H_0), we assert that it is a loaded coin that has a probability of greater than .5 of turning up heads (H_1). Is the probability of the coin's turning up heads .7, .9, or what? We don't know the probability, we only know that it is greater than .5. Generalizing to a population of coin tosses with this coin, we would expect the professor to do a lot of buying if the game continued.

We may summarize the steps in the inference procedure that we have just completed as follows:

1. State H_0 $H_0: P(H) \leq .5$
2. State H_1 $H_1: P(H) > .5$
3. Set α $\alpha = .05$
4. Find P of the test statistic $P(6H) = .0156$
5. Compare P with α $.0156 < .05$
6. Make the decisions   reject H_0, accept H_1

When we test the hypotheses regarding the outcome of observing six heads in a row, we are performing a *one-tailed* or *one-sided test of significance*. The formulation of H_1 designated an experimental outcome in a *specific direction*—getting heads. In addition, when calculating the probability of the test statistic, $P(6H)$, only one tail of the binomial distribution for $N_T = 6$ was applied. In the six heads problem the left tail of Figure 5.2 is used to get the probability. One-tailed tests are proper: (a) when an investigator predicts the *direction* of an outcome from a theory prior to an experiment, or (b) when an investigator is replicating a previous finding. When researchers do not have sufficiently articulated theories to generate directional predictions or they are simply looking for something, whatever its direction, a *two-tailed* or *two-sided test of significance* is appropriate. One-tailed tests possess the advantage of greater statistical power. Their disadvantage is that an experimental outcome in the *opposite* direction to that predicted is evidence for the null hypothesis.

We will now perform a two-tailed test on coin tossing to demonstrate the consequences of this procedure for H_0, H_1, and the probability of the test statistic. Once again imagine that six heads turned up in a row. Our research hypothesis is simply that the coin is loaded. Unlike the preceding example there is no hypothesis that it is loaded to turn up in a specific way. Accordingly, we will accept the alternative hypothesis, H_1, when either too many heads occur or too many tails occur. H_1, under these conditions, is that $P(H) \neq .5$. This H_1 subsumes two possibilities: $P(H) > .5$ and $P(T) > .5$. H_0 is also altered from the one-tailed test. H_0 is $P(H) = .5$. Finally, the probability of the test statistic (6H) is modified. Instead of calculating the $P(6H)$ as before, we will compute the $P(6H \text{ or } 6T)$ because either of these outcomes would cause us to reject H_0. The probability of the test statistic is thus based upon *both* tails of the binomial distribution (Figure 5.2) for $N_T = 6$ and $p = q = .5$. We will again impose an $\alpha = .05$. Here, then, are the steps in doing this two-tailed test:

1. State H_0 H_0: $P(H) = .5$
2. State H_1 H_1: $P(H) \neq .5$
3. Set α $\alpha = .05$
4. Find P of the test statistic $P(6H \text{ or } 6T) = .0312$
5. Compare P with α $.0312 < .05$
6. Make the decisions   reject H_0, accept H_1

Because $P(6H \text{ or } 6T) < \alpha$, the H_0 is rejected and H_1 is accepted. Again, we would decide that the coin is loaded.[3] Note also that if the coin had turned up tails six times in a row the decisions to reject H_0 and accept H_1 would be the same with a two-tailed test.

Let us briefly consider one more coin-tossing example like the last problem. The H_0, H_1, and α will be the same, but the outcome is different: Five heads and one tail occur in six flips of the coin. When performing a two-tailed test the test statistic is $P(5H,1T \text{ or } 6H \text{ or } 5T,1H \text{ or } 6T)$. The $P(5H,1T \text{ or } 6H) = 6/64 + 1/64 = .1094$. Likewise the $P(5T,1H \text{ or } 6T) = 6/64 + 1/64 = .1094$. Remembering that *or* means *addition* of probabilities, the probability of the test statistic is $.1094 + .1094 = .2188$. Since the binomial distributions are symmetrical when $p = q = .5$, the

---

3. Earlier in the chapter we asserted that Type 1 and Type 2 errors are inversely related. If an $\alpha = .01$ had been imposed instead of $\alpha = .05$, then H_0 would not have been rejected. The use of $\alpha = .01$ instead of $\alpha = .05$ *reduces* the chance of making a Type 1 error, but *increases* the chance of not rejecting H_0 and thereby committing a Type 2 error.

probability for one tail, such as $P(5H,1T$ or $6H)$, can be found and then *doubled* for a two-tailed test: $2(.1094) = .2188$. This example demonstrates an important point in testing hypotheses: The probability of a test statistic is the probability of the observed event, such as $P(5H,1T)$, *plus* all other possible, unusual events, such as $P(6H)$. To state this another way, a one-tailed test uses *one particular tail area* of the binomial distribution, and a two-tailed test uses *both tail areas*. In this two-tailed test the probability of the test statistic $(.2188) > (.05)$, so the decision is not to reject H_0. Observing five heads in six tosses is not sufficiently unusual for us to reject the hypothesis of an unbiased coin.

So far we have solved binomial problems by expanding the binomial and calculating the probabilities for various outcomes. A second and simpler method is to read the probability for an outcome from a table of a binomial distribution for the appropriate N_T and value of p. A set of binomial tables can be found in Hollander and Wolfe (1973) and in other statistics books. Another source is to direct a computer to print such tables. For example, the MINITAB statistical package (Ryan, et al., 1981) is programmed to generate binomial tables (see the addendum to this chapter). We will consider an illustration designed to assist you in reading binomial tables. Let us take a section (Table 5.3) from a binomial table (Hollander & Wolfe, 1973, p. 262) for $N_T = 6$ and $p = .5$. Since you are familiar with these probabilities, it will help you to understand the table. The table consists of the first two columns. Here b refers to the outcomes like 3H, 4H, and so on in six tosses. You will recall that $P(6H) = .0156$. What is cP for $b = 5$? It is the probability of getting 5 heads or more. Thus, the probabilities (cP) in column 2 are *one-tailed* and *cumulative*. In column 3 we have inserted the probabilities (P) in *noncumulative* form. Here the P for $b = 5$ is the probability of obtaining *exactly* five heads and one tail in six flips. The probabilities in column 2 are found by accumulating the probabilities from column 3:

TABLE 5.3
Binomial probabilities for $N_T = 6$ and $p = .5$

b	$N_T = 6$ cP	$p = .5$ P
3	.6563	.3125
4	.3438	.2344
5	.1094	.0938
6	.0156	.0156

.1094 = .0156 + .0938; .3438 = .0156 + .0938 + .2344; and
.6563 = .0156 + .0938 + .2344 + .3125. The important points are
the following:

1. When consulting binomial tables you must ascertain whether the probabilities are cumulative or not. Usually, in testing statistical hypotheses, cumulative probabilities are employed. Thus, if the table displays noncumulative values, you will have to accumulate them.
2. The probabilities in column 2 are appropriate for one-tailed tests. For a two-tailed test the tabled values would have to be doubled: the $P(6H) = .0156$ (one tailed); the $P(6H \text{ or } 6T) = .0156 + .0156 = .0312$ (two-tailed.)

TESTING HYPOTHESES WITH THE NORMAL DISTRIBUTION

5.4

The binomial distribution for $p = q = .5$ becomes increasingly like a normal distribution as N_T increases. An implication of this fact is that the normal distribution described in Chapter 4 can be used to obtain approximate binomial probabilities. Ferguson (1981, p. 106) showed that when $N_T \geq 10$ the differences between the exact binomial probabilities and approximate probabilities from the normal distribution are trivial in size. How can we find approximate binomial probabilities from the normal distribution? In Chapter 4 it was stated if a batch of numbers is normally distributed, then their Z scores will conform to the same distribution. Thus, if the outcome of a binomial-type problem (e.g., 6H) is standardized by converting it to a Z score, the probability of the outcome can be determined in Table D for the unit normal distribution. To convert the outcome to a Z score we must know the μ and σ for the particular binomial distribution. If $\mu = N_T p$, then when $p = .5$, $\mu = N_T/2$. If you examine Figure 5.2, you will see that the center is 3 or $N_T/2$. Theoretical statisticians have determined that the σ for a binomial distribution is

$$\sigma = \sqrt{N_T pq}$$

When $p = q = .5$

$$\sigma = \sqrt{\frac{N_T}{4}}$$

The Z score formula when $p = q = .5$ then becomes

$$Z = \frac{X - \mu}{\sigma}$$

$$Z = \frac{\left|X - \frac{N_T}{2}\right| - .5}{\sqrt{\frac{N_T}{4}}}$$

You are probably puzzled over the $-.5$ in the formula. It is a *correction for continuity*. The binomial distribution is discrete; the normal distribution is a continuous function. The $-.5$ corrects for this discrepancy.

Let us try the normal curve method with an $N_T = 10$ to evaluate the adequacy of the approximation. The outcome of $N_T = 10$ coins flips is $X = 10$ tails in succession. The null hypothesis is H_0: $P(H) = .5$ and the alternative hypothesis is H_1: $P(H) \neq .5$. Therefore, a two-tailed test is being performed, and this time we will set $\alpha = .01$. For this problem,

$$Z = \frac{\left|X - \frac{N_T}{2}\right| - .5}{\sqrt{\frac{N_T}{4}}}$$

$$= \frac{\left|10 - \frac{10}{2}\right| - .5}{\sqrt{\frac{10}{4}}}$$

$$= \frac{4.5}{\sqrt{2.5}} = 2.85.$$

In Table D the $P(Z = 2.85)$ is found in column C to be .0022. Because a two-tailed test is being done, the probability must be doubled: $2(.0022) = .0044$. Since $P < \alpha$, H_0 is rejected and H_1 is accepted.

How does this approximate probability compare with the exact probability from the binomial distribution? There is only one way 10 tails can occur in $N_T = 10$ trials and only one way that 10 heads can occur. Thus, $W = 1 + 1 = 2$. By chance the total number of possible outcomes is

$$T = (2)^{N_T}$$
$$= (2)^{10}$$
$$= 1024$$

Accordingly, the exact two-tailed binomial probability is

$$P = \frac{W}{T}$$
$$= \frac{2}{1024}$$
$$= .002$$

By the alternative method the probability is the same:

$$P = p^{10} + q^{10}$$
$$= (.5)^{10} + (.5)^{10}$$
$$= .002$$

The approximate probability derived from the normal curve (.004) differs little from the exact probability (.002) from the binomial distribution. As N_T becomes larger the difference in the two probabilities should become even smaller. The correction for continuity is especially critical when N_T is small as in the last example.

TESTING HYPOTHESES ABOUT CENTERS, SPREADS, AND OTHER STATISTICS

In the cantina examples the outcome or test statistic was the number of heads observed in so many flips of a coin. To decide whether the coin was biased, the probability of the outcome by chance was obtained from a theoretical distribution, the binomial distribution. When the probability of the test statistic was found equal to or less than an arbitrary criterion of significance, α, we concluded that the outcome represented something other than chance. When the probability of the test statistic was found to be greater than α, we did not reject a chance explanation of the outcome.

Instead of dealing with the number of heads, we may be concerned with the centers of two batches of numbers; for example, two means. Again we can ask: Is the difference in the means too large to have occurred by chance? Or, we might want to compare the spreads of two batches, such as two variances. Is the differ-

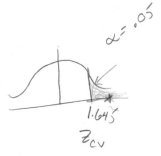

ence in the variances merely chance variation or did something happen to make one spread unusually larger than the other? To answer these questions we will go through the same decision procedure as before. We must state H_0 and H_1 and set α. We must calculate different test statistics and determine their probabilities of chance occurrence by reference to various theoretical distributions like the normal distribution, the t distribution, the F distribution, and so on.

As Hinkle et al. (1979) showed, test statistics often follow a general formula:

$$\text{test statistic} = \frac{\text{a statistic} - \text{a parameter}}{\text{random variability}}$$

For example, the $Z = (X - \mu)/\sigma$ paradigm followed in the last binomial problem fits this general formula. Notice that the denominator, random variability or "noise," is a measure of spread. As was promised, spread is central in statistical inference.

When the probability of a test statistic is compared to α, the object is to reach decisions about the null and alternative hypotheses. Furthermore, we want to go beyond samples and talk about populations. We asserted early that the aim of applied statistics is to make sense of numbers. The numbers are hooked to reality and our ultimate goal is to find out what is happening out there. Statistics is a tool to help us attain this goal and inference plays a major part in the undertaking.

SUMMARY

A research hypothesis is translated into the alternative hypothesis, H_1. The hypothesis of a chance outcome is translated into the null hypothesis, H_0. The probability (P) of the test statistic, for example, the number of heads, is compared to an arbitrary criterion of significance, α. When $P > \alpha$, H_0 is not rejected. When $P \leq \alpha$, H_0 is rejected and H_1 is accepted. The probability (P) of an outcome is defined as W/T, where W is the number of ways an event can occur and T is the total number of possible mutually exclusive and equally likely outcomes.

An examination of the decision-reality matrix reveals that whenever a statistical decision is made there are two ways to err and two ways to be correct. If H_0 is true and the decision is not to reject H_0, a correct decision has been reached. If H_0 is false and the decision is to accept H_1, another correct decision has been reached. If H_0 is true and H_0 is rejected, a Type 1 error of inference has been committed. If H_1 is true and H_0 is not rejected, a Type 2 error has been committed. The probability of a Type 1 error is α; the probability of a Type 2 error, β. The probability of a correct decision when H_0 is true is $1 - \alpha$; the probability of a correct decision when H_1 is true is $1 - \beta$. The latter probability, $1 - \beta$, is also known as statistical power—the chances of finding something when it is there to be found. Type 1 and Type 2 errors are inversely related. The criterion of significance, α, is usually set at .05. If α is relaxed to .10, Type 1 errors will be increased and Type 2 decreased. If α is tightened to .001, Type 1 errors will be decreased and Type 2 errors increased.

Problems with a dichotomous outcome like heads or tails are solved by reference to probabilities obtained from a theoretical distribution, the binomial distribution. The exact probabilities may be obtained by expanding the binomial and by using binomial tables. Approximate probabilities may be found by using the normal distribution. Examples of testing hypotheses for various outcomes of coin tossing were presented. Pascal's triangle was presented as a technique to get coefficients for the terms in the binomial expansion. One- and two-tailed tests of significance were explained. In a one-tailed test an experimental outcome in a specific direction is tested and only one tail of the theoretical distribution is used to find the probabilities. A two-tailed test is nondirectional and both tails of the theoretical distribution provide the probabilities. One-tailed tests are more powerful, suited to testing specific directional predictions from a theory, and appropriate for assessing replications. One-tailed tests pose difficulties when an experimental outcome is in the opposite direction to that predicted. Two-tailed tests are best suited to exploratory or atheoretical research endeavors.

In examples of computing binomial probabilities, two probability rules were described. When A or B happens, the probabilities of A and B are summed—the additive rule of probabilities. When A happens and then B happens, the probabilities of A and B are multiplied—the multiplicative rule of probabilities. *Or* and addition go together; *and* and multiplication go together.

Binomial probabilities can be found from tables in books or generated by computer programs. In reading these tables it is important to ascertain whether the probabilities are one- or two-tailed and whether they are cumulative or noncumulative. Generally, cumulative probabilities, such as the probability of five heads or more, are utilized in testing statistical hypotheses. If the tables include probabilities for one tail of the binomial distribution, they are directly applicable for one-tailed tests of significance; for a two-tailed test of significance, the probabilities must be doubled.

When the number of binomial trials, N_T, is equal to or greater than 10 and the probability of one outcome equals the probability of the other ($p = q = .5$), the standard normal distribution can be used to compute approximate binomial probabilities. The approximate probability is determined by a Z score transformation of a test statistic, such as 10 heads in 10 trials. The proportion of the tail area(s) of the unit normal distribution corresponding to the Z is the approximate probability of the test statistic.

Much of the book will be concerned with testing hypotheses about other statistics—comparing two means, two variances, and so on. The same inference procedure as for binomial problems will be followed except that other test statistics will be computed and they will be related to various theoretical distributions such as the t distribution, F distribution, and so on. Test statistics often follow a general formula: test statistic = (a statistic − a parameter)/random variability. Hypothesis testing and inference from samples to populations are a large part of applied statistics, the aim of which is to make sense of numbers. The numbers represent things and the ultimate goal is to understand phenomena.

PROBLEMS

1. Expand the binomial, $(p + q)^8$. Construct a Pascal's triangle to obtain the coefficients for the expansion. State a rule for obtaining the powers for p and q.
2. For Problem 1 show that T from the triangle for $N_T = 8$ is equal to $(2)^{N_T}$.
3. Assuming an unbiased coin ($p = q = .5$), what is the probability of getting 8H in eight tosses? 7H or more? 8H or 8T? 7H or more *or* 7T or more?
4. Answer the same questions as in Problem 3 for a coin in which the probability of getting a head is $p = .7$.
5. A rat has nine test trials in a T maze. From the rat's prior training the investigator predicts that the rat should turn right more often than chance. Test the prediction with $\alpha = .05$. Go through the six steps in the inference procedure in Section 5.3 for an outcome of seven right turns and two left turns. Repeat the procedure for the outcome of eight right turns and one left turn.

6. A gambler buys a "lucky" coin that is guaranteed to be loaded. He does not know whether it is loaded to turn up head or tails. He tosses the coin 12 times and observes 10 tails and 2 heads. Using $\alpha = .05$, can he reject the hypothesis that $P(H) = .5$? Enumerate the six steps in the inference procedure.

7. A public relations company does a test on the riding comfort of two luxury automobiles. Fifty blindfolded subjects ride in two luxury cars, L and C. After their two rides the subjects report which ride was more comfortable. The company predicts that the L car will give a smoother ride than the C car. The outcome was 32L and 18C. Test the company's hypothesis at $\alpha = .05$ following the six steps in the inference procedure.

8. A self-proclaimed psychic contends that he can "see through" cards. He (a) correctly draws a black card from a standard deck, (b) an ace, (c) a black ace, and (d) the 7 of clubs. What are the probabilities of each of these four events? (The cards are returned to the deck after each draw.)

9. What is the probability of getting either six dots or one dot in a single throw of a die? What is the probability of getting six dots and one dot in that order on two throws of a die?

10. In Problem 8, what is the probability of the "psychic's" doing the four events in succession? Suggest an alternative hypothesis to seeing through the cards.

ADDENDUM: USING COMPUTERS IN STATISTICS

As stated in the chapter, one source for binomial probabilities is a computer-produced table. Here, MINITAB has generated the probabilities for $N_T = 4$ (see 1), $N_T = 6$ (see 2), and $N_T = 10$ (see 3) when $p = .5$. The probabilities in column 2 are presented in noncumulative form. From the chapter you know these are one-tailed values. In column 3 the probabilities have been accumulated. If you have mastered Chapter 5 you should be able to do problems like: What is the probability of getting 10 heads or 10 tails in 10 tosses? What is the probability of getting 9 or more heads in 10 tosses? I get answers of .002 and .0107, respectively. Do you? MINITAB will also generate tables for other values of p.

```
--BINOMIAL PROBABILITIES FOR N=4, P=.5
  BINOMIAL PROBABILITIES FOR N = 4 AND P = 0.500000
    K       P(X = K)       P(X LESS OR = K)
    0       0.0625         0.0625
    1       0.2500         0.3125
    2       0.3750         0.6875                      (1)
    3       0.2500         0.9375
    4       0.0625         1.0000
--BINOMIAL PROBABILITIES FOR N=6, P=.5
  BINOMIAL PROBABILITIES FOR N = 6 AND P = 0.500000
    K       P(X = K)       P(X LESS OR = K)
    0       0.0156         0.0156
    1       0.0938         0.1094
    2       0.2344         0.3438
    3       0.3125         0.6563                      (2)
    4       0.2344         0.8906
    5       0.0938         0.9844
    6       0.0156         1.0000
--BINOMIAL PROBABILITIES FOR N=10, P=.5
  BINOMIAL PROBABILITIES FOR N = 10 AND P = 0.500000
    K       P(X = K)       P(X LESS OR = K)
    0       0.0010         0.0010
    1       0.0098         0.0107
    2       0.0439         0.0547
    3       0.1172         0.1719
    4       0.2051         0.3770
    5       0.2461         0.6230                      (3)
    6       0.2051         0.8281
    7       0.1172         0.9453
    8       0.0439         0.9893
    9       0.0098         0.9990
   10       0.0010         1.0000
```

CHAPTER OUTLINE

6.1 SO YOU HAVE TWO NUMBERS FOR EVERY CASE

6.2 FINDING AN INDEX—TO SUMMARIZE A RELATIONSHIP

6.3 AN EXAMPLE—SELLING ICE CREAM AND LIKING KIDS

6.4 IS OUR RELATIONSHIP JUST A CHANCE AFFAIR?

6.5 SUPPOSE YOU HAVE A PAIR OF RANKS FOR EVERY CASE?

CHAPTER SIX
RHO RHO RHO THE BOAT

SO YOU HAVE TWO NUMBERS FOR EVERY CASE

6.1

Did you ever watch someone stumble, fall, and end up on his or her derrière? What is the first thing the person does? The person looks back and attempts to find out why it happened. Human beings constantly try to relate things—an obstacle or hole and falling, for example. Similarly, in science there is a beginning basic question: "Is there a relationship between the two variables?" (Cook & Campbell, 1979, p. 39).

To this point, we have been mainly concerned with single variables like time estimates or coin tosses. Now it is a new game—we have *two variables*.[1] And for each subject or *case* we have two numbers instead of one. Let us denote the variables X and Y. The two numbers for each case are the *values* for the two *variables*, X and Y. Our new question is this: *Are X and Y related*? What does "related" mean? It means that X and Y covary: As X changes, Y changes.

We have N cases with an X number and Y number for each case. X and Y can be variables measured on the same scale, like the time estimates for trial 1 and trial 2, or more commonly X and Y can be variables measured on different scales, like the Xs being aptitude and the Ys being job performance.

Other parts of the book have stressed the informative value of pictures of numbers. Pictures are also an extremely valuable first step in deciding whether two variables are related. The type of picture for such *bivariate data* is called a *scatter plot* or *scatter diagram*. Table 6.1 contains a hypothetical set of numbers for 20 computer programmers. The Xs are aptitude test scores for the programmers and the Ys are ratings, on a 20-point scale, of overall job performance (higher ratings = better performance).

These numbers can be put into a scatter plot (Figure 6.1) by placing a dot at the intersection of $X = 25$, $Y = 7$, a dot at $X = 21$ and $Y = 5$, etc. The numbers along the X axis increase from left to right; the numbers along the Y axis increase from the bottom to the top. What about the dots? Inspection of the dots reveals that as X *increases, Y increases*. This type of relationship is called a *direct relationship*, and X and Y are said to be *positively correlated*. Figure 6.2 shows a hypothetical relationship between X, the ages of 20 data entry operators, and Y, the ratings of their

1. A variable is something—a property or phenomenon—that varies. That is, a variable can have different values. Height is a variable. Different heights like 5'5", 5'2", and so on are values.

TABLE 6.1
Aptitude (*X*) and performance (*Y*) scores for 20 computer programmers

Case	X	Y	Case	X	Y	Case	X	Y	Case	X	Y
1	25	7	6	32	11	11	36	9	16	35	12
2	21	5	7	38	10	12	43	14	17	45	13
3	46	14	8	40	13	13	31	10	18	47	15
4	28	7	9	23	6	14	39	12	19	34	11
5	37	12	10	30	8	15	33	9	20	29	10

overall job performance. What do the dots reveal here? That is correct, as *X* increases, *Y* decreases. This is called an *inverse relationship*, and *X* and *Y* are said to be *negatively correlated*. Figure 6.3 is something else. In it, increases in *X* are associated with both increases and decreases in *Y* and vice versa. In other words, *X* and *Y* are *not correlated*; *X* and *Y* are *independent*. If you do research in the behavioral sciences, you will encounter quite a bit of this discouraging stuff!

Again, we have observed that turning numbers into pictures can be beneficial. But scatter plots, carefully eyeballed, can tell us even more.[2] Look at Figures 6.4, 6.5, and 6.6. What do they dis-

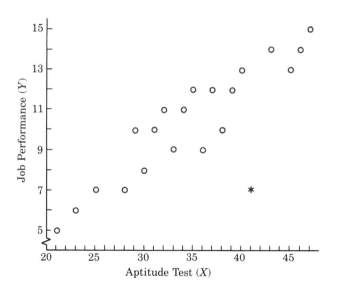

FIGURE 6.1
A scatter plot of the relationship between aptitude (*X*) and job performance (*Y*)

2. Tufte (1983), a specialist in the visual display of numbers, regards scatter plots as "the greatest of all graphical designs."

FIGURE 6.2
A scatter plot of the relationship between age (X) and job performance (Y)

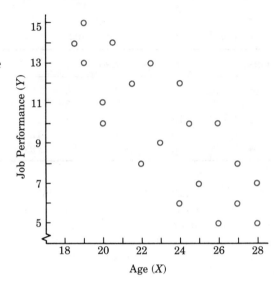

close? These three figures convey information about the *strength of the relationship*. While in all three instances X and Y are positively correlated, what is more notable is that the strength of the relationship increases as we go from Figure 6.4 to Figure 6.6.

Figure 6.7 is both revealing and important. It is a *nonlinear* or *curvilinear relationship*. Here, as X increases, Y increases up to

FIGURE 6.3
A scatter plot showing no relationship between X and Y

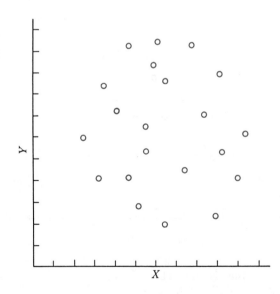

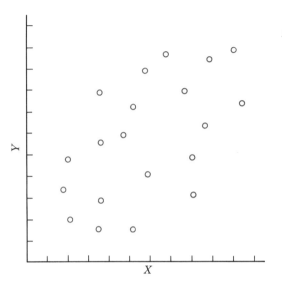

FIGURE 6.4
A scatter plot showing a low relationship between X and Y

a point; but further increases in X are associated with *decreases* in Y. In this book we will only analyze linear relationships between X and Y. Linear relationships are those in which a straight line "fits" the points in a scatter plot (the term *fits* will be defined later). The methods to be presented are entirely inappropriate for scatter plots like Figure 6.7. Although nonlinear relationships can

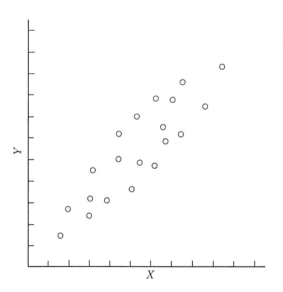

FIGURE 6.5
A scatter plot showing a high relationship between X and Y

FIGURE 6.6
A scatter plot of a perfect positive relationship between X and Y

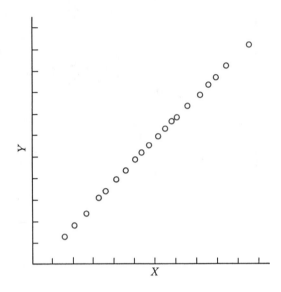

be analyzed, we will not describe the statistical techniques for such relationships. Lastly, eyeballing a scatter plot can disclose another crucial item of information. Look again at Figure 6.1. We have added another case and singled it out with an asterisk instead of a dot. This is an *outlier*. Just as outliers may distort centers and spreads, they can have serious consequences in corre-

FIGURE 6.7
A scatter plot of a curvilinear relationship between X and Y

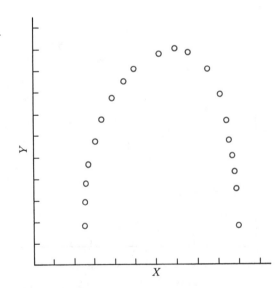

lation. And outliers pose the same troublesome questions of detection, interpretation, and what to do with them.

The new game is a fundamental one in science—finding out if two variables are related. A first step in that search is to put your pairs of numbers into a scatter plot. Note that you are thereby reducing two batches of numbers to one batch of dots—simplifying again. What do we gain from the picture of our data? We obtain eyeball information regarding (a) the direction of the relationship between X and Y—direct or inverse (or none); (b) the strength of the relationship; (c) whether the relationship is linear or curvilinear; and (d) presence of outliers. It pays to plot!

FINDING AN INDEX TO SUMMARIZE A RELATIONSHIP

6.2

Scatter plots provide helpful information. Once more, however, we can profitably go beyond visual inspection and calculate a quantitive index that summarizes the relationship between two variables, X and Y. Moreover, the scatter plots of X and Y numbers are deficient because the two variables are usually measured on different scales. To secure a clearer picture of the relationship and an index of the relationship, the X and Y numbers must be measured on a common scale. In Chapter 4, Z or standard scores were introduced as a device to achieve a common scale with a $\overline{Z} = 0$ and $s_z = 1$. If the X and Y numbers are converted to Z scores, then an index of relationship—*the Pearson product-moment r*—can be calculated. The formula for r, the *correlation coefficient*, is

$$r = \frac{\Sigma Z_x Z_y}{N - 1}$$

where N is the *batch count*. For a population the corresponding parameter is ρ (a small rho) and the formula is

$$\rho = \frac{\Sigma Z_x Z_y}{N_P}$$

where N_P is the number of cases in the population. The coefficient, r, is an *estimate* of ρ based on a sample (Draper & Smith, 1966). In converting X and Y numbers to Z scores s_x and s_y are utilized.[3] Let us demonstrate the limits of r with two examples. In

---

3. Another formula for r is $r = \Sigma Z_x Z_y / N$. If sd_x and sd_y are incorporated to transform X and Y to Z scores, this formula will produce the same values for r as the earlier formula

Table 6.2 X and Y are perfectly related ($Y = 10X$) and the means ($\overline{X}$ and $\overline{Y}$) and standard deviations (s_x and s_y) have been calculated to transform the X and Y numbers into Z scores.

TABLE 6.2
Computing r from standard scores

X	Y	Z_x	Z_y	Z_xZ_y
1	10	−1.3888	−1.3888	1.9288
2	20	−0.9258	−0.9258	0.8571
3	30	−0.4629	−0.4629	0.2143
4	40	0.0000	0.0000	0.0000
5	50	0.4629	0.4629	0.2143
6	60	0.9258	0.9258	0.8571
7	70	1.3888	1.3888	1.9288
$\overline{X}$	$\overline{Y}$	s_x	s_y	$\Sigma Z_x Z_y$
4	40	2.1602	21.6025	6.0004

In column three and four the Z_x and Z_y values are found as follows:

$$Z_x = \frac{X - \overline{X}}{s_x}$$

$$= \frac{1 - 4}{2.1602}$$

$$= -1.3888$$

$$Z_y = \frac{Y - \overline{Y}}{s_y}$$

$$= \frac{10 - 40}{21.6025}$$

$$= -1.3888$$

Using these values,

$$r = \frac{\Sigma Z_x Z_y}{N - 1}$$

$$= \frac{6.0004}{7 - 1}$$

$$= 1.00$$

For a perfect positive relationship $r = 1.00$ and it is one limiting value of r.

If the Y scores are reversed in Table 6.2, Table 6.3 results.

X	Y	Z_x	Z_y	Z_xZ_y
1	70	−1.3888	1.3888	−1.9288
2	60	−0.9258	0.9258	−0.8571
3	50	−0.4629	0.4629	−0.2143
4	40	0.0000	0.0000	0.0000
5	30	0.4629	−0.4629	−0.2143
6	20	0.9258	−0.9258	−0.8571
7	10	1.3888	−1.3888	−1.9288
$\overline{X}$	$\overline{Y}$	s_x	s_y	ΣZ_xZ_y
4	40	2.1602	21.6025	−6.0004

TABLE 6.3
Computing *r* from standard scores

The *X* and *Y* scores are converted to *Z* scores in the same manner as above. In this case of a perfect negative relationship

$$r = \frac{\Sigma Z_x Z_y}{N - 1}$$
$$= \frac{-6.0004}{7 - 1}$$
$$= -1.00$$

Thus, with a perfect inverse relationship the second limiting value of *r* is −1.00. In brief, the limits of *r* are

$$-1.00 \leq r \leq 1.00$$

When the paired Z_x and Z_y values are identical and have the same signs, *r* = 1.00. When the paired Z_x and Z_y values are identical and have the opposite signs, *r* = −1.00. And when the paired Z_x and Z_y values and signs are unsystematically related, *r* = 0.00. As the values and signs become increasingly related, the size of *r* will increase. Since the limits of the Pearson product-moment *r* are 1.00 and −1.00, a computed *r* larger than 1.00 or smaller than −1.00 is an immediate signal for you to search for errors.

The *sign of r* indicates the *direction of the relationship:* + is a direct relationship; − is an inverse relationship. The *size of r* shows the *strength of the relationship*. Consider *r* = .20 versus *r* = .84. Although both signify direct relationships, the relationship is stronger for the *r* = .84. What does an *r* = 0.00 mean? Generally it is said that an *r* = 0.00 means that the variables in question are not related or are independent. If the scatter plot resembles Figure 6.3, then the statement is correct. But a scatter plot like

Figure 6.7, a curvilinear relationship, will also result in an r close to zero. (Applying r to a curvilinear relationship is a data-analysis error.) The r for Figure 6.7 compels us to say that an r of zero means that the two variables are not *linearly* related. This example again demonstrates the importance of making scatter plots.

The Z score formula for r given above is a *rational formula*. If you are a masochist, by all means apply this formula. Two things will probably happen: (a) you will become lost and (b) you will get the wrong answer. Often in correlation work batch sizes are large, perhaps in the hundreds. If you employ the rational formula, you must convert the numbers for each variable to Z scores with their different values and signs, obtain their products, and so on. Don't be fooled by the apparent simplicity of the Z score formula.

If we substitute some raw score formulas into the rational formula, a *computational formula* for r emerges:

$$r = \frac{N\Sigma XY - \Sigma X \Sigma Y}{\sqrt{[N\Sigma X^2 - (\Sigma X)^2][N\Sigma Y^2 - (\Sigma Y)^2]}}$$

Although this computational formula appears somewhat ghastly, it avoids the direct conversion of numbers to standard scores, and it will, in the long run, produce greater accuracy.

AN EXAMPLE—SELLING ICE CREAM AND LIKING KIDS

6.3

Our example is a pilot experiment done by Sam Speedup, an industrial psychologist. "Suggestion Box" Sam, as he is widely known, wanted to select vendors for the Crunchie Ice Cream Bar Company, Inc. Sam thought that liking little kids might be related in *some way* to selling ice cream bars. So he quickly developed and copyrighted a test, ILLK, for I-like-little-kids, and randomly selected 20 of Crunchie's present vendors. Each salesperson took the ILLK test (X) in which high scores = high liking, and Sam found out how many ice cream bars in units of 100 that each sold per day, on the average (Y). The scores for the 20 cases are presented in Table 6.4.

Figure 6.8 is a scatter plot of the numbers. It suggests a fairly strong but inverse relationship between liking kids and

SEC. 6.3 □ AN EXAMPLE—SELLING ICE CREAM AND LIKING KIDS

TABLE 6.4
ILLK (X) and sales (Y) for 20 vendors

Case	X	Y	XY	Case	X	Y	XY
1	4	5.75	23.00	11	14	2.00	28.00
2	5	4.50	22.50	12	15	3.25	48.75
3	7	3.50	24.50	13	16	1.00	16.00
4	8	5.00	40.00	14	17	2.50	42.50
5	9	3.75	33.75	15	17	3.25	55.25
6	9	2.25	20.25	16	18	4.25	76.50
7	11	5.00	55.00	17	19	3.50	66.50
8	12	2.25	27.00	18	20	1.50	30.00
9	13	3.25	42.25	19	20	2.00	40.00
10	14	4.00	56.00	20	22	0.75	16.50

sales. In listing the pairs of numbers in the table, they have been sorted from lowest to highest in terms of their X values. Is *sorting* on one variable helpful for eyeballing relationships? Why?

What now? Find ΣX and ΣX^2 with your calculator by entering the Xs in memory: 4 [M+] 5 [M+] ... 22 [M+]. Then recall ΣX and ΣX^2: [ΣX] 270 [ΣX²] 4170. Repeat for the Ys and recall ΣY and ΣY^2: [ΣX] 63.25 [ΣX²] 235.81. Next find ΣXY: 23 [M+] 22.5 [M+] ... 16.5 [M+]. Recall ΣXY: [ΣX] 764.25. (An alternative procedure

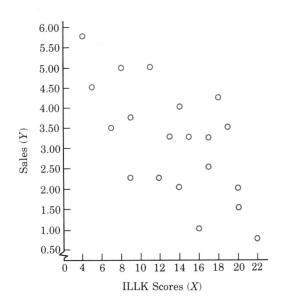

FIGURE 6.8
A scatter plot of the relationship between ILLK scores (X) and sales (Y)

would be to calculate all the *XY*s, put them in memory, and recall ΣXY.) We now have the necessary pieces, and we will substitute them into the computational formula for *r*:

$$r = \frac{N\Sigma XY - \Sigma X \Sigma Y}{\sqrt{[N\Sigma X^2 - (\Sigma X)^2][N\Sigma Y^2 - (\Sigma Y)^2]}}$$

$$= \frac{20(764.25) - 270(63.25)}{\sqrt{[20(4170) - (270)^2][20(235.81) - (63.25)^2]}}$$

$$= \frac{-1792.5}{\sqrt{(10500)(715.6375)}}$$

$$= \frac{-1792.5}{2741.203} = -.65$$

There is, indeed, a fairly strong, *inverse* relationship between liking little kids and sales. The outcome is also a bit surprising: The vendors wearing black hats sell more ice cream bars!

IS OUR RELATIONSHIP JUST A CHANCE AFFAIR?

6.4

There's that old question that lovers have been asking ever since the first furry pair set up housekeeping in a cave. To answer it, we must resort to hypothesis testing. Remember that "Suggestion Box" Sam did *not* predict a relationship between liking kids and selling ice cream bars in a particular direction. Therefore, a two-tailed test of *r* is appropriate. The alternative hypothesis, H_1, is that ρ (the correlation in the population between liking kids and sales of Crunchie bars) is *not* equal to zero. The null hypothesis, H_0, is that ρ is equal to zero, that liking kids and selling Crunchie Bars are not linearly related.

Now *we* come down to the crunch. What is the test statistic? A hypothetical sampling experiment provides an answer to this question. Imagine that we have a population of paired *X* and *Y* numbers that are uncorrelated; that is, ρ = 0. Let us draw 1000 random samples of *X* and *Y* numbers of size *N* from the population and calculate an *r* for each sample. A frequency distribution of these 1000 *r*s is called the *sampling distribution of the correlation coefficient*. If *N* is "very large," the sampling distribution will be a symmetric, bell-shaped normal distribution (Hays, 1963). It consists of infrequent, large negative correlations in the left tail decreasing in size but increasing in frequency to a center *r* of about zero, and then increasing in size but decreasing in frequency to infrequent, large, positive correlations in the right tail. When N is small, the sampling distribution of r is symmetrical but platykurtic. That is, it is pushed down in the center and has fatter tails than a normal distribution. More precisely, the sampling distribution of r follows a known theoretical distribution, the t *distribution*. Like the binomial distribution, the t distribution is a *family of distributions* that becomes increasingly normal as N increases. Since the sampling distribution of r follows the normal distribution only when N is very large, we will use the t distribution to evaluate the statistical hypotheses in the ice cream example.

The sampling distribution of the 1000 rs has a standard deviation. Since it is the standard deviation of a batch of statistics rather than for a sample, it is called the *standard error of the correlation coefficient*, s_r. What is s_r? It is a measure of the *random variability* of r because the rs are the result of drawing 1000 random samples. The test statistic for r follows the general formula suggested earlier for a test statistic:

$$\text{test statistic} = \frac{\text{a statistic} - \text{a parameter}}{\text{random variability}}$$

In this case the test statistic is

$$t = \frac{r - \rho}{s_r}$$

As $\rho = 0$, under the null hypothesis, it drops out of the numerator:

$$t = \frac{r}{s_r}$$

The standard error of r is

$$s_r = \frac{\sqrt{1 - r^2}}{\sqrt{N - 2}}$$

Therefore, the test statistic becomes

$$t = \frac{r\sqrt{N - 2}}{\sqrt{1 - r^2}}$$

We could determine t by putting the $r = .65$ and $N = 20$ into the formula. The two-tailed probability of the obtained t for

$df = N - 2 = 18$ could be found in Table A.[4] Then, the probability of this test statistic could be compared to α, and decisions could be made regarding H_0 and H_1.

Correlation coefficients have been around for a long time, however. We could perform the t test for r, but we do not have to. It has been done and tabled (Table B). Accordingly, we can test H_0 by referring r to Table B. We will set $\alpha = .05$. The probability of $r = \pm.65$ or $r = |.65|$ (recall that Sam's H_1 was a two-tailed hypothesis) can be found in Table B by looking down the left margin for $df = N - 2 = 18$ and across the top at "Level of significance for a two-tailed test." At the intersection of $df = 18$ and .01, the tabled value is $r = .561$. Therefore, our obtained $r = -.65$ has a $P < .01$. We may summarize the steps in testing the hypotheses as follows:

1. State H_0	H_0: $\rho = 0$
2. State H_1	H_1: $\rho \neq 0$
3. Set α	$\alpha = .05$
4. Find P for r	$P(r = -.65) < .01$
5. Compare P with α	$< .01 < .05$
6. Make the decisions	reject H_0, accept H_1

"Suggestion Box" Sam found a significant inverse relationship between sales of ice cream and liking children, and the presumption is that ρ is not zero in the population. Is Sam's r a Type 1 error? Or is he correctly rejecting H_0 when H_0 is false? Prior to mortgaging the farm on the result of this pilot study, wouldn't you like to know more? If another, larger random sample of vendors were tested, would the r reappear? Would the relationship be evident with Good Humor ice cream vendors? Would the relationship generalize to vendors selling other things to kids? The answers to these questions require additional trips down that old *replication* road. Tukey's comments are highly relevant to this situation: "The modern test of significance before which so many editors of psychological journals are reported to bow down, owes more to R.A. Fisher than to any other man. Yet Sir Ronald's standard of firm knowledge was not one very extremely significant result, but rather the ability to repeatedly get results significant at 5% [$P =$

---

4. The term *df* is an abbreviation for *degrees of freedom*. For evaluating rs the $df = N - 2$. Later we will define degrees of freedom in detail.

.05]" (1969, p. 85). To be on more solid ground, Sam must determine that the observed correlation coefficient is repeatedly replicable.

Before leaving the topic of testing hypotheses about correlation coefficients, we need to examine Table B further. Pretend that Sam had a detailed theory that predicted a negative correlation between liking and sales. Then, a one-tailed test of the hypotheses would be appropriate. In accord with the prediction, H_1 would be $\rho < 0$ and H_0 would be $\rho \geq 0$. What would the probability of the test statistic ($r = -.65$) be? Looking in Table B for $df = 18$ and then at the legend, "Level of significance for a one-tailed test" for .01, we observe a tabled value of $r = 0.516$. Notice that this required value is less than the corresponding two-tailed value of $r = .561$. That is why one-tailed tests are more powerful and researchers like one-tailed tests better. Actually, for a one-tailed test $P(r = -.65) < .005$. Why? Because the tabled value for .005 is $r = .561$.

Finally, let us consider one more example. Suppose that Sam tested 102 salespersons and found an $r = -.195$. For $df = 100$ and a two-tailed test, $P(r = -.195) = .05$. For a one-tailed test, $P(r = -.195) = .025$. Thus, as N increases, the size of r required for significance decreases. With large samples very small r values can be judged significant. We will have more to say about this matter in Chapter 7.

One final point on testing rs: To evaluate H_0 and H_1 we compared the probability of the test statistic with α. Equivalently, we could have found the *critical value of the test statistic* for the α level from Table B and compared our test statistic with it.[5] If the obtained r was greater than or equal to the critical value of r, then we reject H_0. If the obtained r is less than critical value of r, then H_0 is not rejected. For $df = 18$ and a two-tailed test at $\alpha = .05$, the tabled value is $r = -.444$. The obtained r of $-.65$ is greater than the critical value of $r = -.444$. Therefore, the decisions would be the same: Reject H_0 and accept H_1.

These alternative procedures for testing hypotheses raise a question that has bothered statisticians: How should significance be reported? Should the investigator report the actual probability of the test statistic, or should the investigator report that the test

5. The critical value of a test statistic is the value of the statistic required for the α level. In the present context, the critical value is r_α for a one- or two-tailed test and $df = N - 2$ as determined from Table B.

statistic is significant at the α level? Sam, for example, could either say that he found an $r = -.65$ ($P < .01$) or that the $r = -.65$ was significant at the .05 level. I lean toward reporting the actual probability of the test statistic. Suppose an investigator did not reject H_0 and $\alpha = .05$. Was $P = .30$, $P = .10$, $P = .06$, or what? If $P = .30$, then I probably would not pursue the problem. If $P = .06$ or $P = .10$, I would be more likely to. More important than this question, however, is the *attitude* taken toward a finding from a single study. The finding may be interesting or even exciting, but until it's been replicated a number of times it isn't too impressive. Given successful replications, the finding becomes more convincing. And the finding becomes even more impressive when it generalizes to other measures of the concepts, populations, and settings. Science, done correctly, is a tough and long row to hoe.

SUPPOSE YOU HAVE A PAIR OF RANKS FOR EVERY CASE?

6.5

In Chapter 1 we said we would analyze three kinds of numbers: frequencies, ranks, and real numbers. In Chapter 5 frequencies were analyzed. Now we will consider the first example in which the numbers are ranks. We still want to know if X and Y are related, but the values of X and Y are the *ranks* from 1 to N for each variable. There are two correlation coefficients for summarizing the relationship between X and Y when the numbers in each variable are ranks: *Spearman's rho coefficient* (r_s) and *Kendall's tau* (r_k). These two coefficients correlate highly. The r_k statistic is more versatile and has a smoother sampling distribution. The r_s is somewhat easier to calculate, it is a Pearson product-moment r for ranks, and tables for testing r_s are more widely available. Despite the advantages of r_k, we will limit the presentation to Spearman's r_s.

Table 6.5 contains two sets of ranks for $N = 6$ cases. One formula for r_s is

$$r_s = 1 - \frac{6\Sigma D^2}{N^3 - N}$$

TABLE 6.5
Ranks in X and Y for six cases

	Cases					
	1	2	3	4	5	6
X	1	2	3	4	5	6
Y	1	2	3	4	5	6

SEC. 6.5 □ SUPPOSE YOU HAVE A PAIR OF RANKS FOR EVERY CASE?

where D is the difference in the ranks, that is, $X - Y$, for each case, 6 is a constant, and N is the number of cases. Here the Ds are all 0, and accordingly, $\Sigma D^2 = 0$.

$$r_s = 1 - \frac{6\Sigma D^2}{N^3 - N} \quad N(N^2-1)$$

$$= 1 - \frac{6(0)}{6^3 - 6}$$

$$= 1 - 0 = 1.0$$

When the ranks for the two variables are in perfect correspondence, as in this example, X and Y are positively correlated, and r_s reaches one maximum value of 1.0.

Table 6.6 shows two other sets of ranks for $N = 6$ cases. Notice that as the ranks in X increase, those in Y decrease. In Table 6.6, D represents $X - Y$. The sum of Ds always equals 0; this offers a good way to check the D scores. $\Sigma D^2 = (-5)^2 + (-3)^2 + (-1)^2 + (1)^2 + (3)^2 + (5)^2 = 70$. Then r_s is

$$r_s = 1 - \frac{6\Sigma D^2}{N^3 - N}$$

$$= 1 - \frac{6(70)}{6^3 - 6}$$

$$= 1 - \frac{420}{210}$$

$$= 1 - 2 = -1.0$$

When there is a perfect negative correlation between the ranks, r_s reaches the other maximum value of -1.0. Thus, the limits of r_s are the same as for r. An $r_s = 0$ again indicates that there is not a linear relation between the ranks of the two variables. The sign of r_s signifies the direction of the relationship, and the size of r_s shows the strength of the relationship.

TABLE 6.6
Ranks in X and Y for six cases

	Cases					
	1	2	3	4	5	6
X	1	2	3	4	5	6
Y	6	5	4	3	2	1
D	-5	-3	-1	1	3	5

If the X and Y numbers are ranks, then r_s can be calculated directly as above. Sometimes investigators will elect to transform their X and Y numbers into ranks and compute r_s. One reason for doing this is that r_s is easy to calculate when N is small. Another reason is to reduce the influence of outliers. Inman and Conover (1983, p. 127) showed that a single outlier may have a more profound effect upon r than upon r_s.[6] We will carry out this rank-transformation procedure on Sam's data. The lowest ILLK test score was 4: Its rank is 1. The score 5 is ranked 2, and so on. Tied numbers are treated by the *midrank method*. The two scores of 9 would normally be ranked 5 and 6; they are given ranks of (5 + 6)/2 = 5.5. The highest score's rank should be N. If it is not, start searching for errors in ranking. The ranking procedure is repeated for Y. The three values of 3.25 would be ranked 9, 10, and 11. By the midrank method their rank is (9 + 10 + 11)/3 = 10. The results of converting the numbers to ranks are shown in Table 6.7.

TABLE 6.7
Difference (Ds) in the ILLK (X) and sales (Y) for 20 vendors

	Cases									
	1	2	3	4	5	6	7	8	9	10
X	1	2	3	4	5.5	5.5	7	8	9	10.5
Y	20	17	12.5	18.5	14	6.5	18.5	6.5	10	15
D	−19	−15	−9.5	−14.5	−8.5	−1	−11.5	1.5	−1	−4.5

	Cases									
	11	12	13	14	15	16	17	18	19	20
X	10.5	12	13	14.5	14.5	16	17	18.5	18.5	20
Y	4.5	10	2	8	10	16	12.5	3	4.5	1
D	6	2	11	6.5	4.5	0	4.5	15.5	14	19

6. The reduced influence of an outlier occurs because an extreme score, no matter how extreme, receives a rank of either 1 or N.

SEC. 6.5 □ SUPPOSE YOU HAVE A PAIR OF RANKS FOR EVERY CASE?

We may observe as a check that $\Sigma D = 0$. The only value needed to calculate r_s is ΣD^2. $\Sigma D^2 = (-19)^2 + (-15)^2 + \ldots + (19)^2 = 2156.5$. Then r_s is

$$r_s = 1 - \frac{6\Sigma D^2}{N^3 - N}$$

$$= 1 - \frac{6(2156.5)}{20^3 - 20}$$

$$= 1 - \frac{12939}{7980}$$

$$= 1 - 1.62 = -.62$$

Thus, r_s is very close to $r = -.65$. A second way to calculate r_s is to employ the ranks in the computational formula for r given earlier. This procedure is applicable because r_s is a Pearson product-moment r for ranks.

Is the $r_s = -.62$ significant? This question evokes a familiar question. What is the *sampling distribution* of r_s? When $\rho_s = 0$, where ρ_s is the rank correlation in the population, and no ties are present in the rankings, the distribution of r_s is symmetrical but a bit on the wild side (Kendall, 1948). For small Ns the exact probabilities of r_s have been worked out. These probabilities are presented in Table C and should be used when N is less than or equal to 14. Beyond $N = 14$ approximate probabilities can be obtained from Table B. In other words, when N is greater than 14 the sampling distribution of r_s follows the t distribution. We can, therefore, test Sam's H_0 and H_1 by following the decision procedure described earlier. Step 4 in the two-tailed test becomes: 4. Find P for r_s: $P(r_s = -.62) < .01$. The P is found in Table B as before, using $df = N - 2$, and the decisions would be the same.

The size of the rank-order correlation coefficient, r_s, is lowered slightly by the presence of ties. These perturbations are unimportant unless the number or extent of ties is large. See Kendall (1948) for the necessary corrections for ties under these conditions.

Spearman's r_s is called a *distribution-free* or *nonparametric* statistic. We will talk more in Chapter 12 about these two labels. Spearman's r_s is especially suitable for instances in which X and Y are in ranks. As we have seen, though, it can also be applied when numbers have been transformed into ranks. We will close with a note of caution: like r, r_s is appropriate *only* for linear relationships between the ranks of X and Y. Nonlinear relationships like the one in Figure 6.7 will yield an $r_s \cong 0$. That is a perfectly ridiculous answer. In Figure 6.7 X and Y are highly related—the relationship just is not a linear one. Good data can be turned into trash by improper analysis, but trash cannot be converted into good data by proper analysis.

SUMMARY

A fundamental question in science is: Are these two variables, X and Y, related? Being related means that X and Y covary. The values for the variables can be on the same scale but are usually on different scales. The relationship between X and Y can be pictured in a scatter plot. Scatter plots impart eyeball information about direction of the relationship (direct or inverse), strength of the relationship, linearity of the relationship, and the presence of outliers.

To obtain an index of relationship, the Pearson product-moment r, the X and Y numbers must be measured on a common scale. The solution to this problem is to convert the X and Y numbers to standard scores (Zs) that have a mean of 0 and a standard deviation of 1. As described in Chapter 4, a Z score is a deviation from the mean divided by the standard deviation.

The correlation coefficient, r, is an index of relationship based upon a sample. The corresponding population parameter is ρ. The correlation coefficient, r, is an "average" of the sum of the products of X and Y when X and Y have been converted to Z scores. A rational formula for r was presented and demonstrated. The obtained r is an estimate of ρ based on a sample. In calculating r the ss for X and Y are employed to convert the X and Y numbers to Zs and an $N - 1$ divisor is used. The coefficient r has a range from -1.00 (a perfect inverse relationship) to 1.00 (a perfect direct relationship). The sign of r indicates the direction of the relationship; the size of r shows its strength. A computational formula for r was presented and its use was recommended. An example of computing r was provided.

Hypotheses, H_0 and H_1, regarding r are evaluated by the same general method of inference described in Chapter 5 for binomial problems. The sampling distribution of r follows a theoretical distribution, the t distribution. The t distribution, a family of curves, is a symmetrical, bell-like distribution that becomes increasingly normal with increasing N. While the probability of r can be ascertained by a t formula, the $P(r)$ can be found more simply using Table B. Examples of one- and two-tailed tests were given, and of comparing $P(r)$ to α or comparing the obtained r to the critical r for the α level to decide the fates of H_0 and H_1. Following Fisher, it was stressed that a significant finding is one that can be reliably replicated rather than a highly significant finding from a single study.

When the X and Y numbers are ranks (or numbers converted to ranks), Spearman's rank-order correlation coefficient, r_s, summarizes the relationship. The coefficient r_s is a Pearson product-moment r for ranks, and like r it ranges from -1.00 to 1.00. It can be calculated by a special formula incorporating the differences in the $X-Y$ ranks for the N cases or by the regular computational formula for r. The sampling distribution for r_s is symmetrical and the exact probabilities of r_s have been determined for small Ns (Table C). For $N > 14$ the t table for r (Table B) can be employed with $df = N - 2$ to find approximate probabilities for r_s. Like r, r_s is only applicable to linear relationships. It was stressed that good data can be ruined by improper analysis, but bad data cannot be saved by proper analysis.

PROBLEMS

1. Forty-two pregnant rats are subjected to eight different intensities (X) of microwave irradiation ($1 = $ low ... $8 = $ high) prior to delivery. The dependent variable (Y) is litter size. Display the data in a scatter diagram on a sheet of graph paper with X on the horizontal axis and Y on the vertical axis.

X	Y	X	Y	X	Y	X	Y
8	4	2	12	7	6	5	10
5	7	8	6	6	9	3	11
3	10	4	8	5	9	8	6
3	10	4	12	4	9	6	7
1	13	2	10	2	11	7	7
6	10	1	11	7	8	3	9
1	13	5	8	7	6	1	12
4	10	4	11	8	6	8	5
2	12	3	12	5	11	8	5
4	12	7	6	1	12	2	13
				1	14	6	9

Use ⊙ for two identical dots; ⊚ for three dots.

2. By inspection, is there a direct or inverse relationship? Is it linear? Are there any obvious outliers?

3. Determine the correlation coefficient between X and Y using the computational formula in Section 6.2.
4. Using Table B, test the researcher's hypothesis of a predicted inverse relationship between irradiation intensity and litter size at $\alpha = .01$. Follow the six steps in the inference procedure.
5. Can the researcher safely conclude that microwave irradiation causes smaller litter sizes?
6. An investigator obtains an r of .381 between locus of control and performance in a stress task. Can she reject H_0: $\rho = 0$ with $N = 37$ and $\alpha = .01$? If the investigator had predicted a positive correlation and employed a one-tailed test at $\alpha = .01$, what would her decision be?
7. Two coaches, A and B, rank 12 defensive backs ($a \ldots l$) from films taken during practices. Do the coaches show a significant degree of agreement ($\alpha = .05$) in their rankings? Perform a one-tailed test.

	Players											
	a	b	c	d	e	f	g	h	i	j	k	l
A	1	3	12	11	9	4	2	7	5	10	8	6
B	2	1	11	12	8	4	3	6	5	10	9	7

8. Given the ranks below for two variables, compute an r_s.

	Cases									
	1	2	3	4	5	6	7	8	9	10
X	3	1	6	4	8	9	7	5	2	10
Y	5	9.5	1.5	3.5	6.5	8	3.5	1.5	6.5	9.5

Why is r_s *inappropriate* for this problem?

ADDENDUM: USING COMPUTERS IN STATISTICS

MINITAB has a nice correlation subprogram. First, the Crunchie data are input, then a scatter plot (1) is made, and finally r (2) is found. Next, the data are ranked and an r (= r_s) is computed (3) by the command, "CORRELATION C5 C6." Note that we obtained an $r_s = -.62$ and MINITAB's answer is r = -.629. This trivial difference probably occurs because of the use of the r formula for r_s and because computers carry numbers out to many decimal places. It will probably shock you to learn that even different statistical packages may produce slightly different values for r, and so on. These differences are well-documented and are

□ **ADDENDUM: USING COMPUTERS IN STATISTICS**

the result of different computing algorithms, the number of decimal places employed, and at what step(s) rounding to fewer decimal places occurs in the program.

```
--SET IN C3
--4 5 7 8 9 9 11 12 13 14 14 15 16 17 17 18 19 20 22
--SET IN C4
--5.75 4.5 3.5 5 3.75 2.25 5 2.25 3.25 4 2 3.25 1                    (1)
--2.5 3.25 4.25 3.5 1.5 2 0.75
--PLOT Y IN C4 VS X IN C3
     C4
   6.1+
      -        *
      -
      -
      -            *       *
   4.7+
      -          *
      -
      -                         *         *
      -                  *
      -            *
   3.3+                             *   *    *        *
      -
      -
      -
      -                                          *
      -              *    *
   1.9+
      -                      *                *
      -                                          *
      -
      -                               *
      -                                              *
   0.5+
        +---------+---------+---------+---------+---------+C3
        0.        5.0      10.0      15.0      20.0      25.0
--CORRELATION C3 C4
                                                                     (2)
    CORRELATION OF    C3 AND C4    =-0.654
--RANK C3 PUT IN C5
--RANK C4 PUT IN C6
--CORRELATION C5 C6                                                  (3)

    CORRELATION OF    C5 AND C6    =-.629
```

CHAPTER OUTLINE

7.1 WHY IT'S A GOOD IDEA TO SQUARE r AND MULTIPLY BY 100

7.2 SPREADS AND r

7.3 THE CRUNCHIE ICE CREAM BAR EXAMPLE REVISITED

7.4 WHY CORRELATION CAN BE "AN INSTRUMENT OF THE DEVIL"

CHAPTER SEVEN
STILL RHOING

7.1 WHY IT'S A GOOD IDEA TO SQUARE r AND MULTIPLY BY 100

Generally, the material we have just discussed is divided into two parts: *correlation* and *regression*. *Correlation* covers the topics in Chapter 6—indices of relationship like r and r_s, testing hypotheses regarding the indices, and so on. *Regression* concerns the *problem of prediction*. Given that two variables, X and Y, are related, it is possible to make quantitative predictions of Y from X or X from Y. This chapter will (a) deal with a couple of odds and ends on correlation, (b) consider prediction, and then (c) return to correlation for a few parting shots. But you won't be able to forget correlation—you will encounter it again later in the book. In teaching, one should never follow a straight path when one can take a tortuous and an adventurous route. Also remember: REAL MEN never go into gas stations and ask for directions when they're lost.

What does a correlation coefficient mean? You now know that the sign of r (or r_s) indicates the direction of the relationship between X and Y, direct or inverse, and that the size of r shows the strength of the relationship. Also, r is a summarizing statistic: It describes the relationship between two variables.

Another interpretation of correlation is called a *variance interpretation*. The X numbers and Y numbers each have a sum of squares and a variance. $SS_x = \Sigma(X - \overline{X})^2$ and $SS_y = \Sigma(Y - \overline{Y})^2$; $s_x^2 = SS_x/(N - 1)$ and $s_y^2 = SS_y/(N - 1)$. When X and Y are transformed to Z scores in calculating r, the s_z^2 for each transformed variable is 1. You will recall that Z scores have an $s_z = 1$; therefore, $s_z^2 = 1$. If we square a correlation coefficient, r^2 is the *proportion of the total variance* in Y that can be explained in terms of X. Conversely, r^2 is the proportion of the total variance in X that can be explained in terms of Y. Instead of the proportion of variance we can think of the total variance in X and in Y as 100%, or 100(1). If a correlation coefficient is squared and multiplied by 100, the result is the *percentage of the total variance* in Y due to X, or the percentage of the total variance in X due to Y.

In Chapter 6 Sam found an r of $-.65$, an inverse relationship between ice cream sales and liking little kids. The proportion of the total variance that can be explained is: $r^2 = (-.65)^2 = 0.4225$. The percentage of the total variance that can be explained is: $r^2(100) = 0.4225(100) = 42.25$. We have an $s^2 = 1$, of which we can account for 0.4225. What about $1 - r^2$ or $1 - 0.4225 = 0.5775$? It is the part of the spread that we cannot explain—the *unexplained variance*. We have a chunk of variance, s^2. It can be divided into two pieces: r^2, the explained portion; and $1 - r^2$, the unexplained portion. Thus: $s^2 = r^2 + (1 - r^2)$ or $1 = 0.4225 + 0.5775$. In percentage terms everything is multiplied by 100; that is, $s^2(100) = r^2(100) + (1 - r^2)(100)$. What is the unexplained variance? It is random variability due to sources like individual differences, intraindividual differences, errors of measurement, and unknown sources. Sam can explain about two-fifths (42%) of the total variance in sales in terms of liking kids, but about three-fifths (58%) of the total variance in sales is due to unknown causes. Variance interpretations of correlation quickly bring investigators back to reality. For example, to account for half of the variance, an r of $\pm.707$ is required.

Earlier we mentioned that an r of $-.195$ was significant ($P = .05$) with a two-tailed test and $N = 102$ cases. In this instance we can explain: $r^2(100) = (-.195)^2(100) = 3.8\%$ of the variance. While the obtained $r = -.195$ is statistically significant, it can only account for a trivial amount of the variance. Therefore, do not stop after calculating rs and testing them for significance; square the rs and multiply them by 100. What will that do for you? It may cause you to despair, but your rs will be brought into the realm of reality.

The view that statistics and significance tests should be supplemented by information regarding accounted-for variance has been promoted by Cohen (1965). I agree strongly with this tactic. Accordingly, you can expect to be doing it for other statistics as well. Like many important ideas, Cohen's idea is a simple one. In a word, he is advocating that we *generalize* the variance interpretation of correlation to other statistics.

SPREADS AND r

7.2

Let us digress by talking about pies. How do pies relate to statistics? We can represent the correlation in the Crunchie Ice Cream Bar example using a couple of pies. "Isn't that amazing?" as they say on those TV ads where they offer to sell you 65 assorted oriental knives, guaranteed for life, for $19.95. What is a pie? It is simply a circle whose area represents the total variability for a variable, say Y. If the Y numbers are standardized or converted to Z scores, then the pie's area is 1. If we multiply the area by 100, the area is 100%. If we have another variable, X, we can represent it by another pie. If X is standardized, its area is also 1 (or 100%). Since both variables have the same total variability as a result of being standardized, we will draw two circles with the same radius to represent them.

FIGURE 7.1
A ballantine representing two unrelated variables

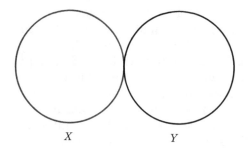

"Where are we now?" you might ask. We have two pies and we can represent a correlation using these two pies. If the pies don't overlap, then X and Y are not correlated (Figure 7.1). If they overlap, then X and Y are correlated. Furthermore, the amount of the overlap is related to the strength of r. More precisely, the overlapped area is r^2. The nonoverlapped area is $1 - r^2$. Figure 7.2 depicts the Crunchie Bar example. The shaded, overlapped area of 0.4225 (or 42.25%) is the explained variance. The unshaded areas of 0.5775 (or 57.75%) represent the unexplained variance. Notice that the pies can be viewed in either of two ways:

1. The Y circle shows the portion of the total variance in Y that is explained by X and the remaining unexplained Y variance.
2. The X circle displays the portion of the total variance in X that is explained by Y and the remaining unexplained X variance.

In more advanced statistics courses, you will be confronted by correlation problems with more than two variables. A pie diagram, termed a *ballantine* or *Venn diagram,* is a valuable picture for illustrating the relationships among a number of variables.

FIGURE 7.2
A ballantine representing the correlation between X and Y

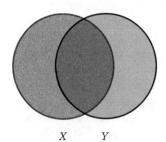

THE CRUNCHIE ICE CREAM BAR EXAMPLE REVISITED

7.3

"Suggestion Box" Sam found a significant r of $-.65$ between the vendors' liking of little kids and their sales of Crunchie Ice Cream Bars. Since the variables are related, Sam can predict one variable from the other.

The Crunchie owners are in the ice cream business. They want to sell ice cream. Accordingly, we will designate sales (Y) as the *criterion* and liking (X) as the *predictor*.[1] How does one predict sales from the liking test scores? To do this, we draw a straight line through the dots in the scatter plot. This line is called a *regression line*. We can then obtain the predictions graphically. Or we can find the equation for the regression line, termed a *regression equation*, and use the equation to generate quantitative predictions.

In this description we have cleverly dodged *the* problem: How do we put the line through the dots in the scatter plot? This problem cannot be avoided. We want the line to pass through the dots so that the predictions will be as accurate as possible. We require, therefore, a *line of best fit*. A line of best fit is one that minimizes errors in prediction. What is an error in prediction? Let Y be the actual score for a vendor with a liking score of X. Let Y' be the predicted score from the regression equation for that vendor. An error in prediction, then, is $Y - Y'$ and the errors summed across all N cases would be $\Sigma(Y - Y')$. But the plan to minimize $\Sigma(Y - Y')$, reasonable as it seems, simply will not work. Some errors will be positive, others will be negative, and when the errors are summed, $\Sigma(Y - Y') = 0$. This equation is thus not helpful. If, however, each error is squared and summed, $\Sigma(Y - Y')^2$, then we can find the errors. Accordingly, a line of best fit is one in which the errors in prediction conform to the *least squares criterion;* that is, the quantity $\Sigma(Y - Y')^2$ is minimal.

The regression equation for predicting Y from X is

$$Y' = b_y X + A_y$$

where Y' is the *predicted value*, b_y is the *regression coefficient*, X is the *score in the predictor variable*, and A_y is the Y intercept

---

1. A criterion is a variable for which the investigator wants to make as accurate predictions as possible. A variable used to achieve this goal is termed a predictor.

(where the regression line crosses Y when $X = 0$). The formula for A_y, the Y intercept is

$$A_y = \overline{Y} - b_y\overline{X}$$

If the right half of the intercept equation is substituted into the regression equation above, another regression equation results:

$$Y' = b_yX + A_y$$
$$= b_yX + \overline{Y} - b_y\overline{X}$$
$$Y' = b_y(X - \overline{X}) + \overline{Y}$$

These regression equations are equivalent. Each equation will generate the same Y' values, but in this discussion we will use the second regression equation.

Although we have talked at length, we still have not put a line through the scatter plot. If you study the last regression equation, you can see that it contains only one mysterious term, b_y. The regression coefficient, b_y, is a "weight" for the X deviation scores $(X - \overline{X})$ that will produce predictions as accurately as possible. It is also the *slope* of the line of best fit. The slope is the *change in Y* divided by the *change in X*; it is the *rise* divided by the *run*. If a vertical line is dropped from the regression line and intersected by a horizontal line drawn from the regression line, a right triangle will be formed. In the triangle the length of the vertical line is the rise, and the length of the base is the run.

FIGURE 7.3
Dividing the data for the ice cream study into three equal batches

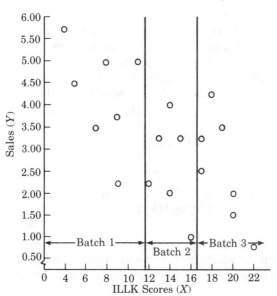

SEC. 7.3 ☐ THE CRUNCHIE ICE CREAM BAR EXAMPLE REVISITED

Next we will describe a method for approximating b_y. This method is presented for two reasons: (a) it should further your understanding of the slope of the line of best fit and (b) it is handy when a quick, approximate value for b_y is sufficient.

Quenouille (1959) suggested the method of putting the dots from the scatter plot into *three roughly equal-size batches* on the basis of the X numbers (see Figure 7.3).

The top and bottom batch should be the same size. An alternative procedure is to select the batches from a table of X and Y numbers for the cases. In the Crunchie Ice Cream problem the X and Y numbers (Table 6.4) are ordered on their X values. The first batch ($N_1 = 7$) consists of cases 1–7. The second batch ($N_2 \doteq 6$) is made up of cases 8–13. The third batch ($N_3 = 7$) includes cases 14–20. The second batch is deleted and sums of X and Y are found separately for batches 1 and 3. Because $N_1 = N_3$ the sums can be used instead of their means. Quenouille's formula (1959, p. 27) is

$$\text{slope} = \frac{\text{rise}}{\text{run}}$$

$$b_y = \frac{\Sigma Y_1 - \Sigma Y_3}{\Sigma X_1 - \Sigma X_3}$$

where the subscripts on X and Y denote batches 1 and 3. Inserting the sums for the Crunchie numbers, b_y is

$$b_y = \frac{29.75 - 17.75}{53 - 133}$$

$$= \frac{12}{-80} = -0.15$$

To repeat, $b_y = -0.15$ is an approximate value for the slope of the regression line for the data in Figure 7.3. It is the regression coefficient for predicting Y from X.

While Quenouille's method is handy for a quick analysis, we will now determine b_y by a more precise method. If you did as you were advised in Chapter 6 and calculated r using the computational formula, you will be in wonderful shape to determine b_y. The computational formula for b_y is

$$b_y = \frac{N\Sigma XY - \Sigma X \Sigma Y}{N\Sigma X^2 - (\Sigma X)^2}$$

where all terms are defined as for r. Because the values for the terms are the same as those for calculating r, b_y is easy to compute:

$$b_y = \frac{20(764.25) - 270(63.25)}{20(4170) - (270)^2}$$

$$= \frac{1792.5}{10500} = -0.17$$

Note that the approximate estimate of b_y by Quenouille's method is very close to the actual b_y. Since we have b_y, we can specify the regression equation for predicting Y from X: $Y' = b_y(X - \overline{X}) + \overline{Y}$. Since $\Sigma X = 270$ and $\Sigma Y = 63.25$,

$$\overline{X} = \frac{\Sigma X}{N}$$

$$= \frac{270}{20} = 13.50$$

$$\overline{Y} = \frac{\Sigma Y}{N}$$

$$= \frac{63.25}{20} = 3.16$$

Substituting $\overline{X}$, $\overline{Y}$, and b_y in the formula for the regression equation, we have

$$Y' = b_y(X - \overline{X}) + \overline{Y}$$

$$= -0.17(X - 13.50) + 3.16$$

Simplifying,

$$Y' = -0.17X + 2.30 + 3.16$$

$$= -0.17X + 5.46$$

This final regression equation is our "working equation" for predicting sales (Y) from scores on the liking scale (X).

We can apply the regression equation to draw the regression line through Sam's scatter plot. The procedure: pick three spread-out values of X, insert them into the equation. and calculate the predicted values, Y'. We will select $X = 8$, $X = 13$, and $X = 18$. For $X = 8$,

$$Y' = -0.17X + 5.46$$

$$= -0.17(8) + 5.46$$

$$= 4.10$$

When $X = 13$, $Y' = 3.25$; $X = 18$, $Y' = 2.40$. We can display these values in Table 7.1 and observe the errors in prediction.

X	Y	Y'	Y − Y'
8	5.00	4.10	0.90
13	3.25	3.25	0.00
18	4.25	2.40	1.85

TABLE 7.1
Errors ($Y - Y'$) in predicting Y from three X scores

It is apparent that for these X values we are making two errors in the predictions, but for this particular r they are the best possible predictions. Plotting the three pairs of X, Y' values in the scatter plot and connecting them by a *solid* straight line produces the regression line for predicting Y from X (see Figure 7.4).

Imagine that a job applicant scored $X = 6$ on the ILLK test. How many ice cream bars would the applicant be expected to sell? An approximate answer can be found graphically by using the regression line in Figure 7.4. Draw a vertical line from $X = 6$ to the regression line. Draw a horizontal line from the intersection of the vertical line and the regression line to the Y axis. The intersection of this horizontal line and the Y axis is Y', which is about 4.5. In other words, the predicted sales of ice cream bars for this applicant with a liking score of 6 would be about 4.5 in units of 100 per day.

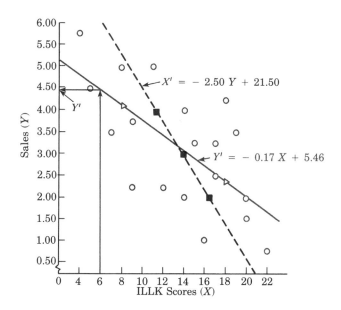

FIGURE 7.4
Regression lines for predicting Y from X (solid line) and X from Y (dashed line)

For a more precise prediction, the job applicant's X score of 6 would be entered in the regression equation:

$$Y' = -0.17X + 5.46$$
$$= -0.17(6) + 5.46$$
$$= 4.44$$

To summarize: (a) the regression equation provides the Y' values needed to draw a regression line through the scatter plot, (b) graphical predictions can be made from this regression line, or (c) the regression equation can be applied to produce quantitative predictions.

Actually, the prediction story is more complicated. So far we've only predicted Y from X. In the beginning of the chapter it was stated that prediction of X from Y is also possible. Accordingly, there are *two regression lines*. If there are two lines, then there are *two regression equations*. In the Crunchie example, Y (sales) was the criterion and X (liking) was the predictor. The Crunchie Co. is in business to make money. So treating Y as the criterion is reasonable. On the other hand, if we had two personality tests, for example, locus of control and introversion/extroversion, then it is not obvious which test is the predictor and which is the criterion. We could either predict Y from X or X from Y.

Instead of predicting ice cream sales (Y) from liking (X), we will now predict the vendors' liking of kids from their sales performance. We need a second regression line and a second regression equation for predicting X from Y. The regression equation is

$$X' = b_x(Y - \overline{Y}) + \overline{X}$$

where X' is the predicted X and b_x is the regression coefficient for doing the best possible job of predicting X.[2] Note carefully that $b_x \neq b_y$. The coefficient b_x is once more related to part of the computational formula for r:

$$b_x = \frac{N\Sigma XY - \Sigma X \Sigma Y}{N\Sigma Y^2 - (\Sigma Y)^2}$$

$$b_x = \frac{(20)(764.25) - (270)(63.25)}{20(235.81) - (63.25)^2}$$

$$= \frac{-1792.5}{715.64} = -2.50$$

2. Earlier the symbol X' was used to denote the midpoint of a class. In the prediction context the symbol X' signifies a predicted value in X from Y.

SEC. 7.3 □ THE CRUNCHIE ICE CREAM BAR EXAMPLE REVISITED

Substituting $\overline{X}$, $\overline{Y}$, and b_x in the regression equation,

$$X' = b_x(Y - \overline{Y}) + \overline{X}$$
$$= -2.50(Y - 3.16) + 13.5$$

Simplifying,

$$X' = -2.50Y + 21.40$$

Again, we may select three Y values of 4, 3, and 2 and insert them successively in the simplified regression equation to generate predicted values (X's) of 11.4, 13.9, and 16.4, respectively. Plotting points in the scatter plot (Figure 7.4) corresponding to these three pairs of Y, X' values and joining them by a *dashed* straight line results in the second regression line. This line may be used to predict graphically X from Y.

Let us talk some more about the two regression lines. If we had a perfect relationship between X and Y with $r = -1.00$ or $r = 1.00$, then there would be only *one* regression line and all dots would fall on the line. Such relationships are almost never seen out in the trenches that people call research. At the other extreme, when $r = 0$, there is no way to put a useful regression line through the dots. Between these extreme rs there are two regression lines, and as the size of r decreases from 1.00 or -1.00 to 0, the two regression lines diverge more widely. Conversely, looking back to Figures 6.4, 6.5, and 6.6, the two regression lines would converge from Figure 6.4 to 6.5 and become a single line in Figure 6.6. It should be obvious that our errors in prediction would decrease from Figure 6.4 to Figure 6.5. In Figure 6.6 a limit is reached; there are no errors in prediction because all the dots fall upon the single regression line. The other extreme (Figure 6.3) is where the dots form a horrible circle—here the errors in prediction reach a maximum. Our best prediction for Ys in this situation is $\overline{Y}$. Then the errors in prediction will be $\Sigma(Y - \overline{Y})^2$ or SS_y because $\overline{Y}$ has replaced Y' in the error formula. If you cut through this verbiage, you'll find a principle: The accuracy of prediction is related to the strength of the relationship.

In the preceding paragraph we attempted to reveal intuitively the relationship between predictive accuracy and the size of r. Now we will consider this relationship quantitatively. The variance in the Y scores is

$$s_y^2 = \frac{\Sigma(Y - \overline{Y})^2}{N - 1}$$

Note that s_y^2 with an $N - 1$ divisor is an unbiased estimate of the population parameter, σ_y^2. When Y is predicted from X and r is less than 1.00 or -1.00, errors $(Y - Y')$ in prediction will occur. These errors have a variance termed the *variance of estimate*, $s_{y \cdot x}^2$ that indicates the accuracy in prediction.[3] What is $s_{y \cdot x}^2$? It is

$$s_{y \cdot x}^2 = \frac{\Sigma(Y - Y')^2}{N - 2}$$

Why is $s_{y \cdot x}^2$ called a variance when Y' isn't a mean? If we regard Y' as a "floating mean" or as a series of "means" for the various Xs, as Runyon and Haber (1984, p. 172) suggested, then the variance label seems more acceptable. Why is the divisor $N - 2$ for $s_{y \cdot x}^2$? Because the $N - 2$ divisor results in a value for $s_{y \cdot x}^2$ that is an unbiased estimate of the population parameter, $\sigma_{y \cdot x}^2$.

Consider the formula for $s_{y \cdot x}^2$ and imagine that we had a study with an $N = 200$. To compute $s_{y \cdot x}^2$ we would need to find the regression equation for predicting Y from X, calculate the 200 Y' values, subtract them from the Ys, square the differences, sum the squared differences, and divide the sum by $N - 2$. But there is an easier way to compute $s_{y \cdot x}^2$, and it is dependent upon the relationship between accuracy in prediction and the size of r. It can be shown that $s_{y \cdot x}^2$ is

$$s_{y \cdot x}^2 = s_y^2(1 - r^2)$$

Taking the square root of this equation, we can find the standard deviation of the numbers about the regression line, $s_{y \cdot x}$:

$$s_{y \cdot x} = s_y \sqrt{(1 - r^2)}$$

The $s_{y \cdot x}$ is commonly termed the *standard error of estimate*. The formulas for $s_{y \cdot x}^2$ and $s_{y \cdot x}$ provide easier ways to calculate the accuracy of prediction and describe precisely the relationship between accuracy of prediction and the size of r.

Suppose that $s_y = 20$. What is the accuracy in prediction for rs of different sizes? If $r = 1.00$, then

$$\begin{aligned} s_{y \cdot x} &= s_y \sqrt{(1 - r^2)} \\ &= 20\sqrt{1 - (1.00)^2} \\ &= 20(0) = 0 \end{aligned}$$

3. The $y \cdot x$ subscript refers to predicting Y from X; $x \cdot y$ to predicting X from Y.

If $r = .67$, then
$$s_{y \cdot x} = 20\sqrt{1 - (.67)^2}$$
$$= 20(.74) = 14.80$$

If $r = .33$, then
$$s_{y \cdot x} = 20\sqrt{1 - (.33)^2}$$
$$= 20(.94) = 18.80$$

If $r = 0$, then
$$s_{y \cdot x} = 20\sqrt{1 - (0)^2}$$
$$= 20(1) = 20$$

These examples demonstrate how the accuracy in prediction decreases ($s_{y \cdot x}$ becomes larger) as r decreases in size. This relationship is unaffected by the sign of r because r is squared in the formula for $s_{y \cdot x}$. An equivalent formula for the standard error of estimate when predicting X from Y is

$$s_{x \cdot y} = s_x \sqrt{(1 - r^2)}$$

To complete this section we need to integrate the variance of estimate $s_{y \cdot x}^2$, with the variance interpretation of correlation in Section 7.1. There we said that the total variance in Y can be divided into two portions: the explained and unexplained portions. When Y is standardized, $s_y^2 = 1$ and

$$1 = r^2 + (1 - r^2)$$

where r^2 is the explained proportion of the total variance and $(1 - r^2)$ is the unexplained proportion. In the present section we have focused upon the *actual* s_y^2 (Y scores not standardized). Relating s_y^2 to r^2, we have

$$s_y^2 = s_y^2 r^2 + s_y^2(1 - r^2)$$

What is $s_y^2(1 - r^2)$? It is $s_{y \cdot x}^2$ or the variance of estimate. Thus, the unexplained variance and variance of estimate are the same. The actual s_y^2 is the variability of the Y scores about $\overline{Y}$. The variance of estimate, $s_{y \cdot x}^2$ or $s_y^2(1 - r^2)$, is the variability of Y scores about Y'. What is the explained portion of s_y^2 or $s_y^2 r^2$? It is the variability of Y' about $\overline{Y}$.

Cohen (1969) has proposed that in behavioral science .10 is a *small* correlation coefficient, .30 is a *medium* coefficient, and .50

is a *large* coefficient. I think these r values are realistic. If you have remembered your lesson, you will immediately square the rs and multiply them by 100. The obtained r^2s indicate that we frequently cannot account for very much variance and that our predictions are not very accurate.

WHY CORRELATION CAN BE "AN INSTRUMENT OF THE DEVIL"

7.4

Hilgard (1955), after analyzing many correlational studies on intelligence and nature/nurture, described correlation as "an instrument of the devil." In this section we will consider some difficulties with correlation. One problem concerns correlation and causality. If smoking and cancer are correlated, can we conclude that smoking causes cancer? If liking little kids is related to the sales of Crunchie Ice Cream Bars, can we assume that liking is responsible for sales?

Causality implies correlation. If smoking causes cancer, then the amount of smoking ought to be correlated with the incidence of cancer. The converse, that *correlation implies causality*, is not defensible. Why? Because the possibility always exists that one or more other variables may be involved. This possibility is often referred to as the *third variable problem*. That is, some variable other than X is responsible for the observed correlation between X and Y. In the ice cream case the vendors who liked children less sold more. Suppose that third variable here was age. As your body and mind slowly sink in the west with age, you may like children less. But older vendors have had more experience and may be more skilled in selling and thus have higher sales. Because of the ever-present possibility of such third variables we must be very cautious about inferring cause-and-effect relationships from correlations. It is possible to control statistically for such third variables, but whenever possible a better technique is to investigate the variables directly in controlled experiments.[4]

In the third variable problem, correlations that are not there are being observed. Another problem is missing correlations that are there. That is, while X and Y are related in the population, a sample fails to reveal the correlation. How can this happen? One possibility is that there is *restriction of the range* of scores in one

---

4. A reader has called attention to the problem of *direction* in causality; that is, does X cause Y or Y cause X? The reader pointed out: "Maybe selling a lot of ice cream makes a vendor hate kids." See Cook and Campbell (1979, pp. 53–54) for a discussion of this problem.

or both variables. We can demonstrate the effects of restriction of range upon r by deleting the last four vendors—those with the four highest X scores—from the Crunchie data. Removing these cases results in the following: $\Sigma X = 189$, $\Sigma X^2 = 2525$, $\Sigma Y = 55.5$, $\Sigma Y^2 = 216.75$, and $\Sigma XY = 611.25$. With the reduced sample,

$$r = \frac{N\Sigma XY - \Sigma X \Sigma Y}{\sqrt{[N\Sigma X^2 - (\Sigma X)^2][N\Sigma Y^2 - (\Sigma Y)^2]}}$$

$$= \frac{16(611.25) - 189(55.5)}{\sqrt{[(16(2525) - (189)^2][16(216.75) - (55.5)^2]}}$$

$$= \frac{-709.5}{\sqrt{(4679)(387.75)}}$$

$$= \frac{-709.5}{1346.953} = -.53$$

We observe that by restricting the range of numbers for the variable, r has fallen from $-.65$ to $-.53$. Restricting range in either or both variables has the effect of reducing the size of r.[5]

Another way in which relationships may be overlooked is to conduct studies with *insufficient statistical power*. One factor affecting power is N, the number of cases. Increasing N increases power, which is the probability of detecting a relationship when it is there. The relationship between N and power may be explicated by considering the standard error of r. In Chapter 6 we said that the sampling distribution of r follows the normal distribution when N is very large and $\rho = 0$. The standard deviation of the sampling distribution of r, that is, the standard error or s_r, under these conditions is

$$s_r = \frac{1}{\sqrt{(N-1)}}$$

Thus, in very large samples r could be tested for significance by:

$$\text{test statistic} = \frac{r - \rho}{\frac{1}{\sqrt{(N-1)}}}$$

5. Conversely, correlations may be increased when only high and low cases are included. For example, $r = -.724$ between liking and sales when the middle cases, 7–15 (Table 6.4), are omitted from the analysis.

If $N = 101$, then $s_r = 0.10$. If $N = 401$, then $s_r = 0.05$. In the second case with a smaller s_r, the test statistic would be larger, and the chances of rejecting H_0 and accepting H_1 would be greater. Accordingly, the statistical power is greater with a larger N. An investigator who does a study in which N is too small may fail to reject H_0 when it is false. That is, although two variables may in reality be correlated, the investigator may fail to detect the correlation. Power is also related to the size of ρ. Detecting a small correlation takes a larger sample than detecting a large one. Before leaving correlation, let us sum up this last section.

1. Causality implies correlation, but correlation does not warrant inferring causation.
2. A correlation between two variables may be a reflection of a third variable.
3. Correlations that are present may be missed because of restricted range in the variables or low statistical power.

SUMMARY

An obtained correlation coefficient should be squared and multiplied by 100. The product, $r^2(100)$, is the percentage of the total variance in Y that is explained in terms of X or the percentage of the total variance in X that is explained in terms of Y. The quantity, $(1 - r^2)(100)$, is the unexplained percentage of the total variance in Y or X. Unexplained variance is random variability due to individual differences, errors of measurement, and so on. These ideas constitute a variance interpretation of correlation. When N is very large, small but significant rs may account for little variance.

Relationships may be represented by pies or circles (a ballantine or Venn diagram). With standardized X and Y variables, the overlapping area between two circles is r^2 (the proportion of explained variance) or $r^2(100)$ (the percentage of explained variance). The nonoverlapped portions of the circles represent $1 - r^2$ (the proportion of unexplained variance) or $(1 - r^2)(100)$ (the percentage of unexplained variance). Nonoverlapping circles denote that X and Y are not linearly related.

Regression is concerned with the problem of prediction: Y from X or X from Y. Prediction is accomplished by putting a line of best fit through the dots in a scatter plot. The straight line, a regression line, must conform to the least-squares criterion: the squared errors in prediction, $\Sigma(Y - Y')^2$ or $\Sigma(X - X')^2$, are minimal. The equation for the regression line is called the regression equation and it includes a "weight," either b_y or b_x, for minimizing errors in prediction. The regression coefficient, b_y, is the slope (rise/run) of the regression line for predicting Y from X. And b_x is the slope of the line for predicting X from Y.

A method by Quenouille for approximating b_y was demonstrated. An exact measure of b_y was calculated and a regression line was plotted through a data set. Examples of predicting from the regression line and from the regression equation were given as well as computing b_x for predicting X from Y. The two regression lines converge as the size of r increases. With $r = \pm 1.00$ there is only a single regression line and all dots fall upon it. Accuracy in prediction is directly related to the size of r. When $r = \pm 1.00$, there are no errors in prediction. When $r = 0$, prediction is at its worst. Accuracy was indexed by the standard error of estimate—the variability of the Y scores about Y'. The variance of estimate, $s^2_{y \cdot x}$, is the unexplained portion of the Y variance.

Causality implies correlation, but correlation does not necessarily imply causality. Smoking and the incidence of cancer should be correlated if smoking causes cancer; however, a correlation between smoking and cancer can be spurious—due to the effects of some other variable. Research also may not detect correlations that do exist. If one or both variables are restricted in range, the correlation between the two variables will decrease in size. Studies with insufficient statistical power may result in overlooking relationships. Power is related to sample size and ρ size. Having a too-small sample lowers power, and trying to detect a small ρ requires a larger sample size than that needed to detect a large ρ.

PROBLEMS

1. A correlation coefficient of $r = .41$ is observed between scores on a test for frustration tolerance and performance on an eight-hour signal detection task. What percentage of the variability in the signal detection scores can be accounted for in terms of frustration tolerance? What percentage of the variability is due to unknown causes?
2. Make a pie diagram summarizing the relationship in Problem 1.
3. Using the data on microwave irradiation and litter size (Problem 1, Chapter 6), calculate the regression equation for predicting litter size from the intensity of irradiation.
4. Using the regression equation, draw a regression line through the scatter diagram obtained earlier (Problem 1, Chapter 6). If a rat were exposed to an intensity of 5, what would be the predicted litter size?

5. For the same data set calculate the regression equation for predicting irradiation intensity from litter size.
6. Find the variance of estimate in litter size for the microwave data. If the correlation between X and Y were smaller than that observed, what would happen to the size of the variance of estimate? Why?
7. Examine a journal in your field (e.g., psychology, sociology, education, etc.) that reports research. Locate studies that present rs. Do these rs suggest that Cohen's definitions of a small, medium, and large correlation are reasonable? Why do published articles present a biased picture of the rs actually found in research?
8. A positive correlation is observed repeatedly between a certain X and a certain Y. Does replicability indicate causality?
9. If correlations can be spurious because of possible third variables, why should correlational research be done?

ADDENDUM: USING COMPUTERS IN STATISTICS

Here MINITAB's regression package is being applied to the Crunchie data. First, the regression equation is output (1). Then, the regression coefficient b_y is tested for significance (2). Testing b_y is equivalent to testing r. The output, "R-SQUARED = 42.8 PERCENT" (3) is really $r^2(100)$. The analysis of variance, which we didn't talk about in the chapter, is another way of testing r for significance. Upon command MINITAB will output all predicted values for all predictor values. In this case MINITAB is presenting only the predicted value for $X = 4$, where the X is a value having large effect (4). Is X a possible outlier? The Durbin-Watson statistic (5) is a test to determine if the residuals (errors in prediction) are independent or not serially correlated. Independence of the residuals is an assumption for r. Unfortunately, the tables for the Durbin-Watson statistic are not widely available. Finally, the regression equation (6) for predicting X and Y is found as well as the other statistics described above.

```
--REGRESS Y IN C4 ON 1 PREDICTOR IN C3
THE REGRESSION EQUATION IS
Y = 5.47 - 0.171 X1                                              (1)

                          ST. DEV.   T-RATIO =
     COLUMN  COEFFICIENT  OF COEF.   COEF/S.D.                   (2)
       --       5.4671     0.6723      8.13
X1     C3      -0.17071    0.04656    -3.67
THE ST. DEV. OF Y ABOUT REGRESSION LINE IS
S = 1.067
WITH (20 - 2) = 18 DEGREES OF FREEDOM                            (3)

R-SQUARED = 42.8 PERCENT
R-SQUARED = 39.6 PERCENT, ADJUSTED FOR D.F.

ANALYSIS OF VARIANCE
  DUE TO      DF     SS      MS = SS/DF
REGRESSION     1   15.300      15.300
RESIDUAL      18   20.484       1.138
TOTAL         19   35.784

      X1       Y    PRED. Y  ST. DEV.
ROW   C3       C4   VALUE    PRED. Y   RESIDUAL   ST. RES.       (4)
  1   4.0    5.750   4.784    0.503     0.966      1.03 X
X DENOTES AN OBS. WHOSE X VALUE GIVES IT LARGE INFLUENCE.
DURBIN-WATSON STATISTIC = 2.25                                   (5)
--REGRESS Y IN C3 ON 1 PREDICTOR IN C4
THE REGRESSION EQUATION IS                                       (6)
Y = 21.4 - 2.50 X1

                          ST. DEV.   T-RATIO =
     COLUMN  COEFFICIENT  OF COEF.   COEF/S.D.
       --      21.421      2.345       9.13
X1     C4      -2.5046     0.6831     -3.67
THE ST. DEV. OF Y ABOUT REGRESSION LINE IS
S = 4.086
WITH (20 - 2) = 18 DEGREES OF FREEDOM

R-SQUARED = 42.8 PERCENT
R-SQUARED = 39.6 PERCENT, ADJUSTED FOR D.F.
```

```
ANALYSIS OF VARIANCE
  DUE TO      DF      SS    MS=SS/DF
REGRESSION    1    224.47    224.47
RESIDUAL     18    300.53     16.70
TOTAL        19    525.00

          X1        Y   PRED. Y   ST. DEV.
ROW       C4       C3    VALUE    PRED. Y   RESIDUAL   ST. RES.
  1     5.75    4.000    7.019     1.990     -3.019    -0.85 X
 20     0.75   22.000   19.542     1.884      2.458     0.68 X
X DENOTES AN OBS. WHOSE X VALUE GIVES IT LARGE INFLUENCE.
DURBIN-WATSON STATISTIC = 1.01
```

NUMBERS IN THE NEWS

Who likes short shorts? The hemline theorists

By Steve Rosen
Star business & financial writer

Ticker tape watching may not be more profitable than girl watching, at least if you're a believer of the hemline theory of stock market forecasting.

To most serious investors, the notion that women's skirt lengths and the stock market go hand in hand is pure hogwash. Their reaction to the theory invariably is laughter.

Lucien Hooper, a 63-year Wall Street veteran and a noted stock market analyst at Thomson McKinnon in New York City, dismisses the hemline theory as an amusing sidelight. "I guess the stock market has a sexy side," he said.

Yet the hemline theory still manages to creep into Wall Street conversation. Part of the reason may be that the theory provides some comic relief. But there also are some investment professionals who believe in the theory, although it is hard to get them to admit it.

As the theory goes, stock market averages tend to rise and fall as hemlines rise and fall. If hemlines are going up, then stock market averages will tend to follow. If skirts are being lowered, then the stock market is in for bad times.

So what's in store this year?

On the fashion front, skirts of all lengths are being offered, says Robert Benham, president of Halls Merchandising Inc., a subsidiary of Hallmark Cards.

That led Bryant Barnes, a vice president of Financial Counselors Inc., a subsidiary of Kidder, Peabody in Kansas City, to conclude that the only correlation between skirt lengths and the stock market in 1981 is that "both are all over the place."

According to Wall Street legend, Ralph Rotnem, a longtime Wall Street analyst who worked at what was then Harris, Upham, is credited with being the first one to come up with the hemline theory of stock market forecasting.

Rotnem, according to the story, was reading a report on the garment industry when it suddenly struck him that changes in skirt lengths matched the ups and downs of the economy.

Next, he set out to see if the same was true of stock market fluctuations, and he found it was true.

"If you go through history, you'll find some correlations," said Robert Kirkpatrick, a retail industry analyst at Waddell & Reed Inc., a Kansas City investment firm.

There is some logic behind the theory. Every market—be it the stock or fashion market—"is a reflection of peoples' moods and the economy that they're living in," said Yale Hirsch, a noted investment adviser in Old Tappan, N.J. When people are in a happy frame of mind, there is a kind of euphoria in the air that can lead to high hemlines as well as high stock prices, Hirsch said. Likewise, he said, when society is in a collectively somber mood, it can be infectious, except in a downward direction.

Early in the century, when skirts were low, the Dow Jones industrial average meandered below 100, Hirsch said, Hirsch, a follower of the hemline theory, said the Dow floated over the then-insurmountable level of 300 in the 1920s—the era of short-skirted flappers.

With the great stock market crash in 1929, both the market and hemlines plunged back down again, Hirsch said. But as hemlines fluttered up to knee level in the late 1940s and early 1950s, the stock market showed periodic signs of liveliness.

But it wasn't until miniskirts and hot pants were introduced in the 1960s that the stock market really took off. In 1966, the market soared above 1000. That also was the year miniskirt hems were at their height, some 24 inches off the floor, Hirsch said.

More recently, with miniskirts a memory and pant suits a big part of a woman's wardrobe, the hemline theory has lost some of its usefulness as a harbinger of stock market trends, Hirsch said.

Detractors of the theory also point out that when skirts become extremely short, the chances are that the public is gripped by a sense of excessive euphoria. The stock market may ride up for a while, they argue, but actually the market is riding for a fall. That was the case in 1929 and again in 1970.

Hooper, however, remains unconvinced that the hemline theory has any value, except as cocktail conversation, "I've seen all these theories," said the veteran stock analyst, "and frankly, I don't see any usefulness in any of them."

Rise in child health problems seen

The Associated Press
SAN FRANCISCO—The rate of debilitating, chronic health problems in children has doubled in the last 25 years, according to annual government surveys of parents.

A researcher analyzing the findings said Monday it was too early to tell what was responsible for the increase.

"We're looking at a wide variety of conditions," said Dr. Peter Budetti, director of the Health Policy Program at the University of California, San Francisco. "Some of them will be related to birth, lots of them will not be."

While there didn't appear to be a large increase of children with mental defects, Dr. Budetti said, the figures indicated a large growth in the number of those with debilitating illnesses such as asthma and bronchitis.

Part of the increase might be due to better surveying techniques, he said. But that "won't turn out to account for all of it," he said.

Dick Leavitt, a spokesman for the March of Dimes in White Plains, N.Y., said statistics compiled by the Centers for Disease Control in Atlanta show no significant change in the incidence of birth defects in the last decade.

He said the number of children surviving with birth defects has increased, although the implications of that are uncertain.

Although medical technology has permitted very small premature infants to survive—possibly increasing rates of cerebral palsy—Mr. Leavitt said it also has reduced the rate of the same condition among somewhat older but still premature infants.

The federal survey, known as the National Health Interview Survey, questions about 110,000 people a year. The surveys indicate that in the last 25 years, the number of children with some limitation of activity due to a medical condition or learning disability has climbed by 500,000, Dr. Budetti said.

But he said the study isn't completed and researchers are "a couple of months away from having answers to any of the really important questions."

Still unanswered, he said, are, "How much of this is the survey getting better? . . . How much of this is parents getting aware of problems their children are having? . . . And how much of his is an actual organic disease?"

"Regardless of the exact numbers, we are seeing real increases in children with some form of handicap, and this is resulting in a substantial burden to society, a burden that will increase with time," Dr. Mary Grace Kovar, an analyst at the National Center for Health Statistics in Hyattsville, Md., said.

U.S. Department of Education statistics show that 4 million children were enrolled in special education programs in the 1981–82 school year, up from 3.5 million in the 1976–77 year. Overall school enrollments were declining at the same time.

One-third of workers in survey admit stealing

The Associated Press
WASHINGTON—One-third of the employees who responded in a survey of 47 corporations reported stealing company property and nearly two-thirds reported taking long lunch breaks, misusing sick leave or using alcohol or drugs while at work.

James K. Stewart, director of the Justice Department's National Institute of Justice, which funded the three-year study, said it shows that "employee pilferage" is costing American business from $5 billion to $10 billion a year.

"This loss is hurting us all because it is passed along to all of us as consumers in the form of higher prices," he said.

The researchers, from the University of Minnesota's Department of Sociology, found that long lunch hours and work breaks were much more prevalent than actual theft, but they also found that to a large extent those engaged in these actions also were doing most of the stealing.

The study said that both problems were "more likely among those employees expressing dissatisfaction . . . with their immediate supervisors and the company's attitude toward the work force."

The researchers said, "We found that those employees who felt that their employers were genuinely concerned with the workers' best interests reported the least theft."

The study found that the highest levels of property theft were reported by unmarried male employees between age 16 and the mid-20s.

The survey was conducted under a $249,967 grant in Dallas-Fort Worth, Minneapolis-St. Paul and Cleveland. The 47 corporations included 16 retail department store chains, 21 hospitals and 10 electronics manufacturing firms, which were asked to respond to a questionnaire.

A total of 9,175 randomly selected employees anonymously filled out the questionnaires. In addition, the researchers conducted in-depth interviews with 256 employees in six firms and with 247 executives.

The report said theft and other misconduct could be minimized through a consistent corporate policy that made clear that employees would be punished for such actions and that high-level company officials and low-level employees would receive the same punishment.

Factories operating at highest capacity since February 1982

The Associated Press
WASHINGTON—U.S. industry operated at 74.5 percent of capacity in June, the highest rate since February 1982, the government said Monday.

Private and government economists said the report bodes well for future investment in plant expansion and modernization—investment spending that would contribute to the economic recovery.

Use of manufacturing, utility and mining facilities in June rose 0.7 percent, the seventh monthly advance in a row, the Federal Reserve Board said in a report that substantially expands on and revises previous reports of industrial activity.

Economic gaps still separate races, study says

The Associated Press

WASHINGTON—Wide economic disparities between whites and blacks have not narrowed since 1960, according to a study released Monday that examined incomes, poverty rates and unemployment rates.

The report by the Center for the Study of Social Policy, a non-profit group that analyzes social trends, used data from the Census Bureau and other government sources to "call attention to the continuing economic inequality between blacks and whites as a fundamental national problem that receives far too little systematic scrutiny."

The data showed that the income of black families has remained fairly constant as a proportion of white family income. The median income for black families in 1960 was 55 percent of the median income for white families. That percentage rose to 61.5 percent in 1975 but fell to 56 percent in 1981, the study found.

Income rose during the period for both blacks and white families. But because black family incomes rose at a slower rate, a larger proportion of black families than white families remained at or near the poverty level, the data showed. In 1981, 28 percent of all white families had incomes of less than $15,000, compared with 55 percent of all black families.

Blacks have been three times more likely to live in poverty than whites, the study showed. About 45 percent of black children under 18 lived in poverty in 1981, compared with 14.7 percent of white children.

The report said the disparities in income could not be attributed solely to differences in the educational levels of blacks and whites because the disparities in this area have narrowed significantly.

For example, in 1960 median schooling was 7.7 years for black males and 8.6 years for black females. The median schooling for both sexes rose to 12.1 years by 1981. By comparison, white males had median schooling of 12.6 years in 1981, and white females had 12.5 years. The study also said the gap in the illiteracy rate between blacks and whites had almost been closed.

Yet in families headed by a high-school graduate, 48 percent of blacks had incomes of less than $15,000, compared with 26 percent of whites. In families headed by college graduates, 9 percent of blacks earned $50,000 or more, compared with 21 percent of whites.

The unemployment rate for blacks consistently has been twice that of whites since 1960, the study found.

Fed expected to revise money-supply targets

The Associated Press

NEW YORK—A financial publication reported Tuesday that Federal Reserve Board Chairman Paul Volcker will announce revisions in the way the central bank calculates money-supply growth, something that could calm fears of another surge in interest rates.

Institutional Investor magazine's publication *Bondweek* said Mr. Volcker would announce the decision today in his semiannual testimony on monetary policy before the House Banking Committee.

Bondweek, quoting unidentified "senior Washington sources familiar with the testimony," said the Fed would change the base for calculating the growth of the basic money-supply measure, M1, to the second quarter of 1983 from the fourth quarter of 1982.

The subject of widespread speculation among credit-market analysts for weeks, such a move would erase months of volatile M1 growth from Fed calculations.

Traders in bond, stock, foreign exchange and commodity markets have been concerned that the rapid growth of M1 would force the Fed to either make credit scarcer, pushing interest rates higher, or risk rekindling higher rates of inflation. But if the Fed adjusted the base to ignore a large portion of M1 growth, the risk of a significant tightening move would subside, the analysts said.

Bondweek also said the Fed would reduce growth targets for broader money-supply measures that have not been growing as fast as M1, which is the sum of cash held by the public, plus traveler's checks and checking accounts.

Joe Coyne, a Fed spokesman in Washington, declined to comment on the report.

Last February the Fed set a target of 4 percent to 8 percent growth in M1 from the fourth quarter of 1982 to the fourth quarter of 1983. But M1 has been growing at an annual rate of about 14 percent since the end of last year, something that prompted the Fed to adopt a slightly more restrictive stance in May.

The Fed's tightening already has pushed short- and long-term interest rates up more than a percentage point to the highest levels since last fall.

The central bank's policy-makers were sharply divided in May. A majority, including Mr. Volcker, said they were concerned that the rapid growth of M1 was a sign of growing inflationary pressures. The minority said M1 growth had been distorted by the flow of funds in and out of checking and saving acounts and into new money-market accounts at banking institutions.

Bondweek said Mr. Volcker would announce that the Fed expected M1 growth to be held to 5 percent to 9 percent for the rest of the year under the revised base.

It also said Mr. Volcker would announce that the Fed would no longer have a formal growth target for M1 but instead would follow the basic money supply's performance through a "monitoring range" of 4 percent to 8 percent.

In addition, *Bondweek* said, Mr. Volcker planned to announce a reduction of one-half percentage point in the target for M2 growth from its present range of 7 percent to 10 percent and M3 growth from its current 6.5 percent to 9.5 percent.

M2 is the sum of M1 plus savings and small-denomination time deposits including shares in money-market mutual funds for individual investors.

M3 is the sum of M2 plus large time deposits, large-denomination term repurchase agreements and shares in money-market mutual funds restricted to institutional investors.

Researchers say smokers' children ill more than those of non-smokers

The Associated Press

WASHINGTON—Children whose parents smoke cigarettes suffer more days of illness than children of non-smokers, according to a new analysis of survey information.

The conclusion was based on data collected in the National Health Interview Survey, which sampled 37,000 American households in 1970. The information was not analyzed until recently.

The authors, Gordon Scott Bonham and Ronald W. Wilson, found that youngsters with smokers in their households suffered more days of restricted activity and bed-disability than their peers in non-smoking households.

"This study, using cross-sectional data, offers no direct proof that adult smoking adversely affects children's health," they said in an article in the March issue of the *American Journal of Public Health*.

But they added, "The data support the findings of others that cigarette smoking by adults adversely affects the health of children in their families."

The data showed that 37.8 percent of children up to 16 years old lived in families with no adult smokers. The remaining 62.2 percent lived in families with at least one smoker—37.4 percent in families with one smoker and 24.8 percent in families with two or more smokers.

Children whose parents were non-smokers averaged 9.1 days a year of restricted activity, while those with two or more smokers in their household averaged 10.2 days.

The difference of 1.1 days a year "supports the hypothesis that family smoking adversely affects the child's health," the authors reported.

There was a clearer correlation between acute respiratory illness and family smoking than other conditions. Such illness, which often is linked with smoking and accounts for many childhood ailments, caused an average of 1.2 more days of restricted activity for children in households with two or more smokers than those in non-smoking families.

There was no difference in the number of restricted activity days because of other conditions.

Children in families with two or more smokers also spent an average of eight-tenths of a day more sick in bed each year than children in non-smoking families.

William Toohey, Tobacco Institute spokesman, discounted the study results, saying, "In any type of study like this there are a lot of other variables that should be looked at."

He also cited at least one study, done by a Yale University professor in 1977, which said that smoking by family members had no measurable effect on children.

This latest study, however, showed that children in families where 45 or more cigarettes were smoked daily suffered 1.5 more restricted activity days than did children where less than one cigarette a day was smoked.

The researchers found that the relationship between family smoking and childhood illness held even when variables like age of the child, number of adults in the household, family income and education were checked.

"In general, the more cigarettes smoked, the greater the days of illness," the study said.

Unemployment declines for third straight month

By The Associated Press

Washington—West Virginia and Michigan lead the nation in unemployment but the picture is improving, according to Labor Department statistics.

According to the department's May survey, 10 percent or more of the work force was without work in 19 states. But the trend was downward in at least 37 states and the jobless rate held steady in three others.

West Virginia's unemployment rate was 18.2 percent, followed by Michigan's 14.7 percent. But even those two high figures were down—from 19 percent in West Virginia, which has been hard hit by mine layoffs, and from 15.5 percent in Michigan, where thousands of workers have suffered along with the auto industry in the 1981–82 recession.

Overall, the national unemployment rate declined from 10.2 percent in April to 10.1 percent in May to 10 percent in June.

The Labor Department often takes more than a month to compile statistics from its monthly surveys.

Residents to weigh arsenic risk from Tacoma, Wash., smelter

By the Associated Press

Tacoma, Wash.—In workshops, hearings and debates this summer, people downwind from a Tacoma copper smelter will be asked how much risk they will accept from airborne arsenic, a suspected source of cancer.

Some residents say they fear the decision will come down to jobs versus health.

The debates and hearings were announced last week by William Ruckelshaus, chief of the Environmental Protection Agency, when he introduced new emission standards for copper smelters and glass plants that produce arsenic. The new standards were ordered last winter by a federal judge ruling in a New York state lawsuit against airborne arsenic from New Jersey.

The standards affect 14 copper smelters around the country, but the Asarco Inc. plant in Tacoma uses copper ore with a particularly high arsenic content. In announcing the standards, Mr. Ruckelshaus said special weight would be given to the views of Tacoma residents in an 81-day public comment period.

"I feel we must involve them directly because the risk we are describing there is high," he said. "We must ask them if they are willing to accept certain risks associated with exposures to low levels of arsenic."

Despite the new hearings, arsenic is an old issue for people living near the 93-year-old

Asarco smelter on a hill above Commencement Bay, a plant that employs 575 workers and produces almost 24 percent of the country's airborne arsenic pollution.

In the Tacoma area arsenic, sulfur dioxide and other chemicals emitted from the plant have for a long time been blamed for the often murky air and raw throats, damaged automobile finishes and ruined gardens.

But the smelter is also defended as an important source of jobs and tax revenue.

Asarco officials, who deny that the airborne arsenic is a danger, say they have spent $45 million since 1972 to reduce pollution. The smelter manager, Larry Lindquist, said that particulate emissions had dropped 85 to 90 percent, and sulfur dioxide has been reduced 35 percent.

Some environmentalists at a meeting with Environmental Protections Agency officials last Thursday said the agency should base its decision on scientific information and not on economic concerns.

"We're very glad they're listening to us," said Frank Jackson, a member of the Vashon Island Community Council. "But if it gets down to a vote, we're concerned about whether we'll be heard."

But a spokesman for the protection agency, Anita Frankel, said, "It's not coming down to jobs versus arsenic."

Arsenic, a byproduct of smeltering, is used in herbicides and insecticides and in the production of glass. The Tacoma plant produces about one-third of the U.S. supply.

CHAPTER OUTLINE

8.1 TESTING A SINGLE MEAN

8.2 t FOR TWO—TESTING THE MEANS OF TWO SEPARATE BATCHES

8.3 t FOR TWINS—TESTING MEANS FOR TWO RELATED BATCHES

8.4 TESTING HYPOTHESES VERSUS CONFIDENCE LIMITS

CHAPTER EIGHT
COMPARING CENTERS

TESTING A SINGLE MEAN

8.1

One thing that researchers do a lot is suffer. Why? Research in reality is not like research in the movies. Research isn't the execution of a single brilliant experiment that produces a spectacular breakthrough. Instead it is a lot of hard work characterized more by breakdowns than breakthroughs. Rats die, roofs leak, equipment explodes, questionnaires are not returned, air conditioners become ice makers, grants dry up, and so on. In short, research is a slice of life.

Another thing that researchers do a lot is compare centers. We said earlier that research creates batches of numbers. Batches of numbers are difficult to comprehend until they are distilled into single numbers like measures of centers and spreads. Since the inference procedure is better developed for means than for other kinds of centers, we will focus on comparing means. Our interest, as always, is not in the batch of numbers, but in the parameters of the population from which the batch came. And don't forget that the *goal* is to understand phenomena by analyzing numbers that have been mapped onto things.

In this chapter we will consider three problems: (a) the mean of a single batch; (b) two means from separate batches in which the test is termed a t for two; and (c) two means from related batches in which the test is termed a t for twins. In Chapter 10 we will analyze k means (where $k > 2$) (a) from separate batches and (b) from related batches.

We will start with a single mean. This problem can be divided into two different situations: (a) when σ is known and N is large (≥ 30) and (b) when σ is unknown and N is any size. The first situation is less common than the second. The population standard deviation, σ, is rarely known. We will investigate the less realistic situation first, then examine the typical situation in which σ is unknown.

Another batch of numbers was acquired from the Crunchie Ice Cream Co.—the IQ scores for $N = 64$ vendors. The mean, $\overline{X}$, of these 64 IQ scores was 105.5. Assume that we wish to know whether this sample mean is a chance deviation *in either direction* from the population mean, μ. "In either direction" should be a signal for you to perform a two-tailed test. Two additional items of information are that the IQ scores for the standardization population of the particular intelligence test had a $\mu = 100$ and a $\sigma = 16$.

The test statistic follows the customary general formula:

$$\text{test statistic} = \frac{\text{a statistic} - \text{a parameter}}{\text{random variability}}$$

We have the statistic, $\overline{X}$, and the parameter, μ, but we lack knowledge of the sampling distribution of the test statistic and its random variability. Suppose that 1000 samples of size N were randomly drawn, then 1000 $\overline{X}$s were calculated and a frequency distribution of the $\overline{X}$s was constructed. The frequency distribution, the *sampling distribution of the mean,* is a *normal distribution* when N is large.[1] When $N \geq 30$, the distribution is approximately normal so that the standard normal distribution described in Chapter 4 is applicable as a test distribution. The standard deviation of the sampling distribution of the mean is called the *standard error of the mean,* $\sigma_{\overline{x}}$. The $\sigma_{\overline{x}}$ is *smaller* than the standard deviation, σ, for the population. Why? Because when random samples are selected and their means are computed, the extreme numbers above and below the means are lost. The standard error of the mean is

$$\sigma_{\overline{x}} = \frac{\sigma}{\sqrt{N}}$$

where σ is the standard deviation of the population and N is the number of cases in a batch.

The test statistic for a single mean is

$$Z = \frac{\overline{X} - \mu}{\sigma_{\overline{x}}}$$

where $\overline{X}$ is the mean of the batch, μ is the population mean, and $\sigma_{\overline{x}}$ is the standard error of the mean. The probability of the test statistic, $P(Z)$, can be determined from the unit normal curve (Table D) in the same way as in Chapter 5 for coin-tossing problems. In using Table D our interest is in the tail area (column C) in which the entries in proportion form are the probabilities of obtaining a specified Z value or larger (column A) by chance. By now you have mastered the steps in the decision process so we

1. This result is in accord with the *central limit theorem.* According to this theorem, the sampling distribution of the mean will become normal as N increases regardless of the shape of the population from which the samples are randomly drawn.

won't go through it by the numbers any more. For practice, we will set $\alpha = .01$. H_1 is that $\mu \neq 100$ (or that $\overline{X} \neq \mu$), and H_0 is that $\mu = 100$ (or that $\overline{X} = \mu$).

The standard error of the mean for the IQ problem is

$$\sigma_{\overline{x}} = \frac{\sigma}{\sqrt{N}}$$

$$= \frac{16}{\sqrt{64}}$$

$$= 2.00$$

The test statistic is

$$Z = \frac{\overline{X} - \mu}{\sigma_{\overline{x}}}$$

$$= \frac{105.5 - 100}{2}$$

$$= 2.75$$

The $P(Z = 2.75) = .006$. Why? Table D is entered in column A for a $Z = 2.75$ and a $P = .003$ is read for the right tail in column C. You will recall that the unit normal curve is a symmetrical, bell-shaped curve with large negative Z values on the left, a center of $Z = 0$, and large positive Z values on the right. Therefore, for the left tail the $P(Z = -2.75) = .003$ also, and for both tails the $P(Z = 2.75$ or $Z = -2.75) = .003 + .003 = .006$. (Remember that a two-tailed test is being performed, and use the additive rule for probabilities.) Since a $P = .006 < \alpha = .01$, the decisions are to reject H_0 and accept H_1. The indication is that the Crunchie vendors came from a population whose $\mu > 100$ IQ points. Or stated another way, their average IQ is significantly higher ($P = .006$) than the mean of the standardization population. The Crunchie vendors are, as Yogi Bear would say, "smarter than the average bear."

If a test statistic is normally distributed or approximately normally distributed, then Table D can be used to determine the probabilities for testing hypotheses. So far three test statistics have followed the normal distribution: (a) binomial outcomes when $p = q = .5$ and $N_T \geq 10$ (Chapter 5); (b) correlation coefficients when $\rho = 0$ and N is very large (Chapter 6); and (c) single means when $N \geq 30$ (this chapter). As other test statistics are also distributed normally, we will dwell upon the normal distribution for awhile. If Table D is entered for $Z = 1.00$, a value of .3413 is observed in column B, .3413 of the total curve's area = 1 falls between the mean of $Z = 0$ and $Z = 1.00$. Another .3413 falls between $Z = 0$ and $Z = -1.0$. Between $Z = -1.0$ and $Z = 1.0$ the area is .6826. If we multiply the total area = 1 by 100, we have percentages instead of proportions. If we think about *percentages of the numbers*, then 68.26% of the numbers would be expected to fall between $Z = -1.0$ and $Z = 1.0$. This logic is the basis for the rough check on the reasonableness of a calculated sample standard deviation proposed in Chapter 4. You will recall it was asserted that in a normal distribution about two-thirds (actually, 68.26%) of the numbers would be expected to fall between the limits: $\overline{X} \pm s$. $\overline{X} \pm s$ corresponds to $Z = \pm 1.0$.

The entries in some standard normal tables are percentages rather than proportions or probabilities. To secure the Ps for testing hypotheses, the percentage entries must be divided by 100. (After a while, going from percentages to proportions and back again becomes automatic.) Because a number of test statistics are normally distributed, it is helpful to store a few frequently used, critical Z values in one of your memory registers. Table 8.1 is a microtable showing the four critical Z values that are employed often in testing hypotheses.

P	Two-Tailed Test	One-Tailed Test
.05	1.96	1.645
.01	2.575	2.33

TABLE 8.1
Critical Z values for one- and two-tailed tests at $\alpha = .05$ and $\alpha = .01$.

If these critical Z values are remembered, then hypotheses regarding test statistics that are normally distributed can be evaluated without looking in Table D. The critical Z values can also be used to construct diagrams of hypothesis testing that many students find illuminating. You roughly draw a normal curve and mark off the critical value(s) of Z that are appropriate for the type of test (one- or two-tailed) and the α level. Figure 8.1 represents the problem we just completed. At or beyond $Z = \pm 2.575$ are the *rejection regions*. That is, if a computed $Z \geq \pm 2.575$, H_0 will be rejected and H_1 accepted. If $Z < \pm 2.575$, then H_0 is not rejected. Our obtained $Z = 2.75$ falls in the rejection region on the right. Thus, H_0 is rejected and H_1 is accepted.[2]

2. Note that the decisions about H_0 and H_1 are being made by comparing the test statistic with the critical value of the test statistic. This procedure, which was described in Chapter 6, is equivalent to comparing the probability of the test statistic to α.

FIGURE 8.1
A normal curve with the regions of rejection for a two-tailed test with $\alpha = .01$

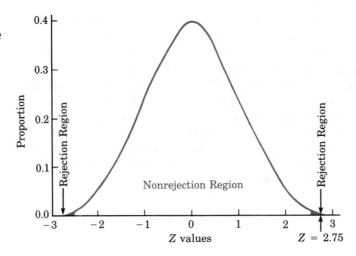

Figure 8.2 shows a normal curve diagram for a one-tailed test at $\alpha = .05$. For example, an obtained $Z = 1.47$ would fall within the *nonrejection* region. Thus, H_0 would not be rejected. It is necessary in one-tailed tests that the obtained Z falls into the rejection region that corresponds with H_1. If H_1: $\mu > 100$, then Z must be positive (right tail). If H_1: $\mu < 100$, then Z must be negative (left tail). Such diagrams can be constructed for hypothesis tests involving other theoretical distributions besides the normal curve.

FIGURE 8.2
A normal curve with the region of rejection for a one-tailed test with $\alpha = .05$

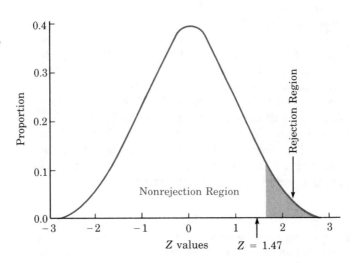

Finally, do not forget that the entries in column C of Table D are one-tailed Ps, and are thus suitable for one-tailed tests of hypotheses. For two-tailed tests the tabled Ps must be doubled. Whenever you apply a theoretical distribution, you need to inquire whether the tabled P values are for a one- or two-tailed test. Some tables like Table A for the t distribution and Table B for r conveniently offer both possibilities.

Next, we turn to the typical situation in which a single mean is to be tested, but σ is unknown and N is any size. We will analyze the time estimates for the women students from Chapter 2 (Section 2.2). Of course, unlike the IQ problem we do not know the value of σ for this batch of numbers. We enter the $N = 71$ numbers in our calculator and recall ΣX and ΣX^2. Next, we calculate $\overline{X}$ and s or call upon the special function keys to output them. I found: $\Sigma X = 7007$; $\Sigma X^2 = 771069$. Do you agree? Then we have

$$\overline{X} = \frac{\Sigma X}{N}$$

$$= \frac{7007}{71} = 98.69$$

and

$$s = \sqrt{\frac{\Sigma X^2 - \frac{(\Sigma X)^2}{N}}{N - 1}}$$

$$= \sqrt{\frac{771069 - \frac{(7007)^2}{71}}{71 - 1}} = 33.71$$

Without knowing σ, we can substitute s as the best estimator of σ. The matter of μ poses another problem. It is unknown. When this happens, a predicted value from a quantitative theory is a possible value for μ. But we do not have such a theory. The actual time interval was 93 seconds. If it is inserted for μ, then we can determine how the average of the women students' time estimates compared with the actual time interval. This appears to be a reasonable research question. The only matter that remains is: What is the sampling distribution of the mean for any size N? It is the t distribution that we employed in Chapter 6 to test correlation

FIGURE 8.3
A series of t distributions demonstrating how t becomes normal as df increases

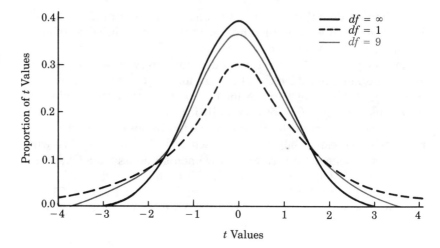

coefficients. Table A is entered for degrees of freedom ($df = N - 1 = 71 - 1 = 70$). The P for a computed t value can be found by comparing it with the tabled t values and with the proper scale for a one- or two-tailed test at the top of the table. As previously stated, the t distribution, unlike the unit normal curve, is a family of symmetrical curves that varies with df and becomes increasingly like a normal distribution as N increases. Since t distributions for small samples have fatter tails than a normal distribution, the critical t values will be *larger* than the Z values in the microtable above. Figure 8.3 shows the thicker tails of the t distributions and how t distributions shift toward normality with increasing degrees of freedom.

Now we will test the research hypothesis that women *overestimate* a time interval. H_1 would be that $\mu > 93$ and H_0 that $\mu \leq 93$. This is a one-tailed test and α is set at .01. The test statistic is

$$t = \frac{\overline{X} - \mu}{s_{\overline{x}}}$$

where $s_{\overline{x}}$ is the standard error of the mean. The standard error of the mean is

$$s_{\overline{x}} = \frac{s}{\sqrt{N}}$$

$$s_{\overline{x}} = \frac{33.71}{\sqrt{71}} = 4.0$$

and

$$t = \frac{98.69 - 93}{4} = 1.42$$

From Table A the $P(t = 1.42)$ for $df = N - 1 = 71 - 1 = 70$ and a one-tailed test lies between .10 and .05, which is symbolized as: $.10 > P > .05$. Since $P > \alpha = .01$, H_0 is not rejected. Although the average of the women's time estimates was higher than the actual time interval, the large variability of the women's estimates put the difference within the limits of chance. Variability is the devil's contribution to experimental research.

Problems are rarely solved by ignoring them. At this point we need to consider a problem that has been ignored—*degrees of freedom (df)*. In Chapter 6 the t table was entered for $df = N - 2$ for testing an r, and we have just entered the t table for $df = N - 1$ for testing a single mean. Why? Degrees of freedom refers to how many numbers in a batch are free to vary. If we have a batch of $N = 5$: 6, 7, 8, 9, 10, then $\Sigma X = 40$ and $\overline{X} = 8$. Earlier it was stressed that $\Sigma(X - \overline{X}) = 0$. The first four $(X - \overline{X})$ are $(6 - 8) = -2$; $(7 - 8) = -1$; $(8 - 8) = 0$; and $(9 - 8) = 1$. The $\Sigma(X - \overline{X})$ for the first four numbers is $(-2) + (-1) + 0 + 1 = -2$. To make the $\Sigma(X - \overline{X}) = 0$ for all five numbers the last number must be $(X - 8) = 2$ or $X = 10$. Four numbers are free to vary; the last number is fully determined. Therefore, the $df = N - 1 = 5 - 1 = 4$ in this example. Likewise, in the case of s, the estimated standard deviation, the rational formula is

$$s = \sqrt{\frac{\Sigma(X - \overline{X})^2}{N - 1}}$$

or:

$$s = \sqrt{\frac{\Sigma(X - \overline{X})^2}{df}}$$

Again, all Xs are free to vary except one.

More generally, the degrees of freedom equal the batch count (N) less the number of restrictions placed upon the data. Students often ask what are *the df*? There is no answer to this question because degrees of freedom vary with the statistical situation. For a t test of single $\overline{X}$ the error term, $s_{\overline{x}}$, involves a sample standard deviation with one restriction (or $df = N - 1$). For a t test of r,

$df = N - 2$. The two restrictions in this instance are (a) estimation of b_y (or b_x) and (b) estimation of a_y (or a_x) (Ferguson, 1981). Throughout the rest of the book values for dfs will be required. Ways to determine the dfs for specific problems will be described.

t FOR TWO—TESTING THE MEANS OF TWO SEPARATE BATCHES

8.2   The second problem is comparing the means of *two separate batches* of numbers. As a mnemonic device (and a little play on words) we will label this a *t-for-two* test. What does separate batches mean? It means that either the cases in the two batches were *randomly drawn* from two populations, or a larger batch of cases was *randomly divided* into two batches. "Randomly" means using coin tosses, a table of random numbers, or a computer procedure to ensure that each case has an equal likelihood of assignment to each batch. It is generally a good strategy, in terms of ease of calculation and soundness of experimental design, to try to have the same number of cases in each batch. Usually, in statistical jargon, separate batches are termed *independent groups*.

When the courageous and foolhardy attempt to teach statistics, they receive a perk: Anyone who has a collection of numbers will bring it in and ask what to do with it. There are two important questions for the consultant to ask the bearer of numbers:

1. What kind of numbers do you have? (Are they numbers, ranks, or frequencies?)
2. How did the cases get into the batches? (Are the numbers from *separate batches,* as defined above, or from *related batches?*) Here we will compare the means of numbers from two separate batches.

Our example will incorporate the time estimates from the women (one batch) and the men (another, separate batch). The research question is: Did the means of the two populations differ in either direction? In this situation we are concerned with a *sampling distribution of the difference in the means*. Conceptually, this would involve randomly drawing about 1000 *pairs of samples,* calculating the two means for each pair, and getting the difference between the two means. The distribution of the 1000 differences in the means would constitute the sampling distribution. As before, this sampling distribution has a standard deviation, *the standard error of the difference in the means*, $s_{\bar{x}_1-\bar{x}_2}$, which is the measure of a random variability for the test statistic.

Sometimes presenting statistics is like a crooked poker game. At this point we will pull a card from under the table. The card says *assumptions*. Statistical tests always depend upon a foundation of assumptions. When correlation was discussed, we emphasized the requirement of *linearity*. Linearity is one assumption—and probably the most critical one—for the Pearson product-moment r and its associated regression procedures. In comparing the means of two separate batches the following assumptions are made about the *populations* from which the batches of numbers were randomly drawn: (a) that both populations of numbers are normally distributed (*normality*), (b) that the variances of the numbers in both populations are equal (*homogeneity of variance*), and (c) that the numbers in the populations are independent of one another (*independence*). The sampling distribution of the difference in the means follows a t distribution with

$$df = (N_1 - 1) + (N_2 - 1) = N_1 + N_2 - 2$$

where N_1 is the number of cases in one batch and N_2 is the number of cases in the other batch. The $df = N_1 + N_2 - 2$ because there is one restriction upon each batch.

The sampling distribution will follow a t distribution if and only if the three preceding assumptions are met. If the assumptions are violated, then the sampling distribution will not conform exactly to a t distribution. For example, the distribution can be skewed instead of symmetrical, or it can be too peaked or too flat in comparison with a t distribution. As a consequence of these departures from a t distribution, the P values from Table A would be wrong. A table value of $P = .05$ might actually be $P = .08$ or $P = .02$.

If the assumption of *homogeneity of variance* is met (the variances are equal), then the variances of the two batches are pooled to obtain a single best estimate (s_p^2) of σ^2:

$$s_p^2 = \frac{(N_1 - 1)s_1^2 + (N_2 - 1)s_2^2}{N_1 + N_2 - 2}$$

Then the standard error of the difference in the means is

$$s_{\bar{x}_1-\bar{x}_2} = \sqrt{s_p^2\left(\frac{1}{N_1} + \frac{1}{N_2}\right)}$$

This standard error is commonly referred to as the *pooled error term*. If the variances are not equal (*heterogeneity of variance*), then the variances are not pooled and the formula for the *non-pooled error term* is

$$s_{\bar{x}_1-\bar{x}_2} = \sqrt{\frac{s_1^2}{N_1} + \frac{s_2^2}{N_2}}$$

The word, equal, does not mean identical; it means that the variances do not differ by more than random fluctuations.

The previously calculated statistics for the men and women are displayed in Table 8.2.

TABLE 8.2
Means and standard deviations of the time estimates for men and women

	Men	Women
$\bar{X}$	91.88	98.69
s	26.95	33.71
N	40	71

The alternative hypothesis is that the population mean of the men (μ_1) differs in either direction from the population mean of the women (μ_2), or $H_1: \mu_1 \neq \mu_2$. The null hypothesis is $H_0: \mu_1 = \mu_2$, and α is set at .05. The test statistic is

$$t = \frac{(\bar{X}_1 - \bar{X}_2) - (\mu_1 - \mu_2)}{s_{\bar{x}_1-\bar{x}_2}}$$

Under H_0, $(\mu_1 - \mu_2) = 0$, so we can omit that term. What about homogeneity of variance? For the time being we will be content to inspect the variances: $s_1^2 = (26.95)^2 = 726.30$ and $s_2^2 = (33.71)^2 = 1136.36$. While s^2s differ somewhat, we will provisionally assume homogeneity of variance and pool the variances:

$$s_p^2 = \frac{(N_1 - 1)s_1^2 + (N_2 - 1)s_2^2}{N_1 + N_2 - 2}$$

$$= \frac{(40 - 1)726.3 + (71 - 1)1136.36}{40 + 71 - 2}$$

$$= \frac{28325.7 + 79545.2}{109}$$

$$= 989.64$$

Then the pooled error term is

$$s_{\bar{x}_1-\bar{x}_2} = \sqrt{s_p^2\left(\frac{1}{N_1} + \frac{1}{N_2}\right)}$$

SEC. 8.2 □ t FOR TWO—TESTING THE MEANS OF TWO SEPARATE BATCHES

$$= \sqrt{989.64\left(\frac{1}{40} + \frac{1}{71}\right)}$$

$$= \sqrt{38.68} = 6.22$$

After omitting $(\mu_1 - \mu_2)$, the test statistic is

$$t = \frac{\overline{X}_1 - \overline{X}_2}{s_{\overline{x}_1 - \overline{x}_2}}$$

$$= \frac{91.88 - 98.69}{6.22} = -1.09$$

From Table A the $P(t = -1.09) > .20$ for a two-tailed test with $df = N_1 + N_2 - 2 = 40 + 71 - 2 = 109$. (There is no entry for $df = 109$, so the nearest value, $df = 120$, is used.) Since

$$P(t = -1.09) > .20 > \alpha = .05$$

H_0 cannot be rejected. There is no evidence that the population means of the men and women differed when a 93-second interval was estimated.

Eyeballs should never be abandoned, but different eyeballs do not always produce the same judgments. Someone else might decide that the variances for the two preceding batches do not look equal. Then, the *nonpooled* formula for $s_{\overline{x}_1 - \overline{x}_2}$ is appropriate. If you do this, you will discover that the same statistical decision about the means will be reached. In Chapter 9 we will show how to test the variances for the two batches statistically.

I am a big booster of demonstrations. How about doing one right now? Suppose we have $N_1 = N_2 = 11$ and $s_1^2 = 5$ and $s_2^2 = 10$. Let us apply both formulas for $s_{\overline{x}_1 - \overline{x}_2}$ and see if there is anything to be learned. The pooled version is

$$s_p^2 = \frac{(N_1 - 1)s_1^2 + (N_2 - 1)s_2^2}{N_1 + N_2 - 2}$$

$$= \frac{(11 - 1)5 + (11 - 1)10}{11 + 11 - 2} = 7.5$$

Then the pooled error term is

$$s_{\overline{x}_1 - \overline{x}_2} = \sqrt{s_p^2\left(\frac{1}{N_1} + \frac{1}{N_2}\right)}$$

$$= \sqrt{7.5\left(\frac{1}{11} + \frac{1}{11}\right)} = 1.17$$

The nonpooled error term is

$$s_{\bar{x}_1 - \bar{x}_2} = \sqrt{\frac{s_1^2}{N_1} + \frac{s_2^2}{N_2}}$$

$$= \sqrt{\frac{5}{11} + \frac{10}{11}} = 1.17$$

What does this living demonstration show? When the Ns for the batches are equal, the two formulas yield the same answer. Therefore, when $N_1 = N_2$, don't wear out your calculator. Always use the simpler, nonpooled formula.

We said, in agreement with Cohen (1965), that significance tests should be supplemented by information regarding the percentage of explained variance. The variance in the time estimation example is the variability of the 111 time estimates. How much of this variability can be accounted for by the sex of the estimator?

There is a Pearson product-moment r for an arrangement in which one variable (Y) is in the form of numbers and the other variable (X) is a dichotomy such as male versus female. It is called a *point-biserial correlation coefficient*, r_{pb}. To find r_{pb}, assign a score of 0 in X for every man and a score of 1 in X for every woman, treat their time estimates as the Ys, and compute a Pearson product-moment r between X and Y for the $N = 111$ cases with the regular computational formula for r in Chapter 6. There is nothing sacred about the 0 and 1 coding for X. If you assign 1 and 0 for the men and women, respectively, it will produce an r_{pb} of the same size but reversed in sign. Or, men and women could be coded 1 and 2 or 12 and 27; the size of r_{pb} will be invariant. If r_{pb} is squared and multiplied by 100, the result will be the percentage of the variance in the time estimates (Y) that can be explained by the sex of the estimator (X). Notice that by coding X, we are converting univariate data (Y) into bivariate data with an X score (the code) and a Y score (the time estimate) for every case.

The numbers for the combined batches are shown in Table 8.3. The quantities needed to calculate r_{pb} include $\Sigma X = 71$, $\Sigma X^2 = 71$, $\Sigma Y = 10682$, $\Sigma Y^2 = 1137040$, $\Sigma XY = 7007$, and $N = 111$. With the regular r computational formula I obtained $r_{pb} = .104$. Why don't you check it? If it is correct, then $r_{pb}^2(100)$ or $(.104)^2 (100) = 1.08\%$. Thus, only a little more than 1% of the variability in the time estimates can be explained in terms of the

SEC. 8.2 □ t FOR TWO—TESTING THE MEANS OF TWO SEPARATE BATCHES

Sex	Case	X	Y	XY
Men	1	0	45	0
	2	0	50	0
	.	.	.	.
	.	.	.	.
	.	.	.	.
	40	0	180	0
Women	1	1	70	70
	2	1	105	105
	.	.	.	.
	.	.	.	.
	.	.	.	.
	71	1	180	180

TABLE 8.3
Group codes (X) and time estimates (Y) for 111 students

estimator's sex. The trivial amount of accounted-for variance is not too surprising in light of the small t value that was observed.[3]

There is an easier way to determine the accounted-for variance. This method is based upon the fact that in the t-for-two situation r_{pb} and t are related:

$$r_{pb} = \sqrt{\frac{t^2}{t^2 + df}}$$

For this problem,

$$r_{pb} = \sqrt{\frac{(-1.09)^2}{(-1.09)^2 + 109}}$$

$$= \sqrt{\frac{1.1881}{110.1881}} = .104$$

When getting r_{pb} with this formula, be careful regarding the df term in the formula: $df = N_1 + N_2 - 2 = 40 + 71 - 2 = 109$. In summary: in the case of two means for separate batches, the percentage of variance accounted for by the grouping variable (here, sex of the estimator) can be determined by computing r_{pb}, squaring it, and multiplying by 100. The coefficient, r_{pb}, can be found either by coding the grouping variable and calculating r_{pb} with the computational formula for r, or more easily by applying a for-

---

3. Usually, when the t test of the means is not significant, the investigator would not be interested in computing the percentage of accounted-for variance.

mula that links r_{pb} and t. The message is: Don't stop after doing a t-for-two test; find the percentage of accounted-for variance, too.

8.3　t FOR TWINS—TESTING MEANS FOR TWO RELATED BATCHES

Lastly, the problem of comparing the means from two *related batches* will be considered. The troublesome word here is related. My mnemonic label for the test statistic in this situation is *t for twins*. Instead of comparing the means from separate batches, as defined above, the contrast is between the means for a set of *paired numbers*. The pairing can occur in two ways: (a) the members of each pair are matched on another variable (appropriately termed the *matching variable*) or (b) the pairs of numbers result from a single batch of cases *tested twice*.

How about a hypothetical example of matching? A researcher believes that performance on a creativity test can be improved by special instructions like urging the test takers to think about novel ways of looking at problems and to imagine unusual solutions to problems. The investigator is worried, however, that if two batches were selected randomly, the two batches might not be equal in intelligence. So a large batch of subjects is first administered an intelligence test, and pairs of subjects are selected who are alike in IQ (the matching variable). For example, say Case 12 had an IQ of 127 and Case 28 had an IQ of 126. These two cases are selected to form one pair. This procedure continues until 20 matched-in-IQ pairs are selected. For each pair a coin is flipped to decide which case undergoes the experimental treatment (special instructions for the creativity test) and which case is exposed to the control treatment (no special instructions). Thus, the selection procedure yields two batches of subjects closely matched in intelligence. Intelligence is thereby ruled out as an explanation for any observed difference in the means of creativity scores for the two batches. To compare the two means, a t-for-twins test is performed (the batches are related by the matching on IQ). Another example would be a study in which the pairs were identical twins. In this study the matching variable would be genetic identity.

The second form of related batches occurs when a batch of cases is *tested twice*. This is labeled in the statistical literature a *repeated-measures design* or "using each subject as his or her own control." In this type of design the matching is perfect—the same case is exposed to both treatments. Previously, we stated that an important question for a consultant to ask a client is: How did the

SEC. 8.3 □ t FOR TWINS—TESTING MEANS FOR TWO RELATED BATCHES

cases get into the groups? The import of that question should now be evident. We are attempting to learn whether: (a) the cases were randomly selected in the two separate batches for which a t-for-two test is appropriate or (b) if matching or repeated testing produced related batches for which the t-for-twins test is appropriate.

Next, we will describe the t-for-twins test. Whether the related batches come from matching cases or from repeated testing of cases is irrelevant—the t-for-twins test is the same for both procedures. A general formula for the standard error of the difference in two means is

$$s_{\bar{x}_1 - \bar{x}_2} = \sqrt{s_{\bar{x}_1}^2 + s_{\bar{x}_2}^2 - 2rs_{\bar{x}_1}s_{\bar{x}_2}}$$

where r is the Pearson product-moment r between the pairs of numbers, $s_{\bar{x}_1}$ is the standard error of the mean in the first batch, and $s_{\bar{x}_2}$ is the standard error of the mean in the second batch. (When there are separate batches, r is assumed to be zero because no pairing is present. Then the formula reduces to

$$s_{\bar{x}_1 - \bar{x}_2} = \sqrt{\frac{s_1^2}{N_1} + \frac{s_2^2}{N_2}}$$

which is the unpooled error term for the t test.) With related batches we could calculate an r for the paired numbers from the related batches, calculate the complicated error term, and compare the means by a t test:

$$t = \frac{\bar{X}_1 - \bar{X}_2}{s_{\bar{x}_1 - \bar{x}_2}}$$

Luckily, a much simpler, equivalent test is possible. The paired numbers in the related batches are correlated. If we subtract the number for each member of a pair in batch 2 from the number for the paired member in batch 1, we end with a set of *difference scores* (Ds) that are free from the effects of the correlation. These D scores are analyzed in the t-for-twins test. How does analysis of the Ds test the means of the related batches? The t test of the Ds does this because $\bar{X}_1 - \bar{X}_2 = \bar{D}$, where $\bar{D}$ is the mean of the D scores. A new test is not required for testing the D scores: Simply treat the D scores as the Xs for a single batch and perform the t test for a single mean described in the first section of this chapter:

$$t = \frac{\bar{X} - \mu}{s_{\bar{x}}}$$

Translating X into D, the formula becomes

$$t = \frac{\bar{D} - \mu_D}{s_{\bar{D}}}$$

where $\bar{D}$ is the mean of the D scores, μ_D is the population mean of the Ds, and $s_{\bar{D}}$ is the standard error of the mean of the Ds. The degrees of freedom for the t test, which are $N - 1$ for the Xs, will be $N_D - 1$ for Ds, where N_D is the number of Ds.

Are you ready for an example? A tire manufacturer, O. Goodie Blimp, wanted to know whether steel-belted radial tires yield more miles per gallon than polyester radials. He did this study. Twenty pairs of new cars were driven from Piscataway, New Jersey to Azusa, California. The cars traveled in a caravan to control for the route, speed driven, and so on. Each pair of cars was a different make and identically equipped except for the tires. For example, pair 1 was two Ford Escorts. (Are you awake? What is the matching variable in this investigation?) The cars in the first batch had new steel-belted radials and those in the second batch had new polyester radials. (Goodie built all the tires.) Goodie is a devotee of steel-belted radials and his prediction was that the cars with the steel-belted tires would get more miles to the gallon. The results are presented in Table 8.4.

Goodie's statistical hypotheses were H_1: $\mu_D > 0$ and H_0: $\mu_D \leq 0$. Observe that forming the hypotheses for the one-tailed test depended on how the D scores are calculated. In the results $D = X_1 - X_2$, so the greater-than sign in H_1 fits Goodie's predic-

TABLE 8.4
Miles per gallon for 20 pairs of cars equipped with steel-belted and polyester radial tires

Pair	Steel	Poly	D	Pair	Steel	Poly	D
1	27.2	26.5	0.7	11	35.6	34.9	0.7
2	17.4	17.5	−0.1	12	27.4	27.9	−0.5
3	33.0	31.9	1.1	13	31.6	30.5	1.1
4	24.6	25.1	−0.5	14	18.7	18.5	0.2
5	12.7	11.9	0.8	15	24.0	23.7	0.3
6	41.4	40.9	0.5	16	16.3	15.3	1.0
7	36.8	36.2	0.6	17	11.7	10.8	0.9
8	25.3	24.8	0.5	18	19.4	19.2	0.2
9	31.2	30.5	0.7	19	16.5	16.5	0.0
10	19.9	19.5	0.4	20	33.1	32.8	0.3
				Means	25.19	24.745	0.445

SEC. 8.3 □ t FOR TWINS—TESTING MEANS FOR TWO RELATED BATCHES

tion of superiority of steel-belted tires. (If $D = X_2 - X_1$, then H_1: $\mu_D < 0$.) There are a few other details to attend to.

1. Does $\overline{X}_1 - \overline{X}_2 = \overline{D}$? If not, start checking.
2. In entering the Ds in your calculator, the *signs* must be included. One way to do this for the score -0.1 is: 0.1 $\boxed{+/-}$ $\boxed{M+}$. The $\boxed{+/-}$ key is a change-of-sign key; in some calculators it is designated as $\boxed{CHS}$.
3. The $D = 0.0$ is a number and must be entered.
4. It does not hurt and often helps to look at your numbers.

There are 20 Ds: Sixteen are positive, three are negative, and one is zero. If chance alone were operating, we would expect 10 positive and 10 negative D scores. Goodie's hypothesis appears promising. Goodie did not want to make a Type 1 error. Therefore, he set $\alpha = .001$. For the D scores my calculator output: $\Sigma D = 8.9$ and $\Sigma D^2 = 8.13$. The standard deviation of the Ds is

$$s_D = \sqrt{\frac{\Sigma D^2 - \frac{(\Sigma D)^2}{N_D}}{N_D - 1}}$$

$$s_D = \sqrt{\frac{8.13 - \frac{(8.9)^2}{20}}{20 - 1}}$$

$$= 0.4685$$

The standard error of the mean of the Ds is

$$s_{\overline{D}} = \frac{s_D}{\sqrt{N_D}}$$

$$s_{\overline{D}} = \frac{0.4685}{\sqrt{20}} = 0.105$$

The t test, then, is

$$t = \frac{\overline{D} - \mu_D}{s_{\overline{D}}}$$

$$= \frac{0.445 - 0}{0.105}$$

$$= 4.24$$

From Table A and $df = N_D - 1 = 20 - 1 = 19$, the $P(t = 4.24)$ is less than .0005. Since $P(t = 4.24) < \alpha$, H_0 is rejected and H_1 is accepted.

The outcome of Goodie's study is instructive. He found what he expected. The cars with steel-belted tires outperformed those with polyester radials. Assuming that his finding could be replicated, should he become ecstatic over it? This question forces us to consider the matter of *effect size,* μ_D. Our best estimate of μ_D is $\overline{D}$. Is a gain of less than a half of a mile per gallon of practical importance? Note also that for three of the car makes the steel-belted tires were possibly less effective. Is the added fraction of a mile worth the additional cost of steel-belted tires? Whatever the answer, this is an important general question in any study: How big is the effect? Is it large enough to make a difference, theoretically or practically?

It is time for the crooked poker game again. What are the *assumptions* for a *t* test with related batches? The assumptions concern the population *difference scores,* not the numbers in the populations for the batches. It is assumed that: (a) the population difference scores are normally distributed (*normality*), and (b) the population difference scores are independent of one another (*independence*). We will discuss normality and independence in greater detail in later chapters.

Let us look again at designs with matched groups. Achieving related batches by matching is a way of controlling for the matching variable. But achieving equivalence of batches by randomization controls for many variables simultaneously.[4] For a related-batches design to be efficient, the matching variable should correlate substantially and positively with the *dependent variable* that the investigator is studying. Then, the *r* in the general formula for the standard error of the difference in two means will also be large and positive and the standard error will be reduced in comparison to the standard error for a separate-batches design. If the *r* between the matching variable and the dependent variable is zero or negative, the standard error will be either unchanged or larger, respectively. In addition, degrees of freedom are lost in matching—the *df* for related batches are one-half of those for separate batches: $N_D - 1$ versus $N_1 + N_2 - 2$.[5] Unless you have evidence of a high positive *r* between the matching variable and the dependent variable, you should be wary of investing a lot of money in a matched-group design.

4. Equalizing batches by randomization is a fundamental principle in experimental design. This principle was contributed by R.A. Fisher.
5. Larger *df* are desirable because smaller critical *t* values are required for larger *df*. Thus, an investigator has a better chance to reject H_0 with larger *df*.

Designs with related batches produced by testing twice also have problems. The effects (learning, fatigue, etc.) from the first test can carry over and influence what happens on the second test. If Goodie had tested 20 cars twice, equipping them with steel-belted tires for the first test and then with polyester radials for the second test, he might have a carry-over problem. The cars would be older, might be less well tuned, or have alignment problems after the first test, and the miles per gallon for the second test could reflect *these* factors as well as the type of tire. It might be dawning on you that I am not a strong advocate of designs involving related batches as a general strategy. You are correct. For openers, they violate a wise and fundamental dictum in research: When in doubt, randomize like crazy.

TESTING HYPOTHESES VERSUS CONFIDENCE LIMITS

8.4

So far in the book we have concentrated on testing statistical hypotheses regarding correlation coefficients, means, and so forth. Another part of applied statistics is concerned with *estimation*. If we know $\overline{X}$ for a batch, what is our best estimate of μ? Usually, the best estimate of μ is $\overline{X}$. $\overline{X}$ is a best estimate of μ in two respects: (a) it is unbiased—the mean of $\overline{X}$s from repeated random samples will converge upon μ; and (b) it is a minimum variance estimator—the standard error of the mean will be smaller than the standard error for other centers like the median or mode. And if we don't know σ, then s from a batch can serve as an estimate of σ. In these instances the values of $\overline{X}$ and s are termed *point estimates*. Other types of estimates are designated as *interval estimates* or *confidence intervals*. Again, we might have the $\overline{X}$ for a batch, but instead of just estimating μ (a point estimate), we might wish to state with a certain degree of confidence between what values or limits μ is likely to fall. Some statisticians are not in favor of hypothesis testing and believe that point and interval estimates should be used instead. They argue that confidence procedures are more informative than hypothesis tests; stating an interval estimate is more informative than simply rejecting or not rejecting an H_0. Other statisticians (e.g., Miller, 1966) see a place for both analytical procedures. Miller (1966) contends that interval estimates are valuable for answering questions about the "apparent magnitude of the effect." Since most behavioral scientists engage in hypothesis testing, we have focused on it. Nevertheless, you should understand interval estimates, so a few examples will be presented.

In the problem on the women's time estimates we observed $\overline{X} = 98.69$, $s_{\overline{x}} = 4.0$, and $N = 71$. Suppose that we want to find two values, the *confidence limits,* for the population mean, μ. Suppose further that we desire two limits such that we could assert with 95% confidence that μ is likely to be contained within them. The confidence limits for the population mean are

$$\mu = \overline{X} \pm t_{.05} s_{\overline{x}}$$

where t is the critical value for t at $\alpha = .05$ (two-tailed) for $df = N - 1 = 71 - 1 = 70$. From Table A, $t_{.05} = 2.00$ for $df = 60$ (the nearest entry to $df = 70$). The confidence limits are

$$\mu = 98.69 - (2.0)(4.0) = 90.69$$
$$= 98.69 + (2.0)(4.0) = 106.69$$

Often the limits are presented as

$$90.69 \leq \mu \leq 106.69$$

We can be 95% confident that these limits contain μ. The *confidence interval,* it should be noted, is the distance between the *confidence limits,* or $106.69 - 90.69 = 16$.

An investigator might want to be more certain. For example, if the investigator desired 99% confidence, then the limits would be

$$\mu = \overline{X} \pm t_{.01} s_{\overline{x}}$$

From Table A, $t_{.01} = 2.66$ for $df = 60$ and the limits would be

$$\mu = 98.69 - (2.66)(4) = 88.05$$
$$= 98.69 + (2.66)(4) = 109.33$$

Or, using the other mode of presentation,

$$88.05 \leq \mu \leq 109.33$$

Comparing these two sets of confidence limits, we can see that the width of the confidence interval increases with the degree of confidence—the interval is wider for 99% than 95%. It is also noteworthy that the test statistic is set at $\alpha = .05$ (two-tailed) for 95% confidence and at $\alpha = .01$ (two-tailed) for the 99% confidence level.

Suppose that we repeated the time estimation study many times. Different $\overline{X}$s would be observed and different sets of confidence limits (say, at $\alpha = .05$) would result. What could we conclude from this hypothetical situation? If one set of confidence

SEC. 8.4 □ TESTING HYPOTHESES VERSUS CONFIDENCE LIMITS

limits were selected at random, then the *probability* is $1 - \alpha = 1 - .05 = .95$ that this particular set would include μ.

The distinction between hypothesis testing and interval estimation becomes a bit murky if we recall that μ, the actual time interval, was 93 seconds. That value falls within both sets of confidence limits above, and in the hypothesis test we failed to reject H_0.

In the second single mean problem regarding vendors' IQ scores we had $\overline{X} = 105.5$, $\sigma_{\overline{x}} = 2$, and $N = 64$. When $N \geq 30$, confidence limits can be calculated with critical values from the normal curve. The 99% limits here are

$$\mu = \overline{X} \pm Z_{.01}\sigma_{\overline{x}}$$

From the microtable for critical Z values (or Table D) $Z_{.01} = 2.575$; accordingly, the limits are

$$\mu = 105.5 - (2.575)(2) = 100.35$$
$$\mu = 105.5 + (2.575)(2) = 110.65$$

The $\mu = 100$ for this problem falls outside the 99% confidence limits and H_0 was rejected in the hypothesis testing. In the hypothesis test the formula for calculating the test statistic was

$$Z = \frac{\overline{X} - \mu}{\sigma_{\overline{x}}}$$

If $\pm Z_{.01}$ (the critical values of Z for a two-tailed test) are substituted for Z, the formula becomes

$$\pm Z_{.01} = \frac{\overline{X} - \mu}{\sigma_{\overline{x}}}$$

And, solving for μ,

$$\mu = \overline{X} \pm Z_{.01}\sigma_{\overline{x}}$$

The last formula is obviously the formula for calculating confidence limits. The distinction between confidence limits and hypothesis testing is confusing. Let us try to cut through the confusion. If confidence limits are calculated and presented, that is *interval estimation*. If the confidence limits are calculated *and* a value is examined to determine whether it falls within or beyond the limits, *interval estimation becomes hypothesis testing*. Proponents of confidence procedures would, at this point, stress that confidence limits permit a data analyst to test many hypotheses in addition to the single hypothesis that $\mu = 93$. Are, for example, $\mu = 95$, $\mu = 91$, and so on contained in the interval?

Can confidence limits be found for other parameters besides a single mean? Yes, all that must be known is the sampling distribution and the standard error. As another example of confidence limits for means we will reconsider the previous problem comparing the mean time estimates for the men and women students. The 95% confidence limits for the difference in the means of separate batches is

$$\mu_1 - \mu_2 = \overline{X}_1 - \overline{X}_2 \pm t_{.05} s_{\overline{x}_1 - \overline{x}_2}$$

Putting the values obtained earlier and $t_{.05} = 1.98$ for $df = 120$ (the nearest entry to $df = 109$) in the formula we have

$$\mu_1 - \mu_2 = 91.88 - 98.69 - (1.98)(6.22)$$
$$= -19.13$$
$$\mu_1 - \mu_2 = 91.88 - 98.69 + (1.98)(6.22)$$
$$= 5.51$$

Or displaying the limits the other way,

$$-19.13 \leq \mu_1 - \mu_2 \leq 5.51$$

If we assert that we are 95% confident that the difference in the population means would fall between -19.13 and 5.51, then we are doing an *interval estimate*. If, however, we checked to see if a hypothesized difference in the means is contained in the confidence limits, then an hypothesis test is being performed. Similarly, confidence limits can be calculated for population correlation coefficients, medians, and so on. If we are cognizant of the sampling distribution and standard error, we are in business.

In the rest of the book the emphasis will continue to be upon hypothesis testing. But do not ignore the fact that a value with some limits around it is more informative than the value alone.

SUMMARY

One activity that researchers do often is compare centers, particularly means. This chapter considered three problems: (a) the mean of a single batch, (b) the means of two separate batches (t for two), and (c) the means of two related batches (t for twins). The single mean problem includes a single batch in which σ is known and $N \geq 30$ and the more typical situation in which σ is unknown and N is any size. In the first instance the sampling distribution

□ SUMMARY   185

of the mean is a normal distribution, the test statistic is Z, and the probabilities for Z are found in the unit normal curve (Table D). In the more common situation, in which σ is unknown, the sampling distribution of the mean follows the t distribution (Table A) for degrees of freedom, df, of $N - 1$. In this situation N can be any size, large or small. The df is the batch count, N, less the number of restrictions on the data. Statistical hypotheses regarding single means were tested by following the same inference procedure employed earlier for binomial and correlation problems.

Next, the problem of testing means from two separate batches was presented. Separate batches occur when the cases in the batches are randomly selected or when a large batch is randomly split into two batches. The sampling distribution of the difference in the means follows a t distribution for $df = N_1 + N_2 - 2$. The t-for-two test assumes that: (a) the numbers in both populations from which the batches were randomly drawn are normally distributed (normality), (b) both populations have equal variances (homogeneity of variance), and (c) the numbers in the populations are independent (independence). If these assumptions are not met, then the sampling distribution of the difference in the means will not follow a t distribution exactly and probability values read from Table A will not be accurate. When the variances of the batches are equal (homogeneity of variance), they may be pooled, and the pooled estimate is used to obtain the standard error of the difference in the means (the pooled error term). When the variances of the batches are unequal (heterogeneity of variance), the variances are not pooled, and a nonpooled error term is employed. If $N_1 = N_2$, the formulas for the pooled and nonpooled error terms give the same answer. Thus, the simpler, nonpooled formula should always be applied when $N_1 = N_2$.

After performing a t-for-two test, the percentage of accounted-for variance due to the treatments should be ascertained. This can be done by computing a point-biserial correlation coefficient (r_{pb}), squaring it, and multiplying by 100. The r_{pb} can be found in two ways: by coding membership in the treatments (X) and calculating an r with the dependent variable numbers (Y); or more easily, by a formula that relates r_{pb} and t.

Comparing the means for two related batches was described next. Related batches arise in two ways: the cases are formed into pairs on the basis of numbers from a matching variable, or the same batch of cases is tested twice (repeated measures). The t-for-twins test is performed upon the difference scores (Ds) from the paired numbers. The test is simply the t test for a single mean from the first section of the chapter. The test of the Ds tells about the difference in the means of related batches because $\bar{X}_1 - \bar{X}_2 = \bar{D}$. The t-for-twins test assumes: (1) normality of the Ds in the population and (2) independence of the Ds. The computational example raised the question of effect size, μ_D. Is the observed effect in any study large enough to be deemed important? Finally, difficulties with designs involving related batches were considered. In the matching procedure the matching variable must be substantially and positively correlated with the variable being investigated (dependent variable) to reduce the error term and to offset the loss in df in comparison with the t-for-two test. Designs with related batches, as a result of repeated measures, may be plagued by carryover effects in which the events of the first test influence performance on the second test.

Another topic in applied statistics is estimation. In point estimation a statistic (e.g., $\bar{X}$) serves as an estimator of a parameter (e.g., μ). In interval estimation the limits within which a parameter is likely to fall are determined. Confidence limits define a confidence interval about which we may have a certain percent of confidence that the interval will contain the parameter. Proponents of confidence methods contend that they are more informative than hypothesis testing. Whenever the sampling distribution and standard error of a statistic are known, confidence limits may be constructed. Greater confidence (e.g., 99%) results in wider confidence intervals than less confidence (e.g., 95%). In computing 95% confidence limits a critical value of a test statistic for $\alpha = .05$ (two-tailed) is employed; for the 99% limits the test statistic for $\alpha = .01$ (two-tailed) is used. When confidence limits are calculated and presented, the data analyst is doing interval estimation. When a value is compared to confidence limits to see whether it falls within or beyond the limits, interval estimation becomes hypothesis testing.

PROBLEMS

1. In a drug study 24 patients who have chronic headaches are randomly and equally assigned to two treatments—a new drug designed to prevent headaches (E) and no drug (C). The patients record the number of headaches they experience over a 90-day period. Given the data below, follow the six-step inference procedure to perform a one-tailed test ($\alpha = .05$) of the hypothesis $H_0: \mu_E \geq \mu_C$.

   E: 10, 13, 12, 17, 11, 11, 15, 14, 17, 18, 14, 14
   C: 16, 16, 20, 19, 16, 18, 13, 13, 19, 14, 15, 12

2. A critic immediately raises the possibility of a placebo effect in the drug study. She substitutes a placebo for the drug and repeats the study exactly with 24 new subjects. Test the same hypothesis as in Problem 1.

   E: 13, 15, 17, 18, 13, 16, 10, 12, 16, 13, 12, 9
   C: 15, 17, 19, 17, 19, 12, 14, 20, 13, 14, 13, 20

3. Considering the results of these two drug studies, what would you conclude? Why?

4. A student executes a computer program to output 50 random observations from a normal population with $\mu = 10$ and $\sigma = 2$. The $\overline{X}$ of the 50 scores is 10.75. Should the student suspect the computer program? Use $\alpha = .05$.

5. In a second execution of the program a $\overline{X} = 9.73$ is found for 50 random observations. Do the means of the two batches differ significantly ($\alpha = .05$, two-tailed)?

6. Seven female executives in a large corporation contend that they are being discriminated against in terms of salaries. Their salaries in thousands of dollars are: 31.3, 29.5, 27.4, 28.7, 29.4, 29.0, 30.7. The $\overline{X} = 31.4$ for the male executives. Use $\alpha = .05$ and make a decision about the women's charge.

7. Find r_{pb}s for Problems 1 and 2.

8. What percentages of the variabilities in the numbers of headaches can be attributed to the treatments in Problems 1 and 2?

9. The scores for 10 students in an advanced math course on the midterm and final examinations are presented here. In both tests 100 points were possible. Did a significant change in the average performance occur in either direction? Use $\alpha = .05$.

Exam	\multicolumn{10}{c}{Students}
	1	2	3	4	5	6	7	8	9	10
Midterm:	47	58	87	79	79	84	77	64	98	78
Final:	51	54	92	82	81	86	75	62	97	80

10. Find the 99% confidence limits for $\mu_1 - \mu_2$ in Problems 4 and 5.
11. Find the 95% confidence limits for the population mean for women in Problem 6. What do the limits mean?

ADDENDUM: USING COMPUTERS IN STATISTICS

MINITAB is handy for comparing centers. First, a one-tailed test of the women's time estimates against the actual time interval of

93 seconds is done. Note that MINITAB tells you the P-value for a $t = 1.422$ and even makes the statistical decision for you (1)! Second, a t-for-two test (2) is done for the men's time estimates versus the women's. The pooled error term is utilized. It is noteworthy that MINITAB automatically outputs the 95% confidence limits on the difference in the means, states the H_0 and H_1 being tested, provides the probability for t, and makes the statistical decisions. If someone would teach MINITAB how to put the data into memory and what commands to issue, it would be a complete statistical slave. Lastly, a t-for-twins test (3) is carried out upon the tire data. As was indicated in the chapter, this test is simply a t test on the difference scores.

```
--TTEST OF MU=93 ALT.=+1 ON C2
   C2    N = 71    MEAN =    98.690    ST.DEV. =    33.7

   TEST OF MU =    93.0000 VS. MU G.T.    93.0000
   T = 1.422
   THE TEST IS SIGNIFICANT AT 0.0797                              (1)
   CANNOT REJECT AT ALPHA = 0.05

--POOLEDT FOR C1 C2
   C1    N = 40    MEAN =    91.875    ST.DEV. =    27.0
   C2    N = 71    MEAN =    98.690    ST.DEV. =    33.7

   DEGREES OF FREEDOM = 109

   A 95.00 PERCENT C.I. FOR MU1-MU2 IS (-19.1447, 5.5144)

   TEST OF MU1 = MU2 VS. MU1 N.E. MU2
   T = -1.096                                                     (2)
   THE TEST IS SIGNIFICANT AT 0.2756
   CANNOT REJECT AT ALPHA = 0.05
--SET IN C7
--0.7 -0.1 1.1 -0.5 0.8 0.5 0.6 0.5 0.7 0.4 0.7 -0.5 1.1 0.2 0.3 1
--0.9 0.2 0.0 0.3
--TTEST OF MU=0 ALT.=+1 ON C7
   C7    N = 20    MEAN =    0.44500    ST.DEV. =    0.468
TEST OF MU = 0. VS. MU G.T. 0.
T = 4.248                                                         (3)
THE TEST IS SIGNIFICANT AT 0.0002
```

CHAPTER OUTLINE

9.1 WHY SPREADS SHOULD BE COMPARED

9.2 TESTING SPREADS FROM TWO SEPARATE BATCHES

9.3 THE *t*-FOR-TWO TEST REVISITED

9.4 TESTING SPREADS FROM *k* SEPARATE BATCHES

9.5 COMPARING SPREADS FROM TWO RELATED BATCHES

CHAPTER NINE
COMPARING
SPREADS

9.1 WHY SPREADS SHOULD BE COMPARED

Why test spreads? First, spreads should be compared because they are an essential characteristic of a batch of numbers. Howell properly summarized the situation: "While statistical analyses generally deal with means and view variability as something which only makes our interpretation of means more difficult, *there is much to be said for the variance as a measure of the effect of an experimental manipulation*" (1982, p. 141, emphasis added). Previously, we indicated that batches of numbers can be simplified and summarized by single numbers that index the centers and spreads of batches. In Chapter 8 we demonstrated how to compare the centers from separate and related batches. Why not test spreads also? If one treatment produces a greater spread of numbers than another treatment, isn't that important? And if it is important, why aren't spreads compared more often? Perhaps the failure to compare spreads is another reflection of the less intuitive nature of spread in contrast with the notion of a center. In summary, comparing spreads is just as legitimate and important as comparing centers and, therefore, we ought to be doing more of it.

Second, there are situations in which spreads—rather than centers—or both spreads and centers are critical. "In the behavioral sciences, we sometimes expect that an experimental condition will cause some subjects to show extreme behavior in one direction while it causes others to show extreme behavior in the opposite direction" (Siegel, 1956, p. 145). When this happens, the means for the experimental condition and the control condition may not differ, but the variability will be greater in the experimental condition. Here is an example. A colleague routinely does a classroom experiment in which two batches of human subjects do the same task. In one batch the subjects are tested under stress; in the other batch the subjects are not. The means of two batches usually do not differ, but the variances do. Presumably, the stress condition taps a personality dimension of "coping with stress." Some subjects respond favorably to stress and others unfavorably so that the performance is more variable in the stress condition.

Individual differences in responsiveness to drugs is a well-documented phenomenon. Some individuals are even *excited* by barbiturates and some are *relaxed* by amphetamines. Because of the range of responsiveness to drugs, knowing the average effect is not sufficient. If two drugs are equally effective on the average, the drug with the less variable effect would be preferred.

Here is a hypothetical example from cognitive psychology in which the analysis of both means and variances would be relevant. Two batches of subjects are tested on the same task. In one batch the instructions induce the subjects to engage in a search process prior to responding; in the second batch the instructions induce the subjects to respond without a search process. Because searching consumes time, it would be anticipated that the mean response time would be greater for the search batch. But searches may be of different durations and involve different strategies. Therefore, the variance of the response times should be greater for the search batch also. If searching reduces errors, then the means and variances of the errors would be in the opposite direction to response times.

Investigators tend to consider how a treatment will affect the mean of a batch. They also should ask whether the treatment might increase or decrease variability. When gross disparities among variances are encountered, they should be examined rather than being regarded as violation of an assumption for testing means. The investigator needs to inquire why a particular treatment produces excessively large (or small) variability. The variability differences could represent something that is more important than the original independent variable.

A third reason for testing spreads is to investigate the assumption of homogeneity of variance. When we described the t-for-two test, it was pointed out that this test assumes equal population variances. Further, it was stressed that the formula for random variability, $s_{\bar{x}_1 - \bar{x}_2}$, depends on homogeneity of variance. The pooled error term is proper for equal variances, and the non-pooled error term for unequal variances. We also indicated that violations of the assumptions for the t-for-two test may generate empirical sampling distributions of the difference in the means that depart appreciably from theoretical t distributions. When such departures from the theoretical t distribution occur, the tabled probabilities will be incorrect. Since these probabilities are the basis for making statistical decisions, then the decisions may be erroneous. It should be acknowledged that we are in an area of some controversy. Most texts today assert that the t-for-two test is a *robust test*. A robust test is a test that is little affected by departures from the assumptions. Accepting the view of t as a robust test implies that comparing spreads to test for homogeneity of variance is unnecessary. However, what I see in the recent statistical literature and particularly in sophisticated statistical packages for computers is an increased interest in tests for homogeneity of variance and in tests for comparing means when heterogeneity of variance is present. Ann Landers has frequently said: "If it isn't broken, don't fix it." We can turn this advice around by asking: "If it isn't broken, then why are many people trying to fix it?" That is, if the t-for-two test is so robust, then why the recent interest in tests for variances and in modifications of the t-for-two test?

Given these reasons for testing spreads, five kinds of variance tests are possible:

1. comparing a single variance versus a specific population value
2. comparing the variances of two separate batches
3. comparing the variances of k separate batches
* 4. comparing the variances of two related batches
* 5. comparing the variances of k related batches

In every instance, as in testing means, the statistical decisions concern population parameters (σ^2s), and statistics based on samples (s^2s) serve as estimators of the parametric values. We will limit the presentation to situations 2, 3, and 4.

Testing a single variance versus a population variance (1) is an unusual procedure, but it can be done (see Hinkle, et al., 1979, pp. 189–191). Why is this test unusual? First, research workers usually have more than one batch of numbers, and second, they seldom know the population variance. The problem of testing k related variances (5) is excluded because it involves matrix algebra and considerable computational labor (Winer, 1971, pp. 594–599). Be forewarned that in trying to cope with one difficulty (homogeneity of variance) we will become entrapped in another problem (normality).

TESTING SPREADS FROM TWO SEPARATE BATCHES

9.2

When the mean time estimates for the women and men were compared in Chapter 8 we provisionally evaluated the assumption of homogeneity of variance by simply inspecting the s^2s for the two batches. Now, we want to compare the variances statistically. The first difficulty that we face is that literally *dozens* of variance tests are available (see Conover, Johnson, & Johnson, 1981). The second difficulty is that the tests for homogeneity of variance are often deficient in two respects: they are profoundly influenced by nonnormality and they lack power to detect differences in variances when differences exist. In a Monte Carlo study with a variety of positively skewed populations, Church and Wike (1976) rediscovered what many researchers have observed: namely, that the Type 1 error rates for certain widely used tests like Bartlett's (1937) and Hartley's (1950) were excessively high. That is, far too many significant differences in variances were found when, in reality, the population variances were equal. On the other hand, when the Bartlett and Hartley tests were applied to batches from near-normal populations, they had excellent Type 1 error rates and power. Other tests, such as the jackknife (Miller, 1968), were less affected by departures from population normality, but lacked power. Once more it is evident that we have a problem. In view of the deficiencies of variance tests, how can variances be tested for equality?

Let us begin by proposing that we test the variances from two separate batches by two tests that perform well with *normal populations*. This proposal raises a troublesome question: How do we know that the populations are normally distributed? With the small batches of numbers that characterize much behavioral research we cannot be certain about normality. But there are three possible courses of action: (a) *use the available knowledge* regarding the numbers in an area of research; (b) do a *residuals analysis* upon the numbers; and (c) do a *Lilliefors test* upon the numbers (Lilliefors, 1967). These three possible approaches are briefly described here.

When researchers investigate a problem area they get numbers, and they look at them. If they observe roughly symmetrical distributions, then they *assume* normality of the populations. If, on the other hand, they continually observe skewed distributions, then they regard population normality as less tenable. For example, distributions of reaction times are commonly observed to be positively skewed. In the latter situation investigators may transform their numbers by converting them to logarithms, reciprocals, and so on (re-expression). We will discuss *re-expression* more in detail in Chapter 14. The purposes of re-expression are to try to make the numbers more normally distributed and to equalize the variances in the batches. In brief, you can draw on previous research in an area to make guesses about normality, or you can re-express your numbers to try to achieve normality and homogeneity of variance.

A second procedure is to compute residuals of the numbers, plot frequency distributions of the residuals for each batch and for all batches combined, and examine the residual plots to see whether the residuals center on zero and appear to be samples from a normal population. *Residuals* are nothing new. They are the deviations of the numbers in a batch from the batch's mean, $(X - \overline{X}_j)$, where $\overline{X}_j$ is the batch mean. For a fuller description of residual analysis see Box, Hunter, and Hunter (1978, pp. 182–186).

Third, the Lilliefors technique (1967) may be applied. The numbers are converted to standard scores (Zs) and an *empirical distribution function (e.d.f.)* of the Zs is plotted on special graph paper. An e.d.f. is constructed by ordering the Z scores by size and plotting the cumulative relative frequencies of the ordered Z scores. By inspecting the Z plot in relation to bounds in the graph, it is possible to see whether or not the plot falls within the limits of a normal distribution for the batch size. The Lilliefors test has been described fully by Inman and Conover (1983, pp. 153–155).

Suppose a researcher, working with a new measure, obtained the following 25 numbers for a batch: 65, 48, 11, 76, 74, 17, 46, 85, 9, 50, 58, 4, 77, 69, 74, 73, 3, 95, 71, 86, 40, 21, 81, 65, 44. For a second separate batch the 25 numbers were 42, 48, 11, 62, 13, 97, 34, 40, 87, 21, 16, 86, 84, 87, 67, 3, 7, 11, 20, 59, 25, 70, 14, 66, 70. Are the numbers in the two batches from normally distributed populations? Because the Lilliefors test requires special graph paper, we will do a residuals analysis. The mean for the first batch is $\overline{X}_1 = 53.68$; for the second batch, $\overline{X}_2 = 45.6$. Next, residuals $(X - \overline{X}_j)$ are found for each batch: $(65 - 53.68) = 11.32$, $(48 - 53.68) = -5.68, \ldots, (44 - 53.68) = -9.68$ and $(42 - 45.6) = -3.6, (48 - 45.6) = 2.4, \ldots, (70 - 45.6) = 24.4$. Frequency distributions of the residuals for each batch are shown in Figures 9.1 and 9.2. Figure 9.3 is a frequency distribution of the 50 residuals for both batches. Box and associates advised inspecting the distributions to determine whether they ". . . have roughly the appearance of a sample from a normal distribution centered at zero" (1978, p. 183). They also stress the value of looking for outliers—residuals that are much larger or smaller than the other residuals in the batches. The distributions in Figures 9.1, 9.2, and 9.3 obviously do not center on zero, do not appear to be samples from normal populations, and have no outliers. Where did the numbers come from? They are actually random numbers from a rectangular distribution (see Figure 2.4).

Having described some approaches to the problem of normality, we will now return to the comparison of variances. To test the variances from two separate batches, use (a) the F test for variances, for unequal-sized batches, and (b) Hartley's F_{max} test, for

SEC. 9.2 □ TESTING SPREADS FROM TWO SEPARATE BATCHES

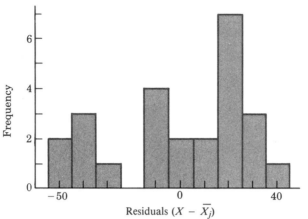

FIGURE 9.1
Histogram of the residuals for batch 1

equal-sized batches. Both tests have good Type 1 error rates and power with normal data, but disintegrate to an incredible degree with skewed data.

In Chapter 8 we found for the time estimates of the men subjects: $s_1^2 = 726.30$, $N_1 = 40$. And for the women: $s_2^2 = 1136.36$, $N_2 = 71$. The F test for variances is

$$F = \frac{\text{larger } s^2}{\text{smaller } s^2}$$

The new theoretical distribution is the F distribution (Table F), a family of asymmetrical curves that vary as a function of the degrees of freedom (df_1) associated with the numerator in the F formula and degrees of freedom (df_2) associated with the denominator in the F formula. The df_1 values are listed across the top of Table F and the df_2 values are down the left side.

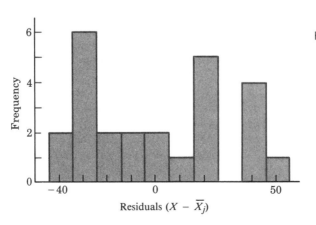

FIGURE 9.2
Histogram of the residuals for batch 2

FIGURE 9.3
Histogram of the residuals for the combined batches

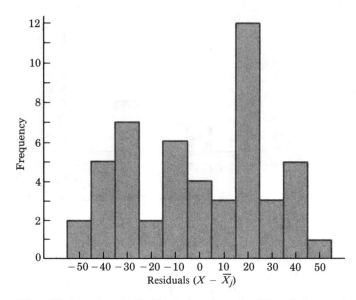

We will do a two-tailed test for the equality of the variances of time estimates. The statistical hypotheses are H_0: $\sigma_1^2 = \sigma_2^2$ and H_1: $\sigma_1^2 \neq \sigma_2^2$. We will employ an $\alpha = .05$. Entering Table F for $df_1 = N_1 - 1 = 71 - 1 = 70$ ($df_1 = 60$ is the nearest tabled value) and $df_2 = N_2 - 1 = 40 - 1 = 39$ ($df_2 = 40$ is the nearest tabled value) and the "upper .025 points" section of the F table, the critical value is: $F_{.05} = 1.80$. Our obtained F is

$$F = \frac{1136.36}{726.30} = 1.56$$

Because the obtained $F = 1.56 <$ critical $F_{.05} = 1.80$, we will not reject H_0. In other words, we will assume that the variances in the two populations do not differ. In addition, this decision on H_0 would legitimize the use of the pooled error term for the t-for-two test in Chapter 8. Why was the F table entered at the "Upper 0.25 points" when $\alpha = .05$? Because when *testing variances for equality* by a two-tailed test, we must *double* the listed P values in Table F.

We reached this statistical decision by comparing the obtained F value with the critical value of F for $\alpha = .05$. Equivalently, we could have determined $P(F = 1.56)$ and compared P with $\alpha = .05$. From Table F, $P(F = 1.56)$ is $.20 > P > .10$. Since this $P > \alpha = .05$, the same decisions would be reached. Did you get this P value from Table F? Don't forget the doubling routine.

With two variances, the *Hartley F_{max} test* is performed in the same manner as the F test for variances. The formula is

$$F_{max} = \frac{\text{largest } s^2}{\text{smallest } s^2}$$

Therefore:

$$F_{max} = F = \frac{1136.36}{726.30} = 1.56$$

Critical values for the F_{max} test are displayed in Table E. Across the top of the table is k, the number of variances in a set (here $k = 2$), and down the left side of the table are the degrees of freedom ($df = N - 1$) that *each variance* is based upon. The F_{max} test can thus only be applied when the Ns are equal in the batches. Two other features of the F_{max} test are that: (a) it can be used for more than two batches and (b) the tabled P values do *not* have to be doubled for a two-tailed test. Since the Ns were unequal in the time estimates for the men and women, the Hartley test was inapplicable.

This section ends with an explanation and elaboration of the recommendation to apply the F and F_{max} tests for comparing the spreads of two separate batches. We tested the preceding variances with the F test and did not reject the H_0 that the population variances were equal. Since the F and F_{max} have good Type 1 error rates with normal distributions, we feel comfortable with the decision not to reject H_0.

Suppose the F test led to the rejection of H_0. This outcome could suggest either that the variances were unequal or that our numbers might have come from skewed populations. If the F test led to rejection of H_0, then we could test the numbers for the two batches for normality by the residuals method or the Lilliefors test. If the normality assumption were satisfied, then we would assert that the variances differed. If the normality assumption were not satisfied, we could try to achieve normality by re-expression and re-test the variances with the F or F_{max} test. Or, failing to normalize with re-expression, we could apply a different variance test that is more robust to nonnormality than the F or F_{max} test. We will describe such a test in Chapter 14.

THE t-FOR-TWO TEST REVISITED

9.3

Imagine that we had found unequal variances and that the numbers in the two batches did not deviate from normality as determined by a residuals analysis or the Lilliefors test. What are the implications of unequal variances for the t-for-two test? Again, a Monte Carlo study by Boneau (1960) can help us. Boneau reported that when $N_1 = N_2$, the sampling distribution of difference in the means follows the t distribution reasonably well even when the σ^2s are unequal. But when the Ns and σ^2s are both unequal, there is trouble. (This arrangement of unequal Ns and σ^2s is a classic statistical problem called the Behrens-Fisher problem.) Boneau found that when a smaller N was associated with a larger variance (e.g., $N_1 = 5$, $\sigma_1^2 = 4$) and a larger N was associated with a smaller variance (e.g., $N_2 = 15$, $\sigma_2^2 = 1$), the empirical sampling distribution of t was symmetrical, but had fatter tails than the theoretical t distribution for $df = 18$. The opposite arrangement ($N_1 = 5$, $\sigma_1^2 = 1$ vs. $N_2 = 15$, $\sigma_2^2 = 4$) produced a symmetrical empirical sampling distribution for t whose tails were thinner than the theoretical t distribution with $df = 18$.

To compensate for the lack of fit between the empirical sampling distributions and the theoretical t distributions either the t values or the df for the t test must be modified. Satterwaite (1946) developed a procedure to adjust the df. Let us assume that the variances of the time estimates for women and men were *unequal* to demonstrate Satterwaite's procedure. First, we must use the *nonpooled* error term to calculate t. Then, we will compute the *Satterwaite df* to evaluate the obtained t. The data are displayed in Table 9.1. You will recall that

$$s_{\bar{x}_1 - \bar{x}_2} = \sqrt{\frac{s_1^2}{N_1} + \frac{s_2^2}{N_2}}$$

$$= \sqrt{\frac{726.30}{40} + \frac{1136.36}{71}}$$

$$= \sqrt{(18.1575 + 16.00507)} = 5.84$$

The *t*-for-two test is

$$t = \frac{\bar{X}_1 - \bar{X}_2}{s_{\bar{x}_1 - \bar{x}_2}}$$

$$= \frac{91.88 - 98.69}{5.84} = -1.17$$

The Satterwaite *df* are

$$df = \frac{(s_{\bar{x}_1}^2 + s_{\bar{x}_2}^2)^2}{\frac{(s_{\bar{x}_1}^2)^2}{N_1 - 1} + \frac{(s_{\bar{x}_2}^2)^2}{N_2 - 1}}$$

where $s_{\bar{x}_1}^2 = s_1^2/N_1$ and $s_{\bar{x}_2}^2 = s_2^2/N_2$. The degrees of freedom for the problem are

$$df = \frac{(18.16 + 16.01)^2}{\frac{(18.16)^2}{(40-1)} + \frac{(16.01)^2}{(71-1)}}$$

$$= \frac{(34.17)^2}{(8.46 + 3.66)} = 96.34$$

which rounded to the nearest integer is 96.

Note two things: 96 is less than the normal $df = N_1 + N_2 - 2 = 40 + 71 - 2 = 109$, and in this problem the larger N is associated with the larger variance. Accordingly, the Satterwaite degrees of freedom should compensate for the thin-tailed sampling distribution that occurs under this N-and-σ^2 arrangement.

To summarize: Variances from two separate batches can be tested using two tests, the F and F_{max}, that work well with normal distributions. On the basis of results from Monte Carlo studies on variance tests, a failure to reject H_0 with the F or F_{max} test should be accepted. If unequal variances are revealed, then tests of normality, residuals analysis or the Lilliefors test, should be performed. If these procedures suggest normality, then the variances would be judged unequal. If the normality tests, on the other hand, indicate nonnormality and the data cannot be normalized by re-expression, then a variance test that is more robust to form should be applied to decide whether the variances differ. Finally, when both the variances and Ns are unequal, the non-pooled error term is appropriate for the t-for-two test and Satterwaite's formula for df is required to compensate for the departures of the empirical sampling distributions of t from the theoretical t distributions.

TESTING SPREADS FROM k SEPARATE BATCHES

This section will be a relatively easy one. Why? Because it will recommend the F_{max} test to compare k variances and the logic of

9.4

	Men	Women
$\overline{X}$	91.88	98.69
s^2	726.30	1136.36
N	40	71

TABLE 9.1
Means and variances of the time estimates for men and women

testing variances will be the same as that proposed for comparing the variances from two separate batches. Therefore, let us go directly to some numbers. The industrial psychologist mentioned earlier was also interested in the effect of background music upon assembly workers' output. He traveled to Hong Kong. There he was able to persuade the manager of a knife factory, Sir Reggie Genuine, to randomly and equally assign 64 knife assemblers to four isolated assembly workshops. All workers assembled the same model of a Genuine Swiss Army Knife.[1] In one workshop there was no background music; the 16 workers in each of the other three workshops heard "elevator" music, Chinese music, and rock, respectively. The results for a two-month period in terms of the *variabilities* in the numbers of Genuine Swiss Army Knives assembled per hour are shown in Table 9.2.

TABLE 9.2
Variances in knife output under four music conditions

Music	N	s^2
none	16	11.6
elevator	16	32.6
Chinese	16	25.7
rock	16	66.2

The statistical hypotheses are H_0: $\sigma_1^2 = \sigma_2^2 = \sigma_3^2 = \sigma_4^2$ and H_1: $\sigma_i^2 \neq \sigma_j^2$, which denotes that at least one variance differs from another. The criterion of significance is set at $\alpha = .05$. Again:

$$F_{max} = \frac{\text{largest } s^2}{\text{smallest } s^2}$$

In this case,

$$F_{max} = \frac{66.2}{11.6} = 5.71$$

We enter Table E for the number of variances in the set $k = 4$ and $df = N - 1 = 16 - 1 = 15$ for each variance. The critical value for $F_{max\ .05} = 4.01$. Since the computed $F_{max} > F_{max\ .05}$, H_0 is rejected and H_1 is accepted—the population variances in knife output are unequal for the background music conditions. At this point, as with two separate batches, it would be appropriate to an-

---

1. I am a big devotee of Swiss Army Knives. With a Swiss Army Knife, a battery charger, and a Roto Rooter machine you can take over the world.

alyze the residuals or to perform Lilliefors tests to check on normality. If these tests suggest normality, then we would conclude that the populations have unequal variances. If these tests suggest nonnormality, then we would try to normalize the data by re-expression and repeat the F_{max} test. If re-expression fails to normalize the numbers, then we would retest the variances with the form-robust test described in Chapter 14.

Assume that the batches of numbers in the knife study were tested, found to be normal, and we concluded that the population variances were unequal. This situation raises a rarely discussed question: What specific variances differ from one another? The reason this question is not considered is that the F_{max} test is usually performed to check on the assumption of homogeneity of variance rather than as a way of comparing spreads. Being unaware of any research on this problem, I would do six pairwise comparisons of the four s^2s with the F_{max} test for $k = 2$ and $df = 15$. This pairwise testing procedure is called a *protected test;* the pairs of variances will be compared *if and only if* the overall F_{max} test is significant. (We will be doing protected, pairwise tests of *means* in Chapter 10.) In Table 9.3 the variances are presented in parentheses for the pairs of batches that are being tested. The table entries are the F_{max} values—$F_{max} = 25.7/11.6 = 2.22$, and so forth.

The critical value of the $F_{max\ .05} = 2.86$ (Table E) for $k = 2$ and $df = 15$. Therefore, the workers listening to rock had more varied work output than those without background music ($F_{max} = 5.71$). The comparison between elevator music and none barely misses significance ($F_{max} = 2.81$).

Lastly, suppose you encountered a variance problem with k separate batches that have unequal Ns in the batches. A test for homogeneity of variance by Bartlett (1937) is suitable for unequal Ns. One Monte Carlo study (Church & Wike, 1976) disclosed that the performance of the Bartlett and F_{max} tests was almost identical when Ns were equal. Actually, *Bartlett's test* can be applied with either equal or unequal Ns, but because the F_{max} test is simpler when Ns are equal the latter was recommended. We will

TABLE 9.3 Variances and comparisons of variances for four music conditions

Music	s^2	Chinese (25.7)	Elevator (32.6)	Rock (66.2)
none	(11.6)	2.22	2.81	5.71
Chinese	(25.7)	—	1.27	2.58
elevator	(32.6)	—	—	2.03

CH. 9 □ COMPARING SPREADS

demonstrate Bartlett's test with the same example. For ease of computation it is useful to list the s^2s, dfs, and other statistics on a worksheet (see Table 9.4).

TABLE 9.4
A worksheet for calculating Bartlett's test

Music	df	s^2	SS	$\log_{10}s^2$	$df \log_{10}s^2$	$1/df$
none	15	11.6	174	1.0645	15.9675	0.0667
elevator	15	32.6	489	1.5132	22.6980	0.0667
Chinese	15	25.7	385.5	1.4099	21.1485	0.0667
rock	15	66.2	993	1.8209	27.3135	0.0667
Sums	60		2041.5		87.1275	0.2668
	A		B		C	D

The SS values in column 3 are the products of s^2 and df. For example, $SS = (15)(11.6) = 174$, and so on.

Bartlett's test follows a new theoretical distribution χ^2 (small chi) with $df = k - 1$ (k = the number of variances being tested). χ^2 (Table J) is a family of asymmetrical distributions that varies with df. The formula for Bartlett's test using the sums at the bottom of Table 9.4 is

$$\chi^2 = \frac{2.3026\,[(A)(E - C)]}{1 + \frac{1}{3(k-1)}\left(D - \frac{1}{A}\right)}$$

where 2.3026, 1, and 1/3 are constants and $E = \log_{10}(B/A)$. In this problem E is

$$E = \log_{10}\left(\frac{2041.5}{60}\right)$$
$$= \log_{10}(34.025) = 1.5318$$

And χ^2 is

$$\chi^2 = \frac{2.3026\,[(60)(1.5318) - 87.1275]}{1 + \frac{1}{3(4-1)}\left(0.2668 - \frac{1}{60}\right)}$$
$$= \frac{(2.3026)(91.9080 - 87.1275)}{1 + (0.1111)(0.2668 - 0.0167)}$$
$$= \frac{11.0076}{1.0278} = 10.71$$

From Table J the critical value of the test statistic, $\chi^2_{.05}$, for $df = k - 1 = 4 - 1 = 3$ is 7.81. Because the computed $\chi^2 > \chi^2_{.05}$, the H_0 is rejected as it was before with the F_{max} test. When Bartlett's test discloses heterogeneity of variance, the batch Ns are unequal, and the tests of normality suggest normality, then protected pairwise tests of the variances may be carried out by the F test for variances.

COMPARING SPREADS FROM TWO RELATED BATCHES

9.5

This problem is not encountered as often as the two previous problems, but we will consider it anyway. Our example is suggested by an experiment by Teel (1981). She was interested in a physiological effect of painful stimulation. The pain, termed ischemic pain, was produced by inflating a blood pressure cuff. The physiological measure was salivary pH, the degree of acidity of the subjects' saliva. Forty-five military personnel were subjected to ischemic pain and their salivary pH was measured twice—just before the painful stimulation and immediately after it. Thus, the design involved related batches with repeated measures on the 45 subjects. The results are presented in Table 9.5.

Don't ever be afraid to look at your data. The s^2 values look very much alike. You might even say that they look so similar you are not going to test them. However, since we're trying to demonstrate a test, we will continue the analysis. But you do not want to ignore what your eyeball suggests to you. If the test for homogeneity of variance for related batches disclosed that the variances were unequal, you should be suspicious.

The test of variances from two related batches is

$$t = \frac{(s_1^2 - s_2^2)\sqrt{N - 2}}{\sqrt{4s_1^2 s_2^2 (1 - r_{12}^2)}}$$

The t statistic follows a t distribution with $df = N - 2$, where N is the number of pairs, 45. The formula contains a term, r_{12}, which is the Pearson product-moment r between the pairs of num-

	Before	After
s^2	0.74	0.67
r_{12}		.75

TABLE 9.5
Variances before and after pain and the correlation between pairs

bers. We will do a two-tailed comparison of the related variances at $\alpha = .05$. The hypotheses are $H_0: \sigma_1^2 = \sigma_2^2$ and $H_1: \sigma_1^2 \neq \sigma_2^2$. Computing t, we have

$$t = \frac{(0.74 - 0.67)\sqrt{(45-2)}}{\sqrt{4(0.74)(0.67)[1 - (.75)^2]}}$$

$$= \frac{0.4590}{0.9315} = 0.49$$

The *df* for t are $N - 2 = 45 - 2 = 43$. But we don't need to look up t in Table A. You will recall from the microtable (Table 8.1) in Chapter 8 that a $Z = 1.96$ is required for a two-tailed test at $\alpha = .05$. Since the critical t values are larger than Z values, we will not reject H_0 when we see that the t is less than one. Sometimes your eyeball may deceive you, but not in this case.

In this experiment the *mean* salivary pH after pain was significantly less than the *mean* salivary pH before pain by a *t*-for-twins test. We have just shown by the *t*-for-related-variances test that the variances of the salivary pH values did not differ on the two tests. Note that the *t*-for-twins test does not assume equal variances. We were curious about equality of variances as a problem in its own right.

According to Snedecor and Cochran (1967), the *t test for related variances* assumes that the paired numbers come from a *bivariate normal population*. Imagine a three-dimensional scatter plot in which the third dimension is the frequencies of the X_1, X_2 points. To be judged bivariate normal the following conditions would have to be fulfilled:

1. The distributions of X_1 and X_2 are normal.
2. Horizontal planes cutting the three-dimensional plot reveal a series of ovals.
3. Vertical planes parallel to the X_1 and X_2 axes produce normal distributions.
4. The lines of best fit are straight lines and the variabilities of the points about the lines are equal. Walker and Lev (1953) point out that lowering r_{12} reduces the size of t. Thus, a restriction of range in either variable or a nonlinear relationship would diminish the chances of rejecting H_0 and finding a difference between the variances.

Elementary statistics books usually do not include a separate chapter on how to compare spreads. There is one here because I think that testing variances is legitimate and important.

It is unfortunate that many tests for spreads are not very form-robust, and the few that are robust lack power. But until these problems are solved, we must use the available techniques. A study can have a variety of outcomes: equal means and variances, unequal means and equal variances, equal means and unequal variances, or unequal means and variances. Why not evaluate all four possible outcomes? Research has been aptly described as "a life of self-inflicted pain." Why throw away half of the information that you have suffered to get?

SUMMARY

Testing spreads is as legitimate and important as testing centers. Testing spreads is also important for evaluating the assumption of homogeneity of variance. Three situations were examined: comparing the variances of two separate batches; comparing the variances of k separate batches; and comparing the variances of two related batches. There are many different variance tests, their performance is generally affected greatly by nonnormality, and they lack power to detect differences in variances.

On the basis of Monte Carlo studies, two tests—the F test for variances and the F_{max} test, which have good Type 1 error rates and power with normal populations—were recommended for testing variances for two separate batches. The F test can be used when $N_1 \neq N_2$; the F_{max} test requires that $N_1 = N_2$. If these tests reveal equality of variance, then this conclusion is accepted. If the tests reveal inequality of variance, then further tests are needed to assess population normality. These procedures include: (a) using the available knowledge from a research area and re-expressing numbers to achieve normality, (b) testing residuals for normality by examining residual plots, and (c) doing Lilliefors graphical tests of Z scores. If these procedures suggest normality, then a finding of inequality of variances would be accepted. If these procedures reveal nonnormality, then re-expression may be tried to achieve normality. If the re-expression is successful, then the transformed numbers are retested with the F or F_{max} test to assess equality of variance. If re-expression fails to achieve normality, then the original numbers can be tested with a more form-robust test for variances (to be described in Chapter 14). A computational example of comparing the variances of two separate batches was provided.

When $N_1 = N_2$ the t-for-two test for means performs reasonably well with unequal variances. When $N_1 \neq N_2$ and $s_1^2 \neq s_2^2$, the empirical sampling distributions of t depart considerably from the theoretical t distributions. Under these latter conditions the nonpooled error term should be used to calculate t and the degrees of freedom for the t test must be adjusted. Satterwaite's procedure was proposed as a way to obtain the adjusted df and an example of this procedure was presented.

In the case of testing k variances from separate batches, the F_{max} test was recommended for equal sizes of batches and Bartlett's test for unequal sizes of batches. The rationale for using these tests was the same as that proposed for testing two variances. If the overall F_{max} or Bartlett test is significant and assumption of normality of the populations appears tenable, then pairwise tests of the variances may be carried out by the F_{max} or F test for variances, respectively. Such pairwise tests, performed *only after* a significant overall test is observed, are termed protected tests. Computational examples of the F_{max} test, Bartlett's test, and pairwise tests of variances were included.

Comparing the variances from two related batches is done by a t test for related variances. The t formula includes r_{12}, the correlation between the N pairs of numbers. A computational example of this t test was given. It was noted again that the t-for-twins test does *not* assume homogeneity of variance. Nevertheless, an investigator might be interested in whether the treatments produced a difference in the variances of two related samples.

An experiment may have four possible outcomes: equal means and variances, unequal means and equal variances, equal means and unequal variances, and unequal means and variances. Spreads should be compared so that all four possible outcomes may be evaluated.

PROBLEMS

1. In a human learning experiment a researcher predicts that subjects in an E condition will be more variable in the number of errors committed than those in a C condition. Twenty-two subjects are randomly and equally assigned to the two conditions. Assume from prior research that the errors are approximately normally distributed in this task. Test the researcher's hypothesis at $\alpha = .05$.

E:	6,	14,	12,	13,	8,	7,	15,	10,	7,	12,	7
C:	15,	16,	12,	14,	16,	15,	13,	14,	15,	16,	15

2. An article in a journal reports the following statistics for two independent batches:

	Group	
	E	C
$\overline{X}$	10.1	12.2
s	4.2	1.8
N	8	12

Assume normality and compare the variances of the two batches. Use $\alpha = .05$ (two-tailed).

3. Using the results of Problem 2, do a two-tailed test of the *means* in Problem 2 ($\alpha = .05$).

4. Do a residuals analysis on the scores in Problem 1 to determine if the batches appear to be samples from normal populations. Are there any obvious outliers?

5. Here are statistics for four separate batches:

	I	II	III	IV
SS	42.3	296.2	121.3	87.6
N	11	11	11	11

Assuming normality from prior research, compare the four population variances for equality ($\alpha = .05$).

6. Perform pairwise tests of the four variances in Problem 5. Use $\alpha = .05$, two-tail. In light of the results of Problem 5, are these pairwise tests appropriate?

7. In a replication of the study in Problem 5 the results were

	I	II	III	IV
SS	50.1	301.5	141.7	77.2
N	8	12	10	9

Assuming normality from prior research, compare the four population variances in the replication study for equality ($\alpha = .05$).

8. Do pairwise tests on the four variances in Problem 7. Use $\alpha = .05$. Are these tests appropriate?

9. Nine pairs of littermate rats were tested in a complex task following their rearing in either an enriched (E) or an impoverished (C) envi-

ronment. Test the variances of the error scores here to determine if the environments had any effect upon variability. Use $\alpha = .05$.

	Pairs								
	1	2	3	4	5	6	7	8	9
E	11	7	15	19	23	16	14	15	17
C	13	11	14	21	27	19	17	19	18

ADDENDUM: USING COMPUTERS IN STATISTICS

MINITAB can be used like a calculator to test homogeneity of variance. In the first example the variances of the time estimates for the men and women are compared by an F test (1). The TWO-SAMPLE program is appropriate for performing a t-for-two test (2) when heterogeneity of variance exists. The nonpooled error term is incorporated into the t test, and "approx. degrees of freedom" is the Satterwaite df (3) described in the chapter. Finally, using MINITAB's arithmetic, an overall F_{max} test (4) is performed on the variances for the four background music conditions. Since the overall F_{max} is significant, pairwise F_{max} tests (5) are then done upon the pairs of music condition variances.

```
--SET IN C11
--27
--SET IN C12
--33.7
--MULT C11 BY C11 PUT IN C13
   ANSWER =          729.0000
--MULT C12 BY C12 PUT IN C14                    (1)
   ANSWER =         1135.6900
--DIVIDE C14 BY C13 PUT IN C15
   ANSWER =            1.5579

--TWOSAMPLE FOR C1 C2
   C1  N = 40  MEAN = 91.875  ST.DEV. = 27.0    (2)
   C2  N = 71  MEAN = 98.690  ST.DEV. = 33.7

   APPROX. DEGREES OF FREEDOM = 96              (3)

   A 95.00 PERCENT C.I. FOR MU1-MU2 IS
   (-18.4203, 4.7900)

   TEST OF MU1 = MU2 VS. MU1 N.E. MU2
   T = -1.166
   THE TEST IS SIGNIFICANT AT 0.2465
   CANNOT REJECT AT ALPHA = 0.05

--SET IN C6
--11.6
--SET IN C7
--25.7
--SET IN C8
--32.6
--SET IN C9
--66.2
--DIVIDE C9 BY C6 PUT IN C10
   ANSWER =            5.7069                   (4)

--DIVIDE C7 BY C6 PUT IN C11
   ANSWER =            2.2155
--DIVIDE C8 BY C6 PUT IN C12
   ANSWER =            2.8103
--DIVIDE C9 BY C6 PUT IN C13
   ANSWER =            5.7069                   (5)
--DIVIDE C8 BY C7 PUT IN C14
   ANSWER =            1.2685
--DIVIDE C9 BY C7 PUT IN C15
   ANSWER =            2.5759
--DIVIDE C9 BY C8 PUT IN C16
   ANSWER =            2.0307
```

CHAPTER OUTLINE

10.1 THE CASE OF k SEPARATE BATCHES—A SIMPLE ANALYSIS OF VARIANCE (ANOVA)

10.2 GENUINE SWISS ARMY KNIVES AND SINGAPORE SLINGS—ANOTHER EXAMPLE

10.3 WHAT TO DO AFTER AN F TEST?

10.4 SEMINASTY PROBLEMS

10.5 THE CASE OF k RELATED BATCHES—A TREATMENTS-BY-SUBJECTS OR REPEATED MEASURES ANOVA

10.6 A STRANGE EXAMPLE—THE WASHINGTON (DC) CONNECTION

10.7 NEW COMPLICATIONS

CHAPTER TEN
A PIE WITH A FEW SLICES— TESTING k MEANS

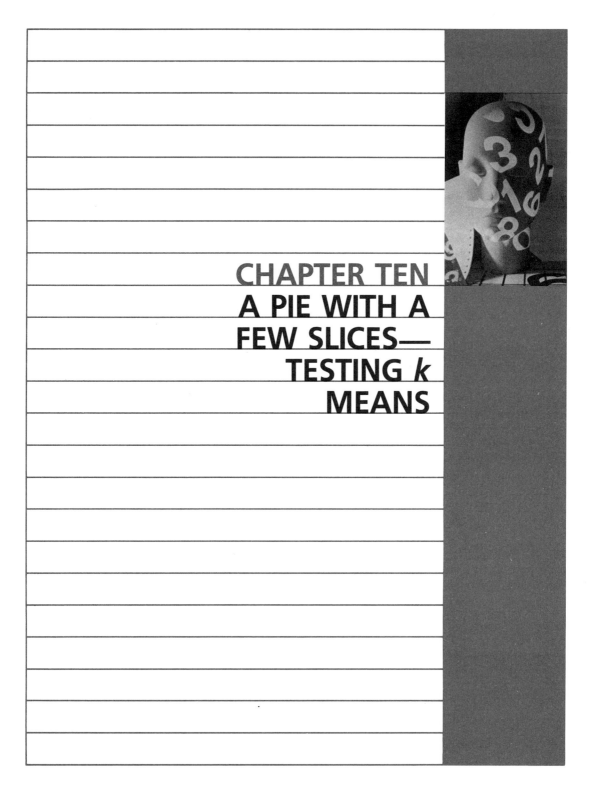

THE CASE OF k SEPARATE BATCHES—A SIMPLE ANALYSIS OF VARIANCE (ANOVA)

10.1

So far we have analyzed some simple cases of one or two batches of numbers. (The sole exception occurred in Chapter 9 in which k variances were compared.) Now we are ready to explore more batches of numbers and more complicated arrangements of batches. In Chapter 8 the means of two separate batches were compared with the t-for-two test and the means of two related batches with the t-for-twins test. In the real world of research investigators often design experiments with k separate batches of numbers or k related batches, where $k > 2$. The definitions of *separate* (random sampling or random assignment) and of *related* (matching or repeated measures) are the same as those for two batches given in Chapter 8. As in the two-batch cases, we want to compare the means, so the new null hypothesis is: H_0: $\mu_1 = \mu_2 = \ldots = \mu_k$. This section will analyze the means from k *separate batches;* later sections will consider the means from k *related batches*. This chapter, then, is simply an extension from two to k batches.

In Chapter 4 we discussed spreads. Let us quickly review relevant parts of that material. One measure of the spread of a batch of numbers and an estimate of σ is the *standard deviation:*

$$s = \sqrt{\frac{\Sigma X^2 - \frac{(\Sigma X)^2}{N}}{N - 1}}$$

The *variance* (s^2) is the *standard deviation* (s) squared. In Chapter 8 we also noted that the $N - 1$ divisor is frequently referred to as *degrees of freedom* (df). We also said that the *sum of squares* (SS) is

$$SS = \Sigma X^2 - \frac{(\Sigma X)^2}{N}$$

Finally, it should be pointed out that s^2 is conventionally, in the context of *analysis of variance* (ANOVA), termed the *mean square* (*MS*) and is defined as

$$MS = \frac{SS}{df}$$

Since the topic is ANOVA at this point, we will be employing the terms *SS, MS,* and *df* throughout. We are just engaged in some relabeling. For example, the MS for a batch of numbers is the s^2 for the batch.

How about a pedogogical law? The law is: Students rarely ask nice questions that instructors know the answers to. Therefore, I'll ask one. To test k means why not do all possible pairwise t-for-two tests among the k batches? Because doing all possible pairwise t tests is tedious and dangerous. Why tedious? Because there are a lot of ts to be run. To be exact, $k(k - 1)/2$ pairwise t tests can be performed. Thus, if $k = 6$, the data analyst could do $6(6 - 1)/2 = 15$ t tests. Why dangerous? In performing pairwise t tests the Type 1 error rate runs wild; the Type 1 error rate greatly exceeds α. Stated otherwise, the use of multiple ts produces far too many rejections of H_0 when, in fact, H_0 is true.

To solve these problems, Sir Ronald Fisher devised a neat, single statistical test—ANOVA—for "looking at" k means simultaneously. Fisher's test permits one to conclude either "Yes, there is at least one significant difference among the k means," or "No, the variability among the k means does not exceed random variability." If H_0 is not rejected (*No*), then your ship has sunk. If H_0 is rejected and H_1 accepted (*Yes*), then you must face the muddled and controversial problem of detecting the difference(s) among the k means. Your ship is afloat, but the problem that was elegantly solved by ANOVA has come back to haunt you.

Here is a description and an intuitive rationale for ANOVA. ANOVA is well-named. ANOVA is simply the partitioning of the total variability of a set of numbers into its parts. If we have k separate batches of numbers with N_G numbers in each batch and the N_Gs are equal, then we would have $k(N_G) = N$ numbers. The total variability of all N numbers is

$$SS_T = \Sigma X^2 - \frac{(\Sigma X)^2}{N}$$

What is SS_T? It is the variability of the N *numbers* about the mean of the N numbers, the *grand mean*.

Another source of variability is

$$SS_G = \frac{(\Sigma X_1)^2}{N_1} + \ldots + \frac{(\Sigma X_k)^2}{N_k} - \frac{(\Sigma X)^2}{N}$$

With equal N_Gs, a *short-cut formula* for SS_G is

$$SS_G = \frac{(\Sigma X_1)^2 + \ldots + (\Sigma X_k)^2}{N_G} - \frac{(\Sigma X)^2}{N}$$

where $\Sigma X_1 \ldots \Sigma X_k$ are batch sums. What is SS_G? SS_G is variability of the *batch means* about the *grand mean* weighted by the number of cases, N_G, in each batch.

Finally, a third source of variability can be found, SS_W. This source of variability may be obtained by subtraction:

$$SS_W = SS_T - SS_G$$

Or SS_W can be directly calculated:

$$SS_W = \Sigma X_1^2 - \frac{(\Sigma X_1)^2}{N_1} + \ldots + \Sigma X_k^2 - \frac{(\Sigma X_k)^2}{N_k}$$

What is SS_W? It is the variability of the *numbers within each batch* about the *mean of the batch,* pooled across all batches. In short, SS_W is an old friend—a measure of random variability. Many statisticians refer to SS_T as the *total variability,* SS_G as the *between-groups variability,* and SS_W as the *within-groups variability.*

To perform an overall test upon the k means we must find MS_G and MS_W. Remember that $MS = SS/df$. Accordingly, after computing the SSs, we need to obtain the degrees of freedom. The SSs are additive:

$$SS_T = SS_G + SS_W$$

Similarly, their respective dfs are additive:

$$df_T = df_G + df_W$$

What is df_T? SS_T is the variability of N numbers about the grand mean. Earlier it was stated that a divisor of $N - 1$ usually generates an optimal estimate of σ^2, and there is one restriction on the N numbers. Therefore,

$$df_T = N - 1$$

SS_G is the variability of k means about the grand mean. Therefore,

$$df_G = k - 1$$

Since the dfs are additive, the df_W may be found by subtraction:

$$df_W = df_T - df_G$$

Accordingly,

$$df_W = (N - 1) - (k - 1)$$
$$df_W = N - k$$

When the N_Gs are equal, we can get df_W in another way. The df for the first batch is $N_G - 1$. Likewise, for the second batch it is $N_G - 1$. For all k batches, when the variabilities within the batches are pooled, it is: $df_W = k(N_G - 1)$. Since $k(N_G) = N$, $df_W = N - k$, as before.

We know SS_G and SS_W and their dfs. Then, $MS_G = SS_G/df_G$ and $MS_W = SS_W/df_W$ or $MS_G = SS_G/(k - 1)$ and $MS_W = SS_W/(N - k)$. To do an overall test upon the k means, the variance of the batch means is divided by the variance due to random variability:

$$F = \frac{MS_G}{MS_W}$$

The test statistic, F (for Fisher), follows the F distribution. F is evaluated in Table F for the degrees of freedom (df_1) associated with the numerator, $k - 1$, listed across the top of the table, and the degrees of freedom (df_2) for the denominator, $N - k$, listed down the left side. If the obtained $F \geq$ critical F_α for $df = k - 1$, $N - k$, then H_0 is rejected and H_1 is accepted. If the obtained $F <$ critical F_α for $df = k - 1$, $N - k$, then H_0 is not rejected. The null hypothesis for ANOVA is H_0: $\mu_1 = \mu_2 = \ldots = \mu_k$. The alternative hypothesis is H_1: $\mu_i \neq \mu_j$, which denotes that at least one population mean, μ_i, differs significantly from another mean, μ_j.

How is it possible that the F test, which is a ratio of variances, tests the differences among k means? The simple answer is that in the F test the variance of the batch means (MS_G) is compared to random variance within the batches (MS_W). A more complicated answer runs along these lines. Underlying ANOVA tests are *statistical models*. One model, a *fixed-effects model*, is appropriate when the *levels* of the independent variable are *arbitrarily selected* by the investigator. In the Genuine Swiss Army Knife example in Chapter 9 the independent variable was the type of background music and the four levels of the independent variable were: none, elevator, Chinese, and rock music. These levels were selected by Reggie Genuine who likes elevator music and detests other music. Another model, a *random-effects model*, is appropriate when the levels of the independent variable are *randomly selected* by the researcher from a large population of levels. The exposition throughout the book will be limited to the fixed-effects model. Under this model, $MS_G = \sigma_G^2 + \sigma_E^2$, where σ_G^2 is the population variance for treatments, σ_E^2 is an estimate of random variability, and $MS_W = \sigma_E^2$ is another independent estimate of random variability. Accordingly, in the F test we have

$$F = \frac{\sigma_G^2 + \sigma_E^2}{\sigma_E^2}$$

Suppose that $\sigma_E^2 = 4$. Suppose further that all k means are identical. In this case $\sigma_G^2 = 0$. Then,

$$F = \frac{0 + 4}{4} = 1.0$$

Now imagine that the k means diverge greatly from one another so that $\sigma_G^2 = 16$. Then,

$$F = \frac{16 + 4}{4} = 5.0$$

These two examples show how the F ratio is affected by the differences among k means. When the levels of a treatment variable are completely ineffectual, F should be 1.0.[1] As the levels of a variable produce differences among the means and the random variability remains unchanged, the F becomes greater than 1.0. Observe, however, that F can be nonsignificant with discrepant means when the random variability is excessive. When this happens, that old demon, random variability, is destroying your research. F cannot be negative, but sometimes $F < 1.0$. An $F < 1.0$ can occur by chance or when the assumptions for ANOVA are violated. We will now turn to some numbers and do an ANOVA.

GENUINE SWISS ARMY KNIVES AND SINGAPORE SLINGS—ANOTHER EXAMPLE

10.2 Actually, the industrial psychologist directed a second experiment on background music and worker output at Sir Reggie Genuine's other knife factory in Singapore. In this study 36 workers were randomly and equally assigned to four isolated workshops with the same four types of background music. In this investigation, however, the investigators were concerned about the effects of background music on the *means* of the output. Table 10.1 presents the numbers of Genuine Swiss Army Knives assembled by the 36 workers. The numbers are means, for some unknown period of work, rounded to the nearest integer for computational simplicity.

Before rushing into the calculations, take a few minutes to eyeball your numbers. What do the spreads look like? Since the N_Gs are alike, we can quickly check the ranges for the four

---

1. More generally, the expected value of $F = df_2/(df_2 - 2)$, where df_2 is the denominator df in the F ratio.

batches. They are, respectively, $13 - 7 = 6$; $18 - 10 = 8$; $12 - 8 = 4$; and $10 - 4 = 6$. They are quite similar. How about the $\overline{X}$s? With equal N_Gs we can examine the $\overline{X}$s by looking at the sums. We observe that the ΣXs differ considerably. It appears, by inspection, that the different types of background music had an effect on the batch $\overline{X}$s. An obtained F value of around 1.0 for these numbers should evoke suspicion. How about the outlier situation? The numbers do not vary much within the batches, and in no instance does a number deviate wildly from its batch mates. We could do more data description, but we will break out our $19.65 calculator and do an ANOVA.

The null hypothesis is $H_0: \mu_1 = \mu_2 = \mu_3 = \mu_4$. The alternative hypothesis is $H_1: \mu_i \neq \mu_j$. Let us set $\alpha = .01$. We will begin by computing SS_T.

---

1. List the formula: $SS_T = \Sigma X^2 - \dfrac{(\Sigma X)^2}{N}$

2. Enter the numbers in memory and sum:
   12 [M+]
   10 [M+]
   . .
   . .
   . .
   8 [M+]

3. Recall ΣX^2: [ΣX^2] 4117
4. Recall ΣX: [ΣX] 369
5. Recall N: [N] 36

6. List formula 1: $SS_T = \Sigma X^2 - \dfrac{(\Sigma X)^2}{N}$
7. Substitute and calculate:

   4117 [−] 369 [x^2] [÷] 36 [=]
   334.75

A quick check on these calculations is to see whether the sum of the treatment (music) sums, ΣX, in the data table $= \Sigma X$ from your calculator: $90 + 125 + 91 + 63 = 369$. The sums check. If that check failed, you might see if the sum of the treatment sums equals the sum of the row sums. If that checked, but the calculator ΣX from step 4 did not equal ΣX in the data table, then you need to enter the 36 numbers in your calculator again.

Next, we will find SS_G with the shortcut formula.

CH. 10 □ A PIE WITH A FEW SLICES—TESTING k MEANS

1. List the formula: $SS_G = \dfrac{(\Sigma X_1)^2 + \ldots + (\Sigma X_k)^2}{N_G} - \dfrac{(\Sigma X)^2}{N}$

2. Enter the numbers in memory and sum:
   90 [M+]
   125 [M+]
   91 [M+]
   63 [M+]

3. Recall ΣX^2: [ΣX^2] 35975
4. Recall ΣX: [ΣX] 369
5. List formula 1: $SS_G = \dfrac{(\Sigma X_1)^2 + \ldots + (\Sigma X_k)^2}{N_G} - \dfrac{(\Sigma X)^2}{N}$

6. Substitute and calculate:
   35975 [÷] 9 [−] 3997.2222
   369 [X²] [÷] 36 [=] 214.97

These calculations require a number of comments.

1. You should notice again that ΣX_1, ΣX_2, and so on are treatment sums.
2. When *treatment sums* are squared, their sum (step 3) does *not* equal the sum of all *numbers* squared in step 3 for SS_T.
3. However, ΣX is the same for both SS_T and SS_G.
4. The short-cut formula for SS_G is applicable when the N_Gs are equal. Instead of repeatedly dividing the squared treat-

TABLE 10.1
Mean number of knives assembled under four music conditions

	None	Background Music Elevator	Chinese	Rock	
	12	11	11	6	
	10	15	9	4	
	13	14	12	7	
	10	14	9	5	
	7	10	8	7	
	9	18	10	7	
	10	16	11	9	
	11	14	12	10	
	8	13	9	8	
Sums	90	125	91	63	$369 = \Sigma X$

ment sums by N_G, that is, $(90)^2/9 + \ldots + (63)^2/9$, it is faster to sum the squared treatment sums and divide once by N_G. If the N_Gs are unequal, then each squared column total *must* be divided by its N_G, i.e., the longer formula for SS_G *must* be implemented.

5. These computational formulas all contain a common term, $(\Sigma X)^2/N$. This term is often called the *correction term*. It is a good idea to calculate it separately to simplify later calculations. Since the correction term is applied over and over, be absolutely certain it is correct.

6. SSs cannot be negative. A *negative SS* is a signal for you to start checking everything, especially your divisors. If you ignore the negative sign, your ANOVA will become a pile of errors.

We found that $SS_T = 334.75$ and $SS_G = 214.97$. Now we can easily obtain SS_W:

$$SS_W = SS_T - SS_G$$
$$= 334.75 - 214.97 = 119.78$$

How about the corresponding *df*s? The $df_T = N - 1 = 36 - 1 = 35$, the $df_G = k - 1 = 4 - 1 = 3$, and $df_W = N - k = 36 - 4 = 32$. Do the *df*s check? The $df_T = df_G + df_W = 3 + 32 = 35$. So the answer is yes. It is both customary and useful to put these various quantities together in an *ANOVA summary table* (Table 10.2). Are we being neurotically compulsive again? Possibly, but a summary table will help you to keep track of your answers, especially in complex designs, and to perform the proper F tests.

After entering the SS and df values in the summary table, the MSs have been calculated (e.g., $MS_G = SS_G/df_G = 214.97/3 = 71.66$). We then find the F statistic:

$$F = \frac{MS_G}{MS_W}$$
$$= \frac{71.66}{3.74} = 19.16$$

Using Table F, the critical value of $F_{.01}$ is determined from the section of the table labeled "Upper .01 points" for the numerator $df_1 = df_G = 3$ and denominator $df_2 = df_W = 32$ ($df_2 = 30$ is the nearest tabled entry).[2] Because the obtained $F = 19.16 > F_{.01} =$

---

2. It is possible to get more precise values of P. See Lindman (1974, pp. 18–19).

TABLE 10.2
ANOVA summary table for the music study

Source of Variation		SS	df	MS	F	$F_{.01}$
Between batches	(G)	214.97	3	71.66	19.16	4.51
Within batches	(W)	119.78	32	3.74		
Total	(T)	334.75	35			

4.51, the decisions are to reject H_0 and accept H_1. The answer is "Yes"—at least one mean, μ_i, differs significantly from another mean, μ_j.

WHAT TO DO AFTER AN F TEST?

10.3

The answer to this question depends on how the F test came out. If you did not reject H_0, a very reasonable thing to do is to become dejected. After this emotional phase, you can do some constructive things. Ask yourself: Was there anything to be found? Did you have sufficient power to detect any differences? The means do not differ, but how about the variances? We will return to the power problem in Chapter 14.

On the other hand, suppose your F test results in the rejection of H_0 and the acceptance of H_1. For openers you should feel elated. But learning that the means differed is not very revealing. Sir Reggie Genuine did not buy it and neither should you. We want to know what *specific* means differed from one another. Regrettably, comparing means after ANOVA is another difficult and controversial problem because different questions can be asked about k means, there are many proposed tests for k means, and different tests yield different results. Here are some questions about k means: Do a few differences between means, from many possible differences, differ? Do the means of the batches differ from a control batch? Do the pairwise differences between the means differ? Do all possible differences among the means differ? To answer the last question not only are pairwise differences assessed, but also all more complicated combinations of means; for example, $(\mu_1 + \mu_2)/2$ versus $(\mu_3 + \mu_4)/2$, and so on are tested. The number of possible comparisons of means performed in this manner is huge—with $k = 4$ means, 25 comparisons of means can be done.

We will restrict the presentation to the pairwise comparison of k means. Here with $k = 4$, there are $k(k - 1)/2 = 4(4 - 1)/2 = 6$ possible comparisons. On the basis of two Monte Carlo studies

(Carmer & Swanson, 1973; Bernhardson, 1975), we recommend a pairwise test that exhibits good Type 1 error rates and power—the *Fisher LSD test* (or the *protected t test*). LSD stands for *least significant difference* and the relevance of the label will be clear in a moment. The Fisher procedure reduces to doing pairwise, modified *t*-for-two tests in *protected* manner. That is, the *t*s will be done *only* when the overall F test is significant. The Type 1 error rates for the *t*s are thus "protected" by the overall F test. It should be noted that multiple-comparison testing is a problem embroiled in controversy. The protected-test procedure that we will employ in this book is a liberal technique—it will reveal more significant differences than more conservative techniques.[3]

The formula for the protected t test when the N_Gs are equal is

$$t = \frac{\overline{X}_i - \overline{X}_j}{\sqrt{\frac{MS_W(2)}{N_G}}}$$

where MS_W *comes from the ANOVA and* N_G *is the number of cases in each batch.* The MS_W is used instead of the s^2s for the batches being contrasted because the MS_W, which is based on all the data, should provide a better estimate of random variability. Instead of actually calculating the *t*s and comparing them to t_α, we can substitute t_α for t and obtain an LSD for the difference in the means, $\overline{X}_i - \overline{X}_j$, required for significance:

$$t = \frac{\overline{X}_i - \overline{X}_j}{\sqrt{\frac{MS_W(2)}{N_G}}}$$

When $t = t_\alpha$, then $(\overline{X}_i - \overline{X}_j) = LSD$:

$$t_\alpha = \frac{LSD}{\sqrt{\frac{MS_W(2)}{N_G}}}$$

$$LSD = t_\alpha \sqrt{\frac{MS_W(2)}{N_G}}$$

3. A host of multiple-comparison tests are available. Two commonly used ones are Tukey's HSD (honestly significant difference) test and the Newman-Keuls test. These procedures, which require a special table for studentized ranges (Q), are described fully in Hinkle et al. (1979).

For the knife problem LSD is

$$LSD = 2.75\sqrt{\frac{3.74(2)}{9}}$$

$$= 2.75(0.9117) = 2.51$$

What does this LSD = 2.51 mean? It means that any differences between pairs of means that are equal to or greater than 2.51 will be declared to be significantly different at $\alpha = .01$. You're probably wondering where the $t_{.01} = 2.75$ came from. It is the $t_{.01}$ (two-tailed) for $df_W = 32$ from Table A ($df = 30$ is the nearest tabled value). Now we will apply the LSD method to compare the means for the different types of background music.

In Table 10.3 the *ordered means* for the kinds of music are enclosed by parentheses and the *differences in the means* are the tabled entries: $\overline{X}_{none} - \overline{X}_{rock} = 10.00 - 7.00 = 3.00$, and so on:[4]

TABLE 10.3
LSD tests among the four music condition means

Music	$\overline{X}$	None (10.00)	Chinese (10.11)	Elevator (13.89)
rock	(7.00)	3.00*	3.11*	6.89*
none	(10.00)		0.11	3.89*
Chinese	(10.11)			3.78*

*Note: P < .01

Five of the six pairwise differences between the means are greater than $LSD = 2.51$, and are, therefore, judged significantly different. The workers hearing elevator music assembled significantly more Genuine Swiss Army Knives than those who heard Chinese music, no music, and rock. The workers with a background of Chinese music or no music had a higher output of knives than the workers under rock music. Notice that when reporting the comparisons of means, you should not only indicate *what differences* are significant, but also the *direction* of the differences. This practice will warm the cruel hearts of journal editors, thesis and dissertation advisors, and your readers.

Before leaving the procedure for pairwise tests of k means, how about a what-if question? What if the N_Gs for the batches are

4. In arranging tables for pairwise comparisons, list the statistics ($\overline{X}$s) from small (at the top) to large in the rows but omit the largest statistic. For the columns, list the statistics from next-to-the smallest statistic (on the left) to the largest. This arrangement will produce the $k(k-1)/2$ comparisons.

unequal? When this happens, a single *LSD* value cannot be computed. Instead, individual t tests must be done, and the obtained ts must then be contrasted with t_α. The modified t formula with unequal N_Gs is

$$t = \frac{\overline{X}_i - \overline{X}_j}{\sqrt{MS_W\left(\frac{1}{N_i} + \frac{1}{N_j}\right)}}$$

where $N_i = N_G$ for the ith batch and $N_j = N_G$ for the jth batch. The *df* for t_α is still df_W. Having equal N_Gs can simplify your computations and your life.

In Chapter 8 we concurred with Cohen's recommendation that a test statistic should be accompanied by a statement regarding the proportion (or percentage) of accounted-for variance. Determining the proportion (or percentage) of accounted-for variance, then, is another procedure that should be done after observing a significant F. A simple way to do this is to play the pie game. When we performed the ANOVA test above, we did not compute MS_T. The MS_T is unnecessary for the F test. $MS_T = SS_T/df_T = 334.75/35 = 9.56$. Are MSs additive? Does $MS_T = MS_G + MS_W$? Definitely not: $9.56 \neq 71.66 + 3.74$. But, as shown previously, the SSs are additive. Therefore, we can find the *proportion* of the total variability in number of knives assembled due to the background music by: $SS_G/SS_T = 214.97/334.75 = 0.64$. Or in *percent*: $100(SS_G/SS_T) = 64\%$. How about the unexplained variability (*residual* or *error*)? It is: $SS_W/SS_T = 119.78/334.75 = 0.36$. Or in percent: $100(SS_W/SS_T) = 36\%$. Of course, we could also determine the proportion of unexplained variability by: $1 - 0.64 = 0.36$ and the percent by: $100 - 64 = 36\%$. Thus, not only do we have a significant F value for the music conditions, but almost two-thirds of the variability in the number of knives assembled can be attributed to these conditions.

We called the comparison of SSs a "simple" way to determine the accounted-for variance. Actually, the quantity, SS_G/SS_T, is a statistic called *eta squared* (η^2) or the *correlation ratio*:

$$\eta^2 = \frac{SS_G}{SS_T}$$

Eta is a general index of relationship that ranges from zero to one. Eta squared is a measure of the proportion of total variability accounted for by the treatments in a *sample*. It tends to overestimate the treatment effect in the population (Cohen & Cohen, 1975). Hays (1963) proposed another measure, *omega squared*

($\hat{\omega}^2$), that is an estimate of the proportion of the total variability accounted for by the treatments in a *population*. (The quantity 100 $\hat{\omega}^2$ is the percentage of accounted-for population variance.) The general formula for $\hat{\omega}^2$ is

$$\hat{\omega}^2 = \frac{SS_{\text{effect}} - df_{\text{effect}}(MS_{\text{error}})}{SS_T + MS_{\text{error}}}$$

In the knife example the proportion of accounted for population variance due to the musical backgrounds is

$$\hat{\omega}^2 = \frac{SS_G - df_G(MS_W)}{SS_T + MS_W}$$

$$= \frac{214.97 - 3(3.74)}{334.75 + 3.74}$$

$$= \frac{203.75}{338.49} = 0.60$$

The answer obtained using Hays' method is very close but smaller than the η^2 method. The similarity in the two values is not surprising, since the first term in the numerator (SS_G) and the first term in the denominator (SS_T) are common to both formulas. $\hat{\omega}^2$ will always be smaller than η^2 because its numerator is reduced by df_{effect} (MS_{error}) and its denominator is increased by MS_{error}. The critical point is that an F test needs to be accompanied by an index of accounted-for variance because an F may be highly significant with a very large N but account for only a small proportion of variance (Keppel, 1982, p. 89).

This brings us to the chapter's title. We can represent the knife experiment's outcome by a pie. The whole pie corresponds to the SS_T, the shaded portion corresponds to the SS_G, and the unshaded portion to SS_W (see Figure 10.1). As in the previous pie pictures, we could think of the whole pie as having an area of 1 (or 100)—the shaded portion as the proportion (or percentage) of the total variability due to the treatments and the unshaded portion as the proportion (or percentage) of the total variability due to random variability.

SEMINASTY PROBLEMS

10.4 What happens to the F test when the assumptions are violated? This question naturally raises the question of the *assumptions* for the F test for k separate means, which are the same as those for the t-for-two test: (a) that the populations of numbers from which

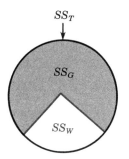

FIGURE 10.1
A pie diagram of the partitioning of SS_T into SS_G and SS_W

the batches were randomly drawn are normally distributed (*normality*); (b) that the variances of the numbers in the populations are equal (*homogeneity of variance*); and (c) that the numbers in the populations are independent (*independence*). The statistical literature on the effects of violating the assumptions upon the F test was carefully reviewed by Glass, Peckham, and Sanders (1972). They concluded that with equal N_Gs: (a) in the case of nonnormality, skewness has "very little effect" and kurtosis has a "slight" effect on the Type 1 error rates; (b) heterogeneous variances have a "very slight effect;" and (c) nonindependence "seriously" affects error rates. When N_Gs are unequal, they concluded that heterogeneous variances and nonindependence produce serious effects. We can reduce the possibility of nonindependence by random selection of cases or random assignment of cases to batches and by experimental controls to prevent effects from spreading across subjects.

The findings by Glass et al. for equal N_Gs have convinced many that the F test is extremely robust. Therefore, they contend that the assumption of homogeneity of variance does not have to be evaluated. This position was stated most eloquently by Box (1953, p. 333): "To make the preliminary test on variances is rather like putting to sea in a rowing boat to find out whether conditions are sufficiently calm for an ocean liner to leave port." What Box said is that the F test is robust (the ocean liner) and homogeneity of variance tests are nonrobust (a rowing boat), so don't waste your time testing variances for equality.

We will reject this position on testing variances for two reasons: (a) as argued earlier, variances ought to be tested just as means are; and (b) the assumption of homogeneity of variance should be evaluated because the F test is affected by heterogeneity of variance with unequal N_Gs and even with equal N_Gs under conditions of rather extreme nonnormality (Bradley, 1980). Wike and Church (1982a and b) replicated Bradley's work and were unable with a variety of techniques and re-expressions of the data to ameliorate the bizarre Type 1 error rates that his distributions produced. Furthermore, it was observed that heterogeneity of variance was the major contributor to the variance in the obtained error rates.

A reader has objected to the idea of testing variances for homogeneity prior to ANOVA. His proposal is: (a) to test for homogeneity with an F or F_{max} test only when the variances look suspicious; and (b) if the variances are unequal, to require a higher level of significance, such as $\alpha = .025$ rather than $\alpha = .05$. I have several objections to this plan. The F and F_{max} tests are extremely sensitive to nonnormality and have excessive Type 1 error rates under nonnormality. Accordingly, α would often be halved unnecessarily and a loss of power would occur in the F test. Halving α is not a universal solution—Norton (Lindquist, 1953) using extremely skewed distributions and Bradley (1980) using L-shaped distributions observed that the F test rejected H_0 *too infrequently*. Here halving α would make a bad situation worse.

In Chapter 9 recommended procedures for testing the variances from k separate batches were offered. We won't repeat the rationale for the recommended procedures, but we will conclude the knife example by comparing the k variances from the study. In the ANOVA we found SS_W by subtraction. An excellent check on an ANOVA is to calculate SS_W or some other error term directly. The direct calculation of SS_W performs two functions: it checks the ANOVA and provides the SSs needed for the equality of variance test. The SS_W is

$$SS_W = \Sigma X_1^2 - \frac{(\Sigma X_1)^2}{N_1} + \ldots + \Sigma X_k^2 - \frac{(\Sigma X_k)^2}{N_k}$$

$$= \left(928 - \frac{(90)^2}{9}\right) + \left(1783 - \frac{(125)^2}{9}\right) + \left(937 - \frac{(91)^2}{9}\right)$$

$$+ \left(469 - \frac{(63)^2}{9}\right)$$

$$= 28 + 46.89 + 16.89 + 28 = 119.78$$

Thus, the SS_W agrees with the SS_W obtained before by subtraction. This increases our faith in the accuracy of the ANOVA.

To test the variances for the k batches we must calculate the s^2s:

$$s^2 = \frac{SS}{df}$$

$$s_1^2 = \frac{28}{8} = 3.5$$

$$s_2^2 = \frac{46.89}{8} = 5.86$$

$$s_3^2 = \frac{16.89}{8} = 2.11$$

$$s_4^2 = \frac{28}{8} = 3.5$$

Parenthetically, it might be pointed out that if you only wanted the s^2s for the batches, there is a faster way to compute them. Enter the numbers for each batch into your calculator, press the $\boxed{\sigma_{n\text{-}1}}$ key, and square the answer. Since the N_Gs are equal, the F_{max} test is appropriate for testing the s^2s:

$$F_{max} = \frac{\text{largest } s^2}{\text{smallest } s^2}$$

$$= \frac{s_2^2}{s_3^2}$$

$$= \frac{5.86}{2.11} = 2.78$$

From Table E the critical value of $F_{max.05}$ for $k = 4$ and $df = 8$ is 7.18. The variances vary, but not enough ($F_{max} < F_{max.05}$) for us to reject H_0. Therefore, in the Singapore knife study the means for the music conditions differed significantly, but their variances did not.

Would your memory crash or go down, as the hackers say, if you heard one more thing? Well, here it is. Suppose we applied the F test for k means to the means from two separate batches? Why not do it and see what happens? We will use the numbers from the none and elevator treatments in the Singapore study. The SS_T is

$$SS_T = \Sigma X^2 - \frac{(\Sigma X)^2}{N}$$

$$= 2711 - \frac{(215)^2}{18}$$

$$= 2711 - 2568.06 = 142.94$$

By the shortcut formula, SS_G is

$$SS_G = \frac{(\Sigma X_1)^2 + (\Sigma X_2)^2}{N_G} - \frac{(\Sigma X)^2}{N}$$

$$= \frac{(90)^2 + (125)^2}{9} - \frac{(215)^2}{18}$$

$$= 2636.11 - 2568.06 = 68.05$$

Then, $SS_W = SS_T - SS_G = 142.94 - 68.05 = 74.89$. $MS_G = SS_G/(k-1) = 68.05/1 = 68.05$ and $MS_W = SS_W/(N-k) = 74.89/(18-2) = 4.68$. Thus:

$$F = \frac{MS_G}{MS_W}$$

$$= \frac{68.05}{4.68} = 14.54$$

The t-for-two formula is

$$t = \frac{\overline{X}_1 - \overline{X}_2}{s_{\overline{x}_1 - \overline{x}_2}}$$

where

$$s_{\overline{x}_1 - \overline{x}_2} = \sqrt{\frac{s_1^2}{N_1} + \frac{s_2^2}{N_2}}$$

Then,

$$s_{\overline{x}_1 - \overline{x}_2} = \sqrt{\frac{3.5}{9} + \frac{5.86}{9}}$$

$$= \sqrt{.39 + .65} = 1.02$$

And t is

$$t = \frac{10 - 13.89}{1.02} = -3.81$$

Notice that if t is squared: $(-3.81)^2 = 14.54$, which is the value of F. Accordingly, when we have two batches, the t-for-two statistic squared is equal to the F for the F test of means. This relationship also holds for critical values of F and t. In Table F the $F_{.01}$ for 1 and 16 $df = 8.53$. In Table A the $t_{.01}$ (two-tailed) for 16 $df = 2.921$; $(2.921)^2 = 8.53$ or $t_{.01}^2 = F_{.01}$. Why not use the F test for the means for both two and k separate batches? You can do it. But would that be fair to the late W.S. Gossett (Student, 1927), who did the pioneering research on small samples at the Guiness Breweries?

10.5 THE CASE OF k RELATED BATCHES—A TREATMENTS-BY-SUBJECTS OR REPEATED MEASURES ANOVA

In Chapter 8 we analyzed the means from two *related batches* by the *t*-for-twins test. Related batches result from: (a) N pairs of cases selected as equivalent on some matching variable (*matching*); or (b) N cases tested twice (*repeated measures*). Now we are confronted with the same situation—matching or repeated measures—except that we have N sets of k matched cases or N cases with k repeated measures, where $k > 2$. We are simply going from two to k related batches. Once more, we could perform multiple *t*-for-twins tests on the k means. But, as with k separate batches, that procedure would be laborious and dangerous for the same reasons offered in Section 10.1. Instead, the means from k related batches will be subjected to an ANOVA. In other books this ANOVA for k related batches may be designated as a *treatments-by-subjects ANOVA* or a *repeated measures ANOVA*. Sometimes, too, statisticians will refer to k related batches as a *two-way ANOVA* and k separate batches as a *one-way ANOVA*.

10.6 A STRANGE EXAMPLE—THE WASHINGTON (DC) CONNECTION

In life one encounters many weird tales. Here is one. A colleague in statistics was sitting in his dingy office. The phone rang. It was the departmental secretary. "There are two men, a Mr. Gltzp and a Mr. Nprst, to see you." "Are they students?" my friend asked anxiously. "I don't think so, Doctor—they're older and they're dressed funny." "OK, send them down," he instructed.

Mr. G and Mr. N were, indeed, dressed funny: they wore suits, ties, and hats. Mr. G said that they were engaged in a "vital mission" for "The Company." They had some valuable numbers from the mission, and they wanted them to be analyzed. Some "cases" had been "tested" five times. They couldn't reveal any more because it would "endanger national security." If my friend agreed in writing not to reveal anything for a period of 10 years, he would be paid well to do the analysis. My friend took the money and ran. He did the job, but he never knew what the numbers represented.

This story provides us with some numbers. We will assume, as the statistician did, that the cases were subjects who were tested five times on the same test. (If the cases took five *different* tests, then a more complicated multivariate statistical procedure would be called for.) The numbers from the vital mission are presented in Table 10.4. Why don't you eyeball the numbers as we did for the example with k separate batches? What did you find out?

Before doing the ANOVA it is prudent to calculate the sums for the cases (taking the signs of the numbers into account), the sums for the tests, and check ΣX for the data table by adding each set of sums separately. Unless ΣX is verified, do not proceed. We will begin the ANOVA in the same manner as before by calculating SS_T and SS_G. To find ΣX^2 enter the 50 numbers into your calculator's memory: 0.5 [+/−] [M+] 1.7 [M+] 1.9 [M+] ... 1.3 [M+]. $\Sigma X^2 = 210.86$ and $\Sigma X = 68.0$. ΣX should agree with the value you found for the table of numbers. SS_T is

$$SS_T = \Sigma X^2 - \frac{(\Sigma X)^2}{N}$$

$$= 210.86 - \frac{(68.0)^2}{50}$$

$$= 210.86 - 92.48 = 118.38$$

TABLE 10.4
Scores for 10 cases tested five times

Cases	1	2	3	4	5	Sums
			Tests			
1	−0.5	0.5	2.7	1.5	1.6	5.8
2	1.7	2.6	0.6	1.4	3.4	9.7
3	1.9	−0.6	2.8	3.6	1.5	9.2
4	−1.6	2.9	−0.5	3.8	2.9	7.5
5	1.1	2.1	−0.7	3.2	3.3	9.0
6	−1.9	−0.6	2.2	1.5	2.1	3.3
7	−1.6	2.2	−0.6	3.3	3.0	6.3
8	1.2	−0.6	2.0	2.9	3.2	8.7
9	1.0	−0.3	2.0	1.3	1.7	5.7
10	−1.3	2.1	−0.4	1.1	1.3	2.8
Sums	0.0	10.3	10.1	23.6	24.0	68.0 = ΣX

SEC. 10.6 □ A STRANGE EXAMPLE—THE WASHINGTON CONNECTION

Note that 92.48 is the correction term which will repeatedly be used. Is it correct? Then:

$$SS_G = \frac{(\Sigma X_1)^2 + \ldots + (\Sigma X_k)^2}{N_G} - \frac{(\Sigma X)^2}{N}$$

$$= \frac{(0)^2 + (10.3)^2 + (10.1)^2 + (23.6)^2 + (24)^2}{10} - 92.48$$

$$= 134.11 - 92.48 = 41.63$$

In an ANOVA for k separate batches we would, at this point, determine SS_W. With k related batches two other sources of variation are obtainable. One is the *sum of squares for cases* or SS_S. It represents the variability of the case means about the grand mean, weighted by the number of trials (N_T) that each case sum is based on. In general, an SS in computational form is a sum for a level (e.g., a case sum) squared and divided by the number of observations for the sum, pooled across all levels of the effect and reduced by the correction term. Since N_Ts are equal, we have

$$SS_S = \frac{(\Sigma X_{S1})^2 + \ldots + (\Sigma X_{S10})^2}{N_T} - \frac{(\Sigma X)^2}{N}$$

where ΣX_{S1} = sum for case 1, and so on. The SS_S is

$$SS_S = \frac{(5.8)^2 + (9.7)^2 + \ldots + (2.8)^2}{5} - 92.48$$

$$= \frac{516.22}{5} - 92.48$$

$$= 103.24 - 92.48 = 10.76$$

Finally, there's a source of variation, SS_E, that is due to *error*. It is a *residual SS* or what is left over:

$$SS_E = SS_T - SS_G - SS_S$$
$$SS_E = 118.38 - 41.63 - 10.76 = 65.99$$

SS_E is also called an *interaction;* here it is a two-factor interaction. The two factors (or independent variables) are *tests* and *cases*.[5] Any two-factor interaction can be computed by the formula:

$$SS_{\text{columns} \times \text{rows}} = SS_{\text{cells}} - SS_{\text{columns}} - SS_{\text{rows}}$$

5. An interaction refers to the combined effect of two or more factors upon the dependent variable. The meaning and analysis of interactions will be considered in detail in Chapter 11.

For our problem the formula is

$$SS_E = SS_{G \times S} = SS_T - SS_G - SS_S$$

We have obtained the SSs, but how about their dfs? The $df_T = N - 1 = 50 - 1 = 49$ and $df_G = k - 1 = 5 - 1 = 4$, as before. The $df_S = c - 1 = 10 - 1 = 9$, but how about df_E? As dfs are additive, df_E can be obtained by subtraction: $df_E = df_T - df_G - df_S = 49 - 4 - 9 = 36$. Or the df for *any* interaction can be determined by the application of a general rule: The df for any interaction is equal to the *product* of the dfs for the interacting factors [e.g., $df_E = (k-1)(c-1) = (4)(9) = 36$]. In Table 10.5 the SSs and dfs have been entered in an ANOVA summary table. The MSs are obtained in the usual manner: $MS_G = SS_G/df_G = 41.63/4 = 10.41$. To evaluate the test means,

$$F = \frac{MS_G}{MS_E}$$

$$= \frac{10.41}{1.83} = 5.69$$

The means for the cases are *not* tested because a valid statistical test is not possible, and the question is meaningless. You do not need a statistical test to tell you that subjects differ. From Table F the critical value for $F_{.05}$ for 4 and 36 ($df_2 = 40$ is the nearest entry) is 2.61. The null and alternative hypotheses are identical to those for ANOVA of k separate batches. Since the obtained $F = 5.69 > F_{.05} = 2.61$, H_0 would be rejected and H_1 accepted. The means of the tests differ significantly. Is that the end of the ANOVA for related batches? No, there is more to be said and it is not especially pretty.

Prior to considering some new difficulties, we will discuss doing protected, pairwise t tests among the k test means. We will not perform the tests because, in dealing with such secret numbers, interpreting the ts would be impossible. But we can sketch

TABLE 10.5
ANOVA summary table for the vital mission study

Source of Variation	SS	df	MS	F	$F_{.05}$
Tests (G)	41.63	4	10.41	5.69	2.61
Cases (S)	10.76	9	1.20		
Error (E)	65.99	36	1.83		
Total (T)	118.38	49			

out the details. We could do $k(k - 1)/2 = 5(5 - 1)/2 = 10$ pairwise t tests on the differences in the means for the five tests. As the N_Gs are equal, computing the LSD value and comparing the differences in means to LSD is the best procedure:

$$LSD = t_\alpha \sqrt{\frac{MS_W(2)}{N_G}}$$

but with related batches we substitute MS_E, the error term for testing the test means, for MS_W and enter df_E into Table A for determining t_α. Thus,

$$LSD = t_\alpha \sqrt{\frac{MS_E(2)}{N_G}}$$

From Table A, $t_{.05}$ (two-tailed) for $df_E = 36$ ($df = 40$ is the nearest entry) is 2.021. Then

$$LSD = 2.021 \sqrt{\frac{1.83(2)}{10}}$$

$$= 2.021\,(.605) = 1.22$$

We could then arrange the ordered test means in a table as was done for k separate batches and compare the differences in the means against the LSD. If $(\overline{X}_i - \overline{X}_j) \geq LSD = 1.22$, the difference would be regarded as significant at the α level. Let us close this section with a warning about two errors that are frequently made when contrasting means after ANOVA. A common error is to insert a wrong value for N_G, which is the number of cases that a single mean is based on when the N_Gs for all batches are equal. A second common error is to do both LSD *and* protected t tests on a set of means. That is redundant and a waste of time. When N_Gs are equal, do *either* LSD tests *or* protected ts but *not both*. (I like LSDs better.) When N_Gs are unequal, however, protected t tests *must* be performed.

NEW COMPLICATIONS

10.7

The new difficulties concern: (a) the assumptions for an ANOVA of k related batches, (b) the consequences of violating the assumptions, and (c) what to do about the problem. We will deal with these matters in order and in an oversimplified manner. Generally it is asserted that the assumption for this type of ANOVA is *compound symmetry*. This assumption means that the variances of the tests are equal and that the covariances—$cov_{ij} = r_{ij}(s_i)(s_j)$—among all possible pairs of tests are equal.[6] Others (Huynh & Feldt, 1970) propose that the variances of the *differences* between all possible pairs of tests must be equal. These two assumptions do not exhaust the possibilities (Huynh & Mandeville, 1979).

If the assumptions are violated, then the general result is that the F test is positively biased so that the F for the repeated measures factor of tests or treatments will be significant too often when the null hypothesis is true. To correct for this positive bias the usual degrees of freedom for the F test for tests are multiplied by a value called epsilon, ϵ:

$$df_G = df_G\epsilon \quad \text{and} \quad df_E = df_E\epsilon$$

or

$$df_G = (k-1)\epsilon \quad \text{and} \quad df_E = (k-1)(c-1)\epsilon$$

Unfortunately, different ϵs have been proposed, and they are difficult to calculate. For one type of ϵ the limiting values are 1 and $1/(k-1)$.

So what should one do? Given these uncertainties, the slightly modified advice of Greenhouse and Geisser (1959) seems reasonable. They proposed a three-step procedure: (a) compare the test means (as we did above) with the *usual dfs* for the F test, (b) do a *conservative F test,* and (c) if a decision cannot be reached by (a) and (b), calculate ϵ and employ it to obtain the *adjusted dfs*. For our example, under step a:

$$df_G = (k-1)\epsilon$$
$$df_G = (5-1)1 = 4$$
$$df_E = (k-1)(c-1)\epsilon$$
$$df_E = (5-1)(10-1)1 = 36$$

That is, with $\epsilon = 1$ the *df*s are the regular *df*s. For step b, the conservative F test, the *df*s include the other limiting value of $\epsilon = 1/(k-1)$. Therefore,

6. In paired-numbers (X and Y) the covariance is a measure of how numbers in X vary with numbers in Y. The $cov_{xy} = (N\Sigma XY - \Sigma X\Sigma Y)/(N-1)$. Note that the numerator is the same as the numerator in the computational formula for r from Chapter 6. The correlation coefficient, r, is the cov_{xy} standardized, that is, divided by s_x and s_y.

$$df_G = \frac{k-1}{k-1}$$

$$= \frac{(5-1)}{(5-1)} = 1$$

$$df_E = \frac{(k-1)(c-1)}{k-1}$$

$$= \frac{(5-1)(10-1)}{5-1} = 9$$

Notice that the conservative F test is *very* conservative—the dfs are reduced from 4 and 36 to 1 and 9. This means a loss in power and is probably an overcorrection. If we apply step b to the vital mission data, $F_{.05} = 5.12$ for 1 and 9 df. Since the obtained $F = 5.69 > F_{.05} = 5.12$, H_0 can be rejected by the conservative F test, also. This is a nice outcome. While we do not know what Mr. G and Mr. N were investigating, we can say that the test means differed. Another clear but unhappy outcome occurs when the F test with the usual df is nonsignificant. Then it is pointless to do the conservative F test. You have had it, but at least you know that you are nowhere.

The third possible outcome is one in which H_0 is rejected by the regular F test and H_0 is not rejected by the conservative F test. Because the conservative F test is such a severe correction, we do not want to leave the situation in this indecisive state. Rather than calculating ϵ as Greenhouse and Geisser suggested for step c, we propose to let a computer do it. Rerun the ANOVA on a computer (e.g., with BMDP). It will output more precise, adjusted dfs that will enable you to resolve the dilemma of the conflicting answers from the regular and conservative F tests. The ebb and flow of ideas in academe is fascinating—Fisher's LSD test, proposed long ago, is back, and 20 years later Rogan, Keselman, and Mendoza (1979) term the Greenhouse and Geisser three-step procedure "an expeditious strategy." I wonder if things like that happen in plumbing, too?

SUMMARY

This chapter was devoted to testing the means of k separate and k related batches of numbers, where $k > 2$. The definitions of separate and related were the same as those given in Chapter 8. The means of k separate batches can be evaluated by Fisher's analysis of variance (ANOVA). In ANOVA the total variability of N numbers, SS_T, is partitioned into the variability due to the treatments, SS_G, and random variability, SS_W. SS_T is the variability of all numbers about the grand mean, the mean of all N numbers, and its $df = N - 1$. SS_G is the variability of the treatment means about the grand mean weighted by the number of cases per mean. The SS_W is the variability of the numbers within a treatment about the treatment mean, pooled for all treatments. The $df_G = k - 1$ and $df_W = N - k$. The F statistic $= MS_G/MS_W$, where $MS_G = SS_G/df_G$ and $MS_W = SS_W/df_W$. Critical values of F_α are found in Table F, the F distribution, for df_G and df_W. If $F \geq F_\alpha$, the null hypothesis (H_0: $\mu_1 = \mu_2 = \ldots = \mu_k$) is rejected and the alternative hypothesis (H_1: $\mu_i \neq \mu_j$) is accepted. If $F < F_\alpha$, H_0 is not rejected. When H_0 is rejected, the difficult and controversial problem of comparing the k means emerges. Fisher's LSD (or protected t tests) was recommended for pairwise tests on the k means. If any $(\overline{X}_i - \overline{X}_j) \geq LSD$, the difference in the means is declared to be significant. A computational example of ANOVA was provided, followed by LSD tests of the means. The F test should be accompanied by an indication of the proportion (or percentage) of the total variance due to the treatments. The proportion for a sample can be found by computing eta squared (η^2)—$\eta^2 = SS_G/SS_T$—or estimated for the population by calculating Hays' omega squared statistic ($\hat{\omega}^2$). For pedagogical purposes the partitioning of variability in ANOVA can be portrayed in a pie diagram.

The assumptions for the F test for k separate batches are the same as for the t-for-two test: normality, homogeneity of variance, and independence. Although the F test is usually viewed as robust, unequal variances with unequal N_Gs and nonindependence with equal or unequal N_Gs can result in empirical sampling distributions that depart from the theoretical F distributions. Even with equal N_Gs heterogeneity of variance can be troublesome when the populations are markedly nonnormal. Accordingly, it was advised that the tests in Chapter 9 for homogeneity of variance should be applied. Lastly, it was shown that the F for an ANOVA of two batches is equal to t^2 from the t-for-two test and that $F_\alpha = t_\alpha^2$.

The ANOVA for k related batches includes the SS_T and SS_G like the ANOVA for k separate batches. Two additional SSs, SS_S and SS_E, are also found. SS_S is the variability of the case means about the grand mean weighted by the number of tests. The SS_E, random variability, is also $SS_{G \times S}$, the two-factor interaction of tests and cases. The $df_S = c - 1$ and $df_E = (k - 1)(c - 1)$. The df for any interaction is the product of the dfs for the interacting factors. The F statistic for evaluating the means of the k related batches is $F = MS_G/MS_E$. If F is significant, LSD tests may be done on the k means. The MS_E replaces MS_W in the LSD formula for separate batches and the df_E is used to get t_α.

The assumptions for the ANOVA for related batches are complicated and a matter of dispute. One proposed assumption is compound symmetry—that the test variances and the covariances among all pairs of tests are equal. The covariance is related to r: $cov_{ij} = r_{ij}(s_i)(s_j)$. Generally, the F test for the k test means is positively biased when the assumptions are violated. Positive bias means that F is found to be significant too often when H_0 is true. To solve the positive-bias problem the usual dfs of $(k - 1)$ and $(k - 1)(c - 1)$ are multiplied by epsilon, ϵ. One type of ϵ has limits of 1 and $1/(k - 1)$. Because of these diverse problems Greenhouse and Geisser's three-step procedure was adopted for testing the test means in ANOVA: (a) do F with the regular dfs or $\epsilon = 1$, (b) do a conservative F test or $\epsilon = 1/(k - 1)$, and (c) if steps a and b do not permit a decision, compute ϵ. (Instead of computing ϵ, we suggested letting a computer do it.) If steps a and b result in the rejection of H_0, the decision is clear—the test means are different. If step a results in the failure to reject H_0, the decision is clear—the test means do not differ. When step a results in rejection of H_0 and step b in the failure to reject H_0, step c is necessary to resolve the dilemma with more precise dfs for the F test.

PROBLEMS

1. In an experiment on the effect of labeled task difficulty upon persistence, 27 subjects were randomly and equally assigned to three conditions. For the E condition the problem was described as "easy." For the A condition the problem was said to be of "average" difficulty. In the H condition the problem was labeled as "hard." Actually, the problem was insoluable and the response measure was the time in minutes that the subject persisted in working. Test the H_0: $\mu_E = \mu_A = \mu_H$ with $\alpha = .05$.

E	A	H
10.7	8.7	9.6
6.4	9.5	15.4
8.3	14.3	19.4
7.6	9.7	8.6
9.8	10.7	10.1
11.6	9.3	13.2
9.4	15.1	14.1
8.9	11.3	11.8
9.5	13.1	16.3

2. If H_0 is rejected in Problem 1, do pairwise tests of the three means at $\alpha = .05$ (two-tailed).
3. Using two different methods, determine the percentage of variability in the persistence scores due to labeled task difficulty.
4. For the data in Problem 1, calculate SS_W directly and compare your answer with the value obtained by $SS_T - SS_B$.
5. Assuming normality, test the H_0: $\sigma_E^2 = \sigma_A^2 = \sigma_H^2$. Use $\alpha = .05$.
6. A verbal discrimination experiment is done in which 30 randomly assigned subjects receive feedback 0, 2, 4, 8, or 16 seconds after responding. The response measure is the number of trials to reach a learning criterion. Test the H_0: $\mu_0 = \mu_2 = \ldots = \mu_{16}$ with $\alpha = .01$.

	Groups			
0	2	4	8	16
3	5	6	8	8
3	2	6	9	6
4	3	3	5	4
1	4	5	8	4
2	6	7	6	6
2	4	4	7	5

7. In view of the outcome for Problem 6, describe what should be done next.
8. In another version of the verbal discrimination experiment, eight subjects underwent all three feedback conditions (0, 4, and 16 seconds.) Again, the number of trials to a learning criterion was the response measure. Test H_0: $\mu_0 = \mu_4 = \mu_{16}$ with $\alpha = .05$.

Subjects	Tests		
	0	4	16
1	4	7	8
2	1	4	5
3	3	2	9
4	3	3	6
5	2	5	7
6	1	6	7
7	2	6	7
8	3	7	9

9. If H_0 is rejected, compare the three means ($\alpha = .05$).
10. In the experiment described in Problem 8 the sequence in which *each* subject received the three feedback conditions was determined by a table of random numbers. Is this a better procedure than giving all subjects the treatments in a 0, 4, 16 sequence? Why?

ADDENDUM: USING COMPUTERS IN STATISTICS

MINITAB's ONEWAY program is appropriate for an ANOVA for k separate batches (1). Besides the ANOVA table, the k means and s values (2) are output. These statistics are convenient for pairwise testing of means and assessing homogeneity of variance. The ANOVA programs always display the confidence intervals (3) for individual means. Means whose confidence limits do not overlap can be regarded as different at $\alpha = .05$.

The TWOWAY program was applied to the "secret data" from the chapter to compare the means of k related batches. Note that the data analyst must compute the F test for tests (C32) with MS_{error} serving as the error term (4). The displays of the confidence limits for the means of cases and tests have been omitted but were available in the output.

```
--SET IN C1
--12 10 13 10 7 9 10 11 8 11 15 14 14 10 18 16 14 13
--11 9 12 9 8 10 11 12 9 6 4 7 5 7 7 9 10 8
--SET IN C2
--9(1),9(2),9(3),9(4)
--ONEWAY ANOVA DATA IN C1 CODES IN C2
```

ANALYSIS OF VARIANCE   (1)

DUE TO	DF	SS	MS=SS/DF	F-RATIO
FACTOR	3	214.97	71.66	19.14
ERROR	32	119.78	3.74	
TOTAL	35	334.75		

LEVEL	N	MEAN	ST. DEV.
1	9	10.00	1.87
2	9	13.89	2.42
3	9	10.11	1.45
4	9	7.00	1.87

(2)

POOLED ST. DEV. =   1.93

INDIVIDUAL 95 PERCENT C. I. FOR LEVEL MEANS
(BASED ON POOLED STANDARD DEVIATION)

```
     +--------+--------+--------+--------+--------+--------+
  1                        I******I******I
  2                                       I******I********I
  3                        I******I******I
  4            I******I******I
     +--------+--------+--------+--------+--------+--------+
    4.0      6.0      8.0     10.0     12.0     14.0     16.0
```
(3)

```
--TWOWAY ANOVA DATA IN C30 CODES IN C31 C32
```

```
ANALYSIS OF VARIANCE                                              (4)

DUE TO     DF        SS      MS=SS/DF
C31         9      10.76       1.20
C32         4      41.63      10.41
ERROR      36      65.99       1.83
TOTAL      49     118.38

OBSERVATIONS
ROWS ARE LEVELS OF C31    COLS ARE LEVELS OF C32
                                                          ROW
             1        2        3        4        5      MEANS
    1     -0.50     0.50     2.70     1.50     1.60     1.16
    2      1.70     2.60     0.60     1.40     3.40     1.94
    3      1.90    -0.60     2.80     3.60     1.50     1.84
    4     -1.60     2.90    -0.50     3.80     2.90     1.50
    5      1.10     2.10    -0.70     3.20     3.30     1.80
    6     -1.90    -0.60     2.20     1.50     2.10     0.66
    7     -1.60     2.20    -0.60     3.30     3.00     1.26
    8      1.20    -0.60     2.00     2.90     3.20     1.74
    9      1.00    -0.30     2.00     1.30     1.70     1.14
   10     -1.30     2.10    -0.40     1.10     1.30     0.56
COL.
MEANS    -0.00     1.03     1.01     2.36     2.40     1.36

POOLED ST. DEV. =        1.35
```

CHAPTER OUTLINE

11.1 WHAT IS A FACTORIAL DESIGN AND WHAT IS SO GREAT ABOUT IT?

11.2 A MILDLY EXCITING EXAMPLE—A CLOSE SHAVE IN SINGAPORE

11.3 AFTER A TWO-FACTOR ANOVA—WHAT NEXT?

11.4 "DEAD RATS BELONG IN THE RESULTS SECTION"

CHAPTER ELEVEN
A PIE WITH MORE SLICES—COMPARING MEANS FROM A FACTORIAL EXPERIMENT

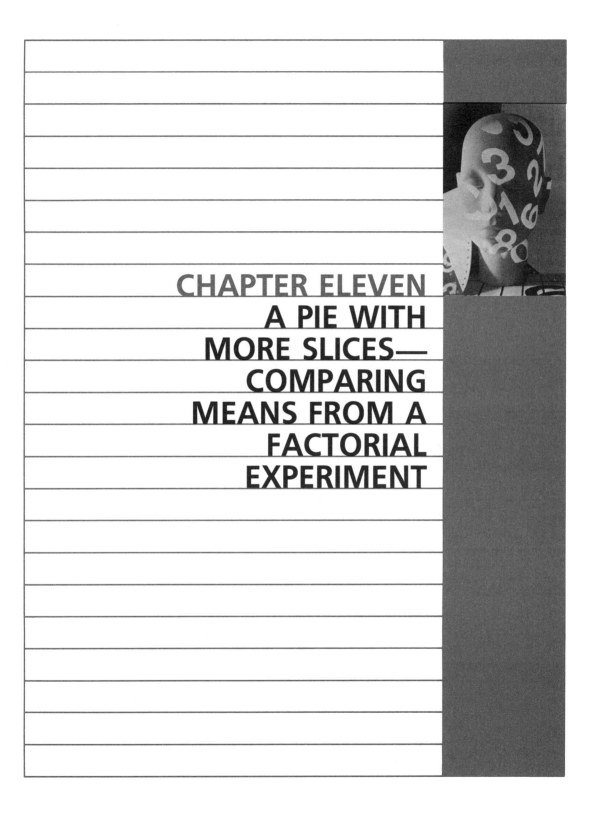

11.1 WHAT IS A FACTORIAL DESIGN AND WHAT IS SO GREAT ABOUT IT?

Research can be done in the style of Hagar The Horrible or it can be done neatly. In the former style, an investigator may manipulate several levels of an independent variable to determine the effect upon some dependent variable. Having done that experiment, the investigator might wonder whether another independent variable influences the same dependent variable and do a second experiment. But there is a tidier way to investigate the two independent variables—do a *single experiment* with a *factorial design*. Let us denote the first independent variable as A and its three levels (a) as A_1, A_2, and A_3. Suppose that B, the second independent variable, also has three levels (b): B_1, B_2, and B_3. Instead of doing an A experiment and a B experiment, A and B may be *crossed*, and the result is termed a 3×3 factorial design. What does crossed mean? It means that every level of A is combined with every level of B to form $(a)(b)$ or nine treatment-combinations. The A and B factors can be either *fixed* (arbitrarily selected levels) or *random* (randomly selected levels). We will only consider the case in which the variables are fixed.

For example, an investigator might be interested in studying the presentation of reading material on a computer terminal. More specifically, suppose the investigator is concerned with the effect of two independent variables or factors upon comprehension of the material (the dependent variable). Factor A is the rate of presentation of the material. The three levels are $A_1 = 300$ words per minute, $A_2 = 500$ words per minute, and $A_3 = 700$ words per minute. Factor B is the difficulty of the reading material. The three levels of B are B_1 = easy, B_2 = medium, and B_3 = difficult. A subject in the A_1B_1 treatment-combination would be presented with easy material at 300 words per minute.

If N_G subjects are randomly assigned to each of the nine treatment-combinations, then the subjects are said to be *nested* in the treatment-combinations. Under the nesting procedure the subjects in the different treatment-combinations are unique—there are *different* subjects in the different treatment-combinations. In another version of a factorial design N subjects undergo *all* treatment-combinations. The subjects are then said to be *crossed* rather than nested. Crossed with what? Crossed with the treatment-combinations. When the subjects are crossed, they are not unique to the treatment-combinations—they are the *same* subjects in all treatment-combinations.

SEC. 11.1 □ WHAT IS A FACTORIAL DESIGN AND WHAT IS SO GREAT ABOUT IT?

By about now the light should be coming on for you. A factorial design with nesting of subjects involves *separate* batches; a factorial design with crossing of subjects involves *related* batches. We will limit the presentation to a factorial design with nested subjects. Often this type of design is referred to as a *between-design* and a design with crossed subjects as a *within-design*. In more advanced courses within-designs and complicated designs incorporating *both* between and within factors are analyzed.

So far then we have asserted that research can be done more neatly by a single factorial design than by a series of independent experiments. Our discussion example will be a 3 × 3 factorial design with $N_G = 10$ subjects who have been randomly and equally assigned to each of the nine cells or treatment-combinations ($N = 90$). The design is depicted in Table 11.1. The 10 subjects in the upper left cell all undergo the A_1B_1 treatment-combination, 10 different subjects in the next cell to the right receive the A_1B_2 treatment-combination, and so on.

Is neatness the only payoff from a factorial design? Decidedly not! A factorial design is also more *efficient* and *informative* than two independent experiments. If an A experiment were done with three levels and an $N = 45$, the estimates of the means for the levels of A would each be based upon 15 numbers and the error term on 45 numbers (with $df_W = N - k = 45 - 3 = 42$). The same would hold true for a B experiment with three levels and $N = 45$. But in a 3 × 3 factorial design with $N_G = 10$ the estimates of the means for the levels of A would each be based upon 30 numbers because in estimating the mean for each A level, pooling occurs across the three B levels. That is, the sum for a mean ($\overline{X}_{A_1}$) is

$$\Sigma X_{A_1} = \Sigma X_{A_1B_1} + \Sigma X_{A_1B_2} + \Sigma X_{A_1B_3}$$

Similarly, the means for the three levels of B are each based on 30 numbers after pooling across the A levels. So the factorial de-

		B		
		B_1	B_2	B_3
A	A_1	$N_G = 10$	$N_G = 10$	$N_G = 10$
	A_2	etc.		
	A_3			$N = 90$

TABLE 11.1
A 3 × 3 factorial design with 10 subjects per cell

sign is more efficient than separate experiments because the estimates of the means for the levels of the factors are based on more observations. The error term for the factorial design is determined from 90 numbers (with $df_W = N - k = 90 - 9 = 81$). Accordingly, the error term should be a better estimate of random variability than the error terms based upon fewer numbers in separate experiments. Thus, a factorial design is more efficient in two respects: It provides better estimates of the effects of the factors and a better estimate of random variability.

How is a factorial design more informative? To answer this question we need to compare the sources of variation from two independent experiments with those from a factorial design. In both situations SS_A and SS_B are found.[1] The variability of the means of the A levels about the grand mean weighted by the number of observations that the means are based on is SS_A; SS_B represents the variability of the means for the B levels about the grand mean. But in the factorial design another unique source of variability is obtainable, the $A \times B$ *interaction*, $SS_{A \times B}$. $SS_{A \times B}$ indicates what happens when the two factors, A and B, are *combined*. If the effects of A and B merely sum or are additive, then the F test for the two-way interaction will not result in rejection of H_0. The interaction will not be significant. If the effects of A and B combine in a more complicated manner, such as a multiplicative relationship, then the F test for the interaction may be significant. We will discuss interactions in more detail in Sections 11.2 and 11.3. The point is that securing information about how the independent variables combine is a decided advantage of employing factorial designs.

A MILDLY EXCITING EXAMPLE—A CLOSE SHAVE IN SINGAPORE

11.2 Have you ever watched the TV commercial for the Remington electric razor in which Victor talks about the electric razor that his wife bought him? "I liked it so much, I bought the company!" he says. Actually, Daphne Genuine bought Reggie Genuine one of those Remington razors, too. These days Sir Reggie says proudly: "I liked it so much, I stole the company!" That is correct, Reggie makes Genuine Remington Electric Razors as well as Genuine Swiss Army Knives. With the assistance of the same industrial

---

1. In the single experiments, the SS_A and SS_B would be found for only one level of B and A, respectively.

SEC. 11.2 □ A MILDLY EXCITING EXAMPLE—A CLOSE SHAVE IN SINGAPORE

psychologist, a factorial experiment was done on the effects of factor A, employee suggestion boxes, and factor B, type of background music, upon how many electric razors were assembled by 64 workers in Reggie's Singapore factory. This time 64 workers were randomly and equally assigned to eight isolated workshops. In the A_1 condition the workshops contained a suggestion box for the workers; in A_2 the workshops did not have a suggestion box. B_1 was no background music, B_2 was Chinese music, B_3 was elevator music, and B_4 was rock. Thus, the experimental design was a 2×4 factorial design. (The workers under A_1 observed that the suggestion boxes were unlocked and emptied daily, but none of their suggestions were actually implemented during the study.) The average numbers of razors assembled for some unknown period of work, which have been rounded to the nearest integer, are displayed in Table 11.2.

TABLE 11.2
Mean numbers of razors assembled under the eight conditions of the 2×4 factorial design

		B (Music)				
		B_1 (None)	B_2 (Chinese)	B_3 (Elevator)	B_4 (Rock)	
		7	4	6	3	
		6	5	7	2	
		4	5	7	3	
	A_1 (present)	5	3	5	1	
		3	5	8	3	
		4	6	9	4	
		5	6	7	2	
		5	7	8	5	
A (Boxes)		Sums 39	41	57	23	$160 = \Sigma X_{A_1}$
		5	4	1	5	
		4	5	3	4	
		3	4	3	5	
	A_2 (absent)	4	2	4	3	
		4	3	4	3	
		2	3	5	3	
		3	4	4	1	
		6	4	6	1	
		Sums 31	29	30	25	$115 = \Sigma X_{A_2}$
		$\Sigma X B_j$ 70	70	87	48	$275 = \Sigma X$

The first task in the ANOVA is, as always, to find the whole pie, SS_T. Before beginning the analysis, the sums for the eight treatment-combinations, the sums for the two suggestion box conditions, and the sums for the four music conditions are determined for the data table. The sum of the suggestion box sums and the sum of the music condition sums must both equal ΣX. Entering all numbers in a calculator, we obtain: $\boxed{\Sigma X^2}$ 1381, $\boxed{\Sigma X}$ 275, $\boxed{N}$ 64. The SS_T is found in the usual way:

$$SS_T = \Sigma X^2 - \frac{(\Sigma X)^2}{N}$$

$$= 1381 - \frac{(275)^2}{64}$$

$$= 1381 - 1181.64 = 199.36$$

Pay particular attention to the correction term $(\Sigma X)^2/N = 1181.64$. If it is incorrect, you are on your way to a disaster. For the moment, forget about the factorial design, and pretend that you have a simple one-way design with eight separate batches. The SS_G for the eight batches (the treatment-combinations) is

$$SS_G = \frac{(\Sigma X_1)^2 + \cdots + (\Sigma X_8)^2}{N_G} - \frac{(\Sigma X^2)}{N}$$

$$= \frac{(39)^2 + \cdots + (25)^2}{8} - 1181.64$$

$$= \frac{(10307)}{8} - 1181.64$$

$$= 1288.38 - 1181.64 = 106.74$$

The SS_G is, of course, the variability of the eight treatment-combination means about the grand mean. By subtraction the random variability, SS_W, is readily found:

$$SS_W = SS_T - SS_B$$

$$= 199.36 - 106.74 = 92.62$$

We have divided the SS_T, total pie (199.36), into two pieces of SS_G (106.74) and SS_W (92.62). So far we have done nothing novel. We have simply obtained the usual SSs for an ANOVA with $k = 8$ separate batches.

Next, we will return to the factorial plan and partition the SS_G part of the pie into three pieces, SS_A, SS_B, and $SS_{A \times B}$:

$$SS_G = SS_A + SS_B + SS_{A \times B}$$

SEC. 11.2 □ A MILDLY EXCITING EXAMPLE—A CLOSE SHAVE IN SINGAPORE

After pooling across B and A, respectively, finding SS_A and SS_B is like computing the SS_Gs for separate batches. SS_A is[2]

$$SS_A = \frac{(\Sigma X_{A_1})^2 + (\Sigma X_{A_2})^2}{N_A} - \frac{(\Sigma X)^2}{N}$$

where N_A is the number of observations that the sums for each level of A are based upon. Then

$$SS_A = \frac{(160)^2 + (115)^2}{32} - 1181.64$$

$$= \frac{38825}{32} - 1181.64$$

$$= 1213.28 - 1181.64 = 31.64$$

Similarly, SS_B is calculated:

$$SS_B = \frac{(\Sigma X_{B_1})^2 + \cdots + (\Sigma X_{B_4})^2}{N_B} - \frac{(\Sigma X)^2}{N}$$

$$= \frac{(70)^2 + (70)^2 + (87)^2 + (48)^2}{16} - 1181.64$$

$$= 1229.56 - 1181.64 = 47.92$$

Remember that we are partitioning SS_G into three additive pieces. Therefore,

$$SS_{A \times B} = SS_G - SS_A - SS_B$$

$$SS_{A \times B} = 106.74 - 31.64 - 47.92 = 27.18$$

This procedure conforms to the formula in Chapter 10 for any two-way interaction. That is,

$$SS_{\text{rows} \times \text{columns}} = SS_{\text{cells}} - SS_{\text{rows}} - SS_{\text{columns}}$$

The completely sliced pie is shown in Figure 11.1.

Having computed the sums of squares, we need to find their respective degrees of freedom: $df_T = N - 1 = 64 - 1 = 63$; $df_W = N - k = 64 - 8 = 56$. The $df_A = a - 1$, where a is the number of levels of A, and $df_B = b - 1$, where b is the number of levels of B.

2. As shown in Table 11.2, the sum ΣX_{A_1} is the total for the 32 subjects who had suggestion boxes (i.e., pooling is across the four music conditions). The sum ΣX_{B_1} is the total for the 16 subjects in the none condition of music (i.e., pooling is across the two suggestion-box conditions).

FIGURE 11.1
A pie diagram of the partitioning of SS_T for a factorial design into SS_A, SS_B, $SS_{A \times B}$, and SS_W

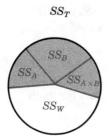

Thus, $df_A = 2 - 1 = 1$ and $df_B = 4 - 1 = 3$. The $df_{A \times B}$ can be found by the rule, stated in Chapter 10, that the degrees of freedom for any interaction are the *product* of the degrees of freedom for the interacting factors:

$$df_{A \times B} = (df_A)(df_B) = (a-1)(b-1) = (2-1)(4-1) = 3$$

Note that in this ANOVA, SS_G is only obtained to permit the calculation of SS_W and $SS_{A \times B}$. The sums of squares and degrees of freedom are listed in an ANOVA summary table (Table 11.3). The MSs are computed in the regular way:

$$MS_A = \frac{SS_A}{df_A} = \frac{31.64}{1} = 31.64$$

and so on. MS_W serves as an error term for testing the so-called *main effects*, A and B, and the $A \times B$ interaction.[3] For example,

$$F = \frac{MS_A}{MS_W} = \frac{31.64}{1.65} = 19.18$$

and so forth.

TABLE 11.3
ANOVA summary table for the 2 × 4 factorial design

Source of Variation		SS	df	MS	F	$F_{.01}$
Suggestion box	(A)	31.64	1	31.64	19.18	7.08
Music types	(B)	47.92	3	15.97	9.68	4.13
Interaction	(A × B)	27.18	3	9.06	5.49	4.13
Within	(W)	92.62	56	1.65		
Total	(T)	199.36	63			

3. This is true when A and B are fixed effects. If A, B, or both are random effects, different error terms are used.

Formulating the null hypotheses becomes cumbersome in complex designs, but we will state them here to facilitate an understanding of the interaction effect. There are *three* H_0s. For the main effect of A, H_0: $\mu_{A_1} = \mu_{A_2}$. For the main effect of B, H_0: $\mu_{B_1} = \mu_{B_2} = \mu_{B_3} = \mu_{B_4}$. The interaction H_0 pertains to the *treatment-combination means,* and more specifically to the *differences* in these means. For the $A \times B$ interaction,

$$H_0: (\mu_{A_1B_1} - \mu_{A_2B_1}) = (\mu_{A_1B_2} - \mu_{A_2B_2}) = (\mu_{A_1B_3} - \mu_{A_2B_3})$$
$$= (\mu_{A_1B_4} - \mu_{A_2B_4})$$

Or equivalently for the $A \times B$ interaction,

$$H_0: (\mu_{A_1B_1} - \mu_{A_1B_2} - \mu_{A_1B_3} - \mu_{A_1B_4})$$
$$= (\mu_{A_2B_1} - \mu_{A_2B_2} - \mu_{A_2B_3} - \mu_{A_2B_4})$$

When α is set at .01, the critical F values are located in Table F (60 was the nearest tabled entry for df_2), and the critical $F_{.01}$ values are entered in the summary table. Since all three obtained Fs are greater than $F_{.01}$s in Table 11.3, the H_0s for three effects, the main effects of A and B and the $A \times B$ interaction, may be rejected.

AFTER A TWO-FACTOR ANOVA—WHAT NEXT?

11.3

Before confronting this question, we can enumerate the eight possible outcomes for a two-factor design in Table 11.4, in which *sig* denotes that an effect is significant and *ns* denotes that an effect is nonsignificant. What to do after an ANOVA depends upon the nature of the outcome. Outcome 8 is a dead duck. We talked about this outcome briefly in Chapter 10 and will consider it further in Chapter 14. Outcomes 2, 4, and 6 with one or more significant main effects are not troublesome. The logical next step in these cases is to compare the means for the *levels* of significant factor(s). If a factor has two levels (such as A, suggestion boxes), the F test in the ANOVA compares the two means. If a factor has more than two levels (e.g., B, the music conditions), then the k means for the levels may be contrasted by protected, pairwise t tests. In computing the *LSD* (a) MS_W is obtained from the ANOVA, (b) the df_W is used to find t_α, and (c) N_G is the *number of cases* that *each level mean is based on.* Regarding (c), a place where errors are commonly made, let us consider an example. If we were contrasting the four music conditions means, N_G would be 16.

Finally, we can lump outcomes 1, 3, 5, and 7 together, and tag them TROUBLE. While their results might be important and

TABLE 11.4
Possible outcomes for a two-factor design

Outcome	A	B	A × B
1	sig	sig	sig
2	sig	sig	ns
3	sig	ns	sig
4	sig	ns	ns
5	ns	sig	sig
6	ns	sig	ns
7	ns	ns	sig
8	ns	ns	ns

exciting for an investigator, they pose data-analysis problems. Notice in each of these four outcomes the interaction is significant. What should we do with the interaction? And if one or more main effects are also significant (outcomes 1, 3, and 5), what should be done with the main effect(s)? How do researchers behave when they are confronted by a significant interaction and one or more significant main effects? I think that researchers follow three different approaches to this confusing problem: (a) report the significant interaction but analyze the main effects, (b) analyze *both* the interaction and main effects, and (c) analyze the interaction and report the main effects. Faced with a confusing situation researchers do different things.

In general, a significant interaction means that the effects of A and B are not additive. If the investigator asserts that a significant $A \times B$ interaction implies that "the effects of A depend upon B," that is a true statement but a decidedly vague one. The presence of a significant interaction forces an investigator to focus upon the interaction as the first order of business. It seems evident that outcomes 1, 3, 5, and 7 *demand* an examination of the $A \times B$ interaction.

A helpful first step in this examination is to graph the interaction. Since the interaction reflects the treatment-combination means, a graph of these cell means is called for. Figure 11.2 shows a plot of the means for the eight treatment combinations in our example. The A_1 means have been connected by one line and the A_2 means by another line. You will notice that the *differences* between A_1 and A_2 are *unequal* for the four B levels. These differences, $\overline{X}_{A_1} - \overline{X}_{A_2}$, for $B_1 \ldots B_4$ are, respectively, $4.88 - 3.88 = 1.00$, $5.12 - 3.62 = 1.50$, $7.12 - 3.75 = 3.37$, and $2.88 - 3.12 = -0.24$. The presence of such unequal differences is the *mark* of an interaction. Figures 11.3 and 11.4 show plots of some hypothetical means. Here the effects of A and B are additive because the dif-

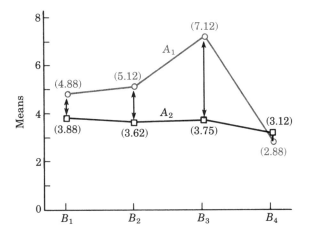

FIGURE 11.2
A graph of means for the simple main effects of A at B

ferences in the A_1 and A_2 means for all levels of B are *equal*. There is *no interaction* between A and B. Figure 11.2 depicts what are termed the *simple main effects* of A at B. A simple main effect is the variability of the means for different levels of a factor (e.g., A) at one level of another factor (e.g., B).

We could also look at the $A \times B$ interaction in a second way. In Figure 11.5 four lines have been drawn connecting A_1B_1 and $A_2B_1, \ldots,$ and A_1B_4 and A_2B_4. Now our attention is directed toward the differences between the means for the B levels (i.e., $\overline{X}_{B_1} - \overline{X}_{B_2}, \overline{X}_{B_2} - \overline{X}_{B_3},$ and $\overline{X}_{B_3} - \overline{X}_{B_4}$ at A_1 and the same three differences in the B means at A_2. The differences at A_1 are $4.88 - 5.12 = -0.24$, $5.12 - 7.12 = -2$, and $7.12 - 2.88 = 4.24$. The differences at A_2 are $3.88 - 3.62 = 0.26$, $3.62 - 3.75 = -0.13$, and

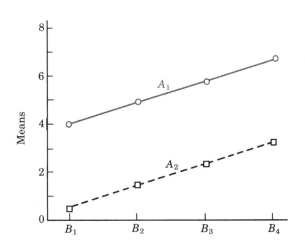

FIGURE 11.3
A graph of means for a factorial design in which the $SS_{A \times B}$ interaction is nonexistent

FIGURE 11.4
A graph of means for a factorial design in which the $SS_{A \times B}$ interaction is nonexistent

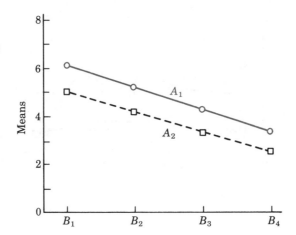

$3.75 - 3.12 = 0.63$. The differences in the B means differ greatly at the A_1 level, but they differ little at the A_2 level. These comparisons are called the *simple main effects of B at A*. If you reexamine the two possible null hypotheses for the preceding interaction, you can see how the simple main effects of A fit the first H_0 and the simple main effects of B fit the second H_0.

The procedure for analyzing an interaction entails four steps.

1. Select a set of simple main effects for analysis.
2. Calculate the variabilities for the simple main effects.
3. Test the simple main effects for significance.
4. Compare the means of the significant simple main effects.

Which set of simple main effects should be investigated? This should be decided by the investigator's interest. Sir Reggie Genuine was interested in the effects of B, the type of background music, upon worker output. He was paying the freight. Therefore, the *simple main effects of B at A* will be tested. How is this done? A sum of squares for the B means at A_1 is found. Another sum of squares for B means at A_2 is found. Then these two SSs are converted to MSs and are evaluated by F tests with the MS_W from the original ANOVA as the error term. The first simple main effect is

$$SS_{B \text{ at } A_1} = \frac{(\Sigma X_{B_1})^2 \cdots + (\Sigma X_{B_4})^2}{N_G} - \frac{(\Sigma X)^2}{N}$$

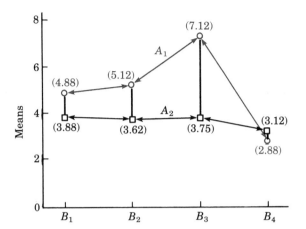

FIGURE 11.5
A graph of means for the simple main effects of B at A

In these calculations only the *top half of the data* in Table 11.2 is included so $\Sigma X_{B_1} = 39$, $\Sigma X = \Sigma X_{A_1} = 160$, $N = 32$, and so on. Then

$$SS_{B \text{ at } A_1} = \frac{(39)^2 + \cdots + (23)^2}{8} - \frac{(160)^2}{32}$$

$$= \frac{6980}{8} - \frac{(160)^2}{32}$$

$$= 872.5 - 800 = 72.5$$

The $SS_{B \text{ at } A_2}$ is for the *bottom half of the data* in Table 11.2:

$$SS_{B \text{ at } A_2} = \frac{(31)^2 + \cdots + (25)^2}{8} - \frac{(115)^2}{32}$$

$$= \frac{3327}{8} - \frac{(115)^2}{32}$$

$$= 415.88 - 413.28 = 2.6$$

We can put these *SS*s for the simple main effects into a summary table (Table 11.5) and test them. For the simple main effects, the degrees of freedom are

$$df_{B \text{ at } A_1} = df_{B \text{ at } A_2} = b - 1 = 4 - 1 = 3$$

The F tests reveal that the music conditions significantly affected worker output when suggestion boxes were present, B at A_1, but did not when suggestion boxes were absent, B at A_2.

TABLE 11.5
A summary table for the simple main effects of B at A

Source of Variation	SS	df	MS	F	$F_{.01}$
B at A_1	72.5	3	24.17	14.65	4.13
B at A_2	2.6	3	0.87	<1	4.13
W	92.62	56	1.65		

To complete the analysis of the $A \times B$ interaction, the means for B, the four music conditions, at the A_1 level (suggestion boxes present) are contrasted by protected, pairwise t tests. The

$$LSD = t_\alpha \sqrt{\frac{MS_{W(2)}}{N_G}}$$

The $t_{.01}$ (two-tailed) for $df_W = 56$ ($df = 60$ is the closest entry) from Table A is 2.66.

$$LSD = 2.66 \sqrt{\frac{1.65(2)}{8}} = 1.71$$

The ordered means for the music conditions when the suggestion boxes were present are shown in parentheses in Table 11.6. The table entries are the differences in the means. All differences in the means exceed the $LSD = 1.71$ except for the difference between the means for no music and Chinese music. Elevator music produced a significantly greater output of assembled electric razors than no music, Chinese music, and rock; Chinese music and no music led to a higher output than rock.

We have just analyzed the interaction in the way that was of interest to the investigator. The industrial psychologist was interested in the effects of the suggestion boxes, however. Without telling Reggie, he examined the interaction the other way. He evaluated the *simple main effects of A at B*. The four sums of squares now are

$$SS_{A \text{ at } B_1} = \frac{(\Sigma X_{A_1})^2 + (\Sigma X_{A_2})^2}{N_G} - \frac{(\Sigma X)^2}{N}$$

$$\vdots$$

$$SS_{A \text{ at } B_4} = \frac{(\Sigma X_{A_1})^2 + (\Sigma X_{A_2})^2}{N_G} - \frac{(\Sigma X)^2}{N}$$

In these computations successive *quarters of the data* or *columns* are analyzed going from left to right in Table 11.2:

TABLE 11.6
LSD tests on music condition means when suggestion boxes were present

Music	$\overline{X}$	None (4.88)	Chinese (5.12)	Elevator (7.12)
rock	(2.88)	2.00	2.24	4.24
none	(4.88)		0.24	2.24
Chinese	(5.12)			2.00

$$SS_{A \text{ at } B_1} = \frac{(39)^2 + (31)^2}{8} - \frac{(70)^2}{16} = 310.25 - 306.25 = 4.00$$

$$SS_{A \text{ at } B_2} = \frac{(41)^2 + (29)^2}{8} - \frac{(70)^2}{16} = 315.25 - 306.25 = 9.00$$

$$SS_{A \text{ at } B_3} = \frac{(57)^2 + (30)^2}{8} - \frac{(87)^2}{16} = 518.62 - 473.06 = 45.56$$

$$SS_{A \text{ at } B_4} = \frac{(23)^2 + (25)^2}{8} - \frac{(48)^2}{16} = 144.25 - 144 = 0.25$$

After putting these SSs into a summary table (Table 11.7), they are evaluated by F tests. The $df = 1$ for the simple main effects of A because the sums of squares at each level of B are the variability of two means about their grand mean. This way of analyzing the interaction reveals that only one comparison was significant: The presence of a suggestion box resulted in a significantly larger output of razors only under the condition of elevator music.

The outcome of the razor experiment was one in which both main effects, A and B, and the $A \times B$ interaction were found to be significant in the original ANOVA. We have just analyzed the $A \times B$ interaction in two ways and drawn conclusions about the treatment-combination means. What are we to do with the significant main effects, A and B? In view of the nature of the interaction, the main effects should *not* be analyzed. Why not? Consider the main effect for A, suggestion boxes, for example. Across all

TABLE 11.7
Summary table for the simple main effects of A at B

Source of Variation	SS	df	MS	F	$F_{.01}$
A at B_1	4.00	1	4.00	2.42	7.08
A at B_2	9.00	1	9.00	5.45	7.08
A at B_3	45.56	1	45.56	27.61	7.08
A at B_4	0.25	1	0.25	<1	7.08
W	92.62	56	1.65		

music conditions the 32 workers with suggestion boxes present assembled significantly more razors than the 32 workers without suggestion boxes. *But* we have just seen in the second interaction analysis that the statement about enhanced output with suggestion boxes is true only for the workers under the condition of elevator music. In the case of the significant main effect of B, music conditions, the first interaction analysis revealed differences among the means for music conditions only when suggestion boxes were present. Thus, for both main effects the presence of the $A \times B$ interaction forces us to qualify greatly any statements we might make about the main effects.

A second reason for not examining the means for the main effects is that the variability (SSs) of the main effects is included when the interaction is analyzed. This assertion requires further explanation. The sum of SSs for B at A is $72.5 + 2.6 = 75.1$, and the sum of their dfs is $3 + 3 = 6$. The $SS_{A \times B} = 27.18$ with $df_{A \times B} = 3$. Why is there a discrepancy in the SSs (75.1 vs. 27.18) and in the dfs (6 vs. 3)? Because the interaction analysis also includes $SS_B = 47.92$ with $df_B = 3$. Note that $SS_{A \times B} + SS_B = 27.18 + 47.92 = 75.1$ and $df_{A \times B} + df_B = 3 + 3 = 6$. Likewise, the sum of the SSs for A at B is $4 + 9 + 45.56 + 0.25 = 58.81$ with $df = 4$, which equals (within the limits of rounding) $SS_{A \times B} + SS_A = 27.18 + 31.64 = 58.82$ and $df_{A \times B} + df_A = 3 + 1 = 4$. In a sense, then, analyzing the main effects for A and B would be redundant since their variability was part of the variability that was partitioned in the interaction analyses.

In this particular example we would complete the ANOVA with the interaction analyses and do nothing further with the main effects. This proposed solution in the razor example to the what-to-do-after-ANOVA problem brings to mind a tale from the Great War (WW II). A torpedo bomber crashed on an aircraft carrier. It ended up half overboard and half on board. Half the crew wanted to deep-six it; the other half wanted to hoist it onto the deck. Finally, the wisdom of a veteran Chief was sought. He quickly solved the problem. "Just paint the damned thing gray and let it hang there," he advised. Faced by an interaction that forces qualifications of possible statements about the main effects, we are painting the main effects gray. There may be other examples in which the interaction, while significant, is trivial in magnitude and does not force a qualification of statements about the main effects. Then, the means of the significant main effects should be contrasted. Under these conditions the gray paint is being applied to the interaction.

SEC. 11.3 □ AFTER A TWO-FACTOR ANOVA—WHAT NEXT?

In Figure 11.6 the means for a 2 × 4 factorial study are shown. Let us assume that an ANOVA disclosed a very large and significant main effect for A and a small but significant *divergent* $A \times B$ interaction. Assume further that the simple main effects of A at B are significant throughout. Here the A main effect does not need to be qualified, and the interaction should be painted gray. Figure 11.7 shows a *cross-over interaction*. Even if the main effect of A were significant, it should be painted gray because A_1 is smaller than A_2 at B_1, A_1 equals A_2 at B_2, and A_1 is larger at B_3 and B_4. Plots of the means for interactions are essential to help investigators decide what to do when one or more main effect and the interaction are significant.

Only three matters are left to be considered: (a) the proportion or percentage of accounted-for variance, (b) the assumptions for a two-factor design, and (c) factorial designs with more than two factors. The simple formula (η^2) for the proportion of accounted-for sample variance is

$$\eta^2 = \frac{SS_{\text{effect}}}{SS_T}$$

For the main effect A it is

$$\eta_A^2 = \frac{31.64}{199.36} = 0.16$$

By Hays' formula for the estimated proportion of accounted-for variance in a population, it is

$$\hat{\omega}^2 = \frac{SS_{\text{effect}} - df_{\text{effect}}(MS_W)}{SS_T + MS_W}$$

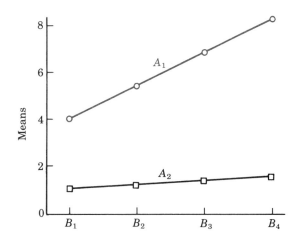

FIGURE 11.6
A graph of means with a large main effect of A and a small divergent $A \times B$ interaction

For A,

$$\hat{\omega}_A^2 = \frac{31.64 - (1)(1.65)}{199.36 + 1.65} = 0.15$$

After calculating the accounted-for variance for B and $A \times B$ by both methods, they are summarized in Table 11.8. Inspection of these proportions discloses that (a) the two methods agree closely in this example, (b) the proportions from the $\hat{\omega}^2$ formula as expected are always a little smaller, and (c) roughly a tenth to a quarter of the total variance is accounted for by the various effects. It should be noted that on occasion $\hat{\omega}^2$ will be a negative number. When this happens, $\hat{\omega}^2$ should be regarded as zero.

We can quickly state the *assumptions* for a factorial design with nested subjects. The three assumptions are the same as those for the t-for-two test and the F test for k means from separate batches—*normality* of the treatment-combination populations from which the batches were randomly drawn, *homogeneity of variance* of the treatment-combination populations, and *independence of observations* in the populations. Violations of these assumptions have the same consequences as were described earlier for the F test for means of separate batches. Nevertheless, as argued before, we believe that the variances should be evaluated for equality as a check on the homogeneity assumption and as a legitimate analysis in its own right. After finding the s^2s for the eight treatment-combinations with a calculator, they are shown in Table 11.9. Despite their obvious similarity, the s^2s will be subjected to an $F_{\max}$ test.

FIGURE 11.7
A graph of means showing a crossover $A \times B$ interaction

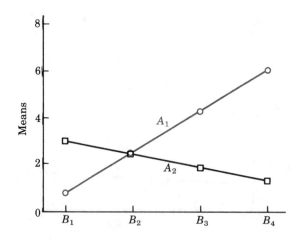

TABLE 11.8
Proportions of accounted-for variance by both methods in the razor study

Effect	η^2	$\hat{\omega}^2$
A	0.16	0.15
B	0.24	0.21
$A \times B$	0.14	0.11

$$F_{max} = \frac{\text{largest } s^2}{\text{smallest } s^2}$$

$$= \frac{2.411}{0.839} = 2.87$$

From Table E, $F_{max\,.01} = 22$ for $k = 8$ and $df = N_G - 1 = 8 - 1 = 7$. There is no reason to question the assumption of homogeneity of variance.

It was proposed in Chapter 10 that a useful check on an ANOVA is to calculate *directly* some source of variation that was obtained by subtraction. One spinoff of testing the variances for equality is getting the SSs for eight treatment-combinations. These SSs when combined should equal the value for SS_W found previously by subtraction. In Table 11.9 we have the variances (s^2s) for the eight treatment-combinations. Since $s^2 = SS/df$, the SSs can be obtained by multiplying the s^2s by df. The $df = N_G - 1 = 7$ and the resulting SSs when summed are

10.88 + 10.88 + 10.88 + 10.88 + 10.88 + 5.87 + 15.50 + 16.88 = 92.65

Within the limits of rounding, their sum is equal to $SS_W = 92.62$ in the ANOVA summary table.

Finally, factorial designs can be extended to include combinations of between and within factors and more factors. Whenever within factors (repeated measures) are involved, the problem of the complicated assumptions described in Chapter 10 arises again. In designs with more than two factors the problem of evaluating interactions becomes more complex. For example, with a three-factor design four interactions ($A \times B$, $A \times C$, $B \times C$, and $A \times B \times C$) result. We have already seen that two-factor interac-

TABLE 11.9
The variances for the eight treatment-combinations

	B_1	B_2	B_3	B_4
A_1	1.554	1.554	1.554	1.554
A_2	1.554	0.839	2.214	2.411

tions are difficult to analyze; when three or more factors interact, the analysis becomes extremely difficult. I think researchers tend to include too many factors in factorial designs. This tactic can produce a nightmare when a four- or five-factor interaction is found to be significant.

11.4 "DEAD RATS BELONG IN THE RESULTS SECTION"

This classic statement was written by the late Harry Harlow (1962) when his long tenure as a journal editor ended. It answered, once and for all times, the authors' age-old question: Where in an article should I put the description of the rats that failed to make the experimental journey?

Subject mortality in research is a formidable methodological problem. Rats do, indeed, die. Human subjects can die as they do in cancer research. Human subjects can participate in part of a study and fail to complete the rest of it. The subjects may live but the apparatus may die. A faithful, underpaid research assistant may bungle the research protocol and thereby lose subjects. And if the experimental treatments cause subject loss, the research is in serious difficulty. A million such sad tales emanate from the lab.

Suppose that you designed an elegant, two-factor study with equal N_Gs but some subjects are lost due to equipment failure and they cannot be replaced. You began with equal N_Gs and now they are unequal. If the unequal N_Gs are *proportional*, there is no computational problem. Proportionality is present when the ratios of cell counts in corresponding cells are identical.

Table 11.10 is a 2 × 3 factorial design with proportional N_Gs. The N_Gs are proportional because at B_1, $A_1/A_2 = 16/36 = 4/9$; at B_2, $A_1/A_2 = 8/18 = 4/9$; and at B_3, $A_1/A_2 = 12/27 = 4/9$. With proportional N_Gs the ANOVA can be calculated in the usual way except that the appropriate Ns must be employed throughout. The ANOVA, itself, is unaffected when proportionality exists. These considerations suggest a therapeutic device. Perhaps by *randomly* discarding a few subjects you can achieve proportionality. If you have many subjects, randomly discarding subjects could be tried as a possibility.

TABLE 11.10
A 2 × 3 design with proportional N_Gs

	B_1	B_2	B_3
A_1	16	8	12
A_2	36	18	27

Now let us turn to the situation in which *disproportionality* occurs. Furthermore, imagine that there is no way to achieve proportionality by discarding a few cases. Here, unfortunately, the ANOVA is perturbed—now the main effects and interaction are not independent. Ordinarily introductory statistics books do not deal with factorial design with disproportional N_Gs. We will, however, examine the problem because subject loss and disproportionality are facts of life in research, and one of the possible solutions to disproportionality, termed an *unweighted means analysis,* is not a difficult procedure.

Table 11.11 contains some hypothetical numbers for a 2 × 3 factorial design with disproportional N_Gs.[4] We will demonstrate the unweighted means method with this data set. First, a simple ANOVA will be done upon the *numbers* in the *six separate batches*. The sole goal of this analysis is to obtain SS_W. Second, a *factorial analysis* will be performed on the *means* of the batches. Third, the resulting SSs for A, B, and $A \times B$ will be *weighted equally,* and their MSs tested for significance against MS_W.

For the numbers in Table 11.11,

$$SS_T = \Sigma X^2 - \frac{(\Sigma X)^2}{N}$$

$$= 1344 - \frac{(202)^2}{34}$$

$$= 1344 - 1200.12 = 143.88$$

TABLE 11.11
A factorial design with disproportionality

		A_1			A_2		
B	B_1	B_2	B_3	B_1	B_2	B_3	
	3	4	7	6	6	3	
	2	5	8	8	7	2	
	4	6	8	8	6	5	
	4	7	9	9	5	4	
	5	6		10		4	
		6		8		4	
		8		7			
				8			
N_G	5	7	4	8	4	6	34 = N
$\overline{X}$	3.6	6.0	8.0	8.0	6.0	3.67	

4. The disproportionality can be verified by constructing a table of cell counts like Table 11.10. At B_1, $A_1/A_2 = 5/8$; at B_2, $A_1/A_2 = 7/4$; and at B_3, $A_1/A_2 = 4/6$.

The SS_G must be calculated with the longer formula, since the N_Gs are unequal:

$$SS_G = \frac{(\Sigma X_1)^2}{N_1} + \cdots + \frac{(\Sigma X_k)^2}{N_k} - \frac{(\Sigma X)^2}{N}$$

$$= \frac{(18)^2}{5} + \frac{(42)^2}{7} + \frac{(32)^2}{4} + \frac{(64)^2}{8} + \frac{(24)^2}{4} + \frac{(22)^2}{6} - \frac{(202)^2}{34}$$

$$= 1309.47 - 1200.12 = 109.35$$

Then

$$SS_W = SS_T - SS_G$$
$$= 143.88 - 109.35 = 34.53$$

And $df_W = N - k = 34 - 6 = 28$. Of these three SSs, only the SS_W is important for the unweighted means procedure. Parenthetically, it might be noted that the analysis so far is an example of an ANOVA for k separate batches with *unequal* N_Gs.

Next, an ANOVA is done on the *cell means* for the factorial design. The sum of the six squared means is $\Sigma X^2 = (3.6)^2 + \ldots + (3.67)^2 = 226.43$. The sum of the six means is $\Sigma X = 3.6 + \ldots + 3.67 = 35.27$. Then

$$SS_T = \Sigma X^2 - \frac{(\Sigma X)^2}{N}$$

$$= 226.43 - \frac{(35.27)^2}{6}$$

$$= 226.43 - 207.33 = 19.1$$

and

$$SS_A = \frac{(\Sigma X_{A_1})^2 + (\Sigma X_{A_2})^2}{N_G} - \frac{(\Sigma X)^2}{N}$$

where ΣX_{A_1} = sum of the three A_1 means.

$$SS_A = \frac{(17.6)^2 + (17.67)^2}{3} - 207.33$$

$$= 207.33 - 207.33 = 0.00$$

Then,

$$SS_B = \frac{(\Sigma X_{B_1})^2 + (\Sigma X_{B_2})^2 + (\Sigma X_{B_3})^2}{N_G} - \frac{(\Sigma X)^2}{N}$$

$$= \frac{(11.6)^2 + (12)^2 + (11.67)^2}{2} - 207.33$$

$$= 207.37 - 207.33 = 0.04$$

The SS_T for a factorial design with one observation (the mean) per batch corresponds to SS_G for a typical two-factor design with N_G observations per batch. Therefore, $SS_T = SS_A + SS_B + SS_{A \times B}$ for one observation per batch is like $SS_G = SS_A + SS_B + SS_{A \times B}$ for N_G observations per batch.[5] Thus,

$$SS_{A \times B} = SS_T - SS_A - SS_B$$
$$= 19.1 - 0 - 0.04 = 19.06$$

Now we must weight the three SSs from the means' analysis by the *harmonic mean* of the N_Gs of the treatment-combinations. The harmonic mean of the N_Gs is

$$N_h = \frac{ab}{\frac{1}{N_1} + \frac{1}{N_2} + \cdots + \frac{1}{N_k}}$$

Where a = the number of levels of the A factor and b = the number of levels of the B factor. Then

$$N_h = \frac{(2)(3)}{\frac{1}{5} + \frac{1}{7} + \frac{1}{4} + \frac{1}{8} + \frac{1}{4} + \frac{1}{6}}$$
$$= \frac{6}{1.1345237} = 5.29$$

The weighted SSs are then $N_h SS_A$, $N_h SS_B$, and $N_h SS_{A \times B}$. (To me the name *equally weighted* would seem more descriptive than *unweighted* since the SSs are multiplied by N_h.) These quantities are computed (e.g., $N_h SS_A = 5.29(0.00) = 0.00$, etc.) and entered in the ANOVA summary table (Table 11.12) along with the SS_W obtained in the first analysis. What can we conclude? The H_0 would not be rejected for the main effects, A and B, but would be rejected for the interaction. What should you do next? That is correct, go after that interaction. Plot it and then analyze it. Since the H_0s for the main effects were not rejected, nothing further needs to be done with them.

Since the N_Gs are unequal and ANOVA is troubled by unequal Ns and variances, we would want to test the variances to check the assumption of homogeneity of variance. With unequal N_Gs the recommended method for an overall test of equal variances is Bartlett's test (which was demonstrated in Chapter 9).

---

5. Since there is only one number ($\overline{X}$) in each cell, an SS_W cannot be obtained from the ANOVA of the means.

TABLE 11.12
ANOVA summary table for an unweighted means analysis

Source of Variation	SS	df	MS	F	$F_{.05}$
A	0.00	1	0.00	< 1	4.20
B	0.21	2	0.11	< 1	3.34
A × B	100.83	2	50.42	40.99	3.34
W	34.53	28	1.23		

We have completed our brief survey of the ANOVA of *numbers*. Chapter 12 will consider the ANOVA of *ranks*.

SUMMARY

A factorial design involves crossing two (or more) factors or independent variables. Crossing consists of combining all levels (a) of one factor, A, with all levels (b) of another factor, B, to form k treatment-combinations, where $k = (a)(b)$. We can randomly assign N_G subjects to each of the treatment-combinations, a procedure termed nesting. Or N subjects can undergo all treatment-combinations, which is termed crossing. A factorial design with nested subjects involves separate batches (treatment-combinations). A factorial design with crossed subjects involves related batches. Only the former design was considered. A single experiment with a factorial design, in contrast to a series of separate experiments, yields (a) better estimates of the effects of the independent variables, (b) a better estimate of random variability, and (c) information regarding the interaction of the factors, such as $SS_{A \times B}$, an indication of the joint effect of A and B.

A computational example of a 2 × 4 factorial design was provided. The total variability, SS_T, was partitioned into the variability between the treatment-combination means, SS_G, and the pooled variability, SS_W, within the treatment-combinations. The SS_G was then partitioned into three parts: SS_A, the variability of the means for the levels of A (the main effect of A); SS_B, the variability of means for the levels of B (the main effect of B); and $SS_{A \times B}$, which is related to the variability of the treatment-combination means. Null hypotheses regarding A, B, and $A \times B$ were evaluated by F tests with the MS_W serving as the measure of random variability.

What is done after an ANOVA of a factorial experiment depends upon the outcome of the experiment. If one or more main effects are significant, then the means for the significant main effects are contrasted. If a factor has only two levels, the F test in the ANOVA compares the two means. If a factor has more than two levels, then the k means are contrasted by pairwise, protected t tests.

If the outcome of a factorial experiment is that only the $A \times B$ interaction is significant or that the $A \times B$ interaction and one or both of the main effects are significant, the $A \times B$ interaction must be explored first. The treatment-combination means may be plotted for an eyeball examination. Unequal differences between the means for the levels of one factor at different levels of the other factor characterize an interaction. An $A \times B$ interaction can be evaluated quantitatively by doing simple main effects tests, either A at B or B at A or both. What simple main effects are tested should be dictated by the investigator's interest. When the $A \times B$ interaction is sizable and demands qualified statements about the main effects, the ANOVA is completed by analyzing the simple main effects. If the $A \times B$ interaction is small in magnitude and does not qualify statements about the main effects, then the means for the significant main effects should be compared, as described above, to complete the ANOVA.

Besides reporting the outcome of F tests and follow-up tests, the proportion of accounted-for variance for the effects, the A and B main effects, and the $A \times B$ interaction should be determined. The simple procedure (eta squared) for computing the proportions of the accounted-for sample variance is: $\eta^2 = SS_{\text{effect}}/SS_T$. Hays' omega squared yields estimates of the accounted-for population variance:

$$\hat{\omega}^2 = \frac{SS_{\text{effect}} - df_{\text{effect}}(MS_W)}{SS_T + MS_W}$$

The assumptions for a two-factor ANOVA are: (a) random sampling from normally distributed treatment-combination populations, (b) homogeneity of variance of the treatment-combination populations, and (c) independence of observations in the populations. These assumptions are the same as those for the t-for-two test and the F test for k separate batches. Violations of these assumptions have the same consequences as those described in Chapter 10 for the F test. It was recommended that the assumption of homogeneity of variance should be assessed by an $F_{\max}$ test. A by-product of this procedure is obtaining the SSs for the treatment-combinations. The accuracy of the ANOVA computations is suggested when the sum of these SSs is equal to the SS_W, which was obtained in the ANOVA by subtraction.

Subject loss in experiments constitutes a serious methodological problem. If the N_Gs in the treatment-combinations are proportional, the ANOVA is valid and can be done as usual except for inserting the proper Ns. However, if the N_Gs are disproportional, the ANOVA is disturbed and remedies are required. Sometimes disproportional N_Gs can be made proportional by randomly discarding a few cases. One simple analytic solution to disproportionality with a factorial design is to do an unweighted means ANOVA. A simple ANOVA on the numbers in the k treatment-combinations is performed to obtain SS_W. An ANOVA is then done upon the means in the factorial design to obtain SS_A, SS_B, and $SS_{A \times B}$. These sums of squares for the means are weighted by the harmonic mean of the N_Gs (e.g., $N_h SS_A$), and the resulting MSs are tested against MS_W to evaluate the H_0s for the factorial design. Because unequal N_Gs and unequal variances affect the sampling distribution of F, it was recommended that the s^2s for the treatment-combinations be tested for equality by Bartlett's test.

PROBLEMS

1. A researcher does a two-factor experiment in which one factor is sex (male vs. female) and the second factor is achievement motivation (low vs. average vs. high). If 12 subjects were randomly assigned to each treatment-combination, what would be the sources of variation and their degrees of freedom in the ANOVA summary table?

2. Fifty-four adults with severe deficits in short-term memory are assigned randomly to nine cells of a 3 × 3 factorial design. Factor A is dosage—low, medium, and high—and factor B is type of drug—1, 2, and 3. Following a month-long regimen of the drug the subjects are given a test of short-term memory. The scores here are the number of items remembered correctly on the test for each subject:

		Drug		
		1	2	3
Dosage	high	6, 5 7, 7 4, 8	2, 1 3, 1 2, 2	5, 4 4, 4 6, 5
	medium	5, 6 3, 5 7, 4	3, 1 1, 2 2, 1	7, 8 6, 8 7, 5
	low	4, 5 3, 4 6, 2	0, 2 3, 1 2, 1	5, 3 4, 4 5, 2

Did the drugs, dosages, or their interaction have any effect ($\alpha = 5.0$) upon memory?

3. In light of the results of Problem 2, do whatever follow-up tests are appropriate. What would you conclude from the study? (Parenthetically, it might be pointed out that $\overline{X} = 1.6$ was found for the whole group of subjects on an equivalent form of the memory test which was administered prior to drugs.)

4. Determine the percentages of variability in the memory test scores due to drugs, dosages, and their interaction.

5. In a replication of the study in Problem 2 an investigator included four treatment-combinations: drug 1—medium; drug 1—high; drug 3—low; and drug 3—medium. Evaluate the design of the replication study.

6. A researcher does an experiment with six separate, equal-sized batches (A_1B_1, A_1B_2, A_1B_3, A_2B_1, A_2B_2, A_2B_3). The researcher then performs a one-way ANOVA, finds a significant difference between the six batches, and compares the batch $\overline{X}$s by pairwise LSD tests. Comment upon the analysis.

7. Given the following data for a 2×3 factorial design:

	B_1	B_2	B_3
A_1	13, 10 14, 9 12, 13	9, 10 10, 11 9, 12	8, 7 8, 6 9, 10
A_2	16, 14 15, 16	12, 12 11, 9	9, 10 11, 11

Are the N_Gs proportional? Test the null hypothesis for A, B, and $A \times B$ ($\alpha = .05$).

8. Perform any follow-up tests on the means in Problem 7 that are appropriate. What would you conclude from the study?

9. In another study on memory deficits, 42 new subjects were exposed to medium dosages of drugs 1 and 3 for 8, 16, and 32 days prior to the short-term memory test. Did the type of drug, length of drug treatment, or their interaction affect memory test performance? Use $\alpha = .05$.

	Length		
	8	16	32
Drug 1	3, 1 2, 1 1, 2 0, 3	1, 4 3, 2 2	6, 5 5, 3 6
Drug 3	2, 1 3, 2 1, 2	3, 2 2, 3 0, 1 2, 3 1, 0	5, 7 6, 8 8, 7 7, 5

10. In light of the results of Problem 9, perform whatever follow-up procedures are necessary and summarize the results of the study.

ADDENDUM: USING COMPUTERS IN STATISTICS

MINITAB's TWOWAY program has been applied here to the factorial design described in the chapter (1). The slight differences between the hand calculations and the computer output are the result of rounding differences—where they occur and what rules are followed. Again the data analyst must complete the F tests. The TWOWAY program requires equal N_Gs. However, MINITAB can be adapted for an unweighted means analysis. First, a one-way ANOVA is performed with the AOVONEWAY program to obtain the error term, MS_{error}, and the cell means (2). Second, a TWOWAY ANOVA is done upon the cell means (3). Third, the analysis would be completed by calculating the harmonic mean, N_h, and then following the procedure described in the chapter for calculating the F tests.

```
--TWOWAY ANOVA DATA IN C6 CODES IN C7 C8
ANALYSIS OF VARIANCE                                       (1)
DUE TO      DF         SS
C7           1       31.64    MS=SS/DF
C8           3       47.92      31.64
C7 * C8      3       27.17      15.97
ERROR       56       92.63       9.06
TOTAL       63      199.36       1.65
```

```
CELL MEANS
ROWS ARE LEVELS OF C7   COLS ARE LEVELS OF C8
                                        ROW
              1      2      3      4    MEANS
       1    4.88   5.13   7.13   2.88   5.00
       2    3.88   3.63   3.75   3.13   3.59
COL.
MEANS       4.38   4.38   5.44   3.00   4.30

POOLED ST. DEV. = 1.29

--AOVONEWAY DATA IN C10-C15

ANALYSIS OF VARIANCE

DUE TO      DF      SS     MS=SS/DF   F-RATIO
FACTOR       5   109.35     21.87     17.73
ERROR       28    34.53      1.23
TOTAL       33   143.88

LEVEL     N    MEAN    ST. DEV.
C10       5    3.60      1.14
C11       7    6.00      1.29
C12       4    8.00      0.82
C13       8    8.00      1.20
C14       4    6.00      0.82
C15       6    3.67      1.03

POOLED ST. DEV. = 1.11

--READ C20-C22
--3.6 1 1
--6 1 2
--8 1 3
--8 2 1
--6 2 2
--3.67 2 3
--TWOWAY ANOVA DATA IN C20 CODES IN C21 C22

ANALYSIS OF VARIANCE

DUE TO     DF      SS     MS=SS/DF
C21         1    0.00      0.00
C22         2    0.05      0.02
ERROR       2   19.05      9.53
TOTAL       5   19.10
```

(2)

(3)

CHAPTER OUTLINE

12.1 STATISTICS (NONPARAMETRIC) THAT DO NOT TAKE SO MUCH FOR GRANTED

12.2 TESTING THE CENTER OF A SINGLE BATCH

12.3 COMPARING CENTERS FOR TWO SEPARATE BATCHES—BIG *T* FOR TWO

12.4 COMPARING CENTERS FOR TWO RELATED BATCHES—A TIRED EXAMPLE

12.5 COMPARING THE CENTERS OF *k* SEPARATE BATCHES OF RANKS—THE SINGAPORE STORY ONE MORE TIME

12.6 COMPARING CENTERS FOR *k* RELATED BATCHES—A REANALYSIS OF THE SECRET MISSION NUMBERS

CHAPTER TWELVE
MEETING RANKS AND SIGNS AT HIGH NOON

STATISTICS (NONPARAMETRIC) THAT DO NOT TAKE SO MUCH FOR GRANTED

12.1

In this chapter we will examine some statistics that apply to *ranks* and *signs*. In Chapter 13 statistics suitable for *frequencies* will be considered. These techniques are sometimes termed *nonparametric*. Other writers call them *distribution free*. And often the two names are treated as synonyms. The term *nonparametric* implies that population parameters are not estimated. But there are exceptions to this implication. Distribution-free implies that *no* assumptions are necessary for the population(s). As we shall see, this implication is not always true, either. The terms nonparametric and distribution-free will be employed interchangeably in this chapter to refer to statistical tests whose *assumptions* are *less restrictive* than those for parametric tests.[1] For example, the *t*-for-twins test, a parametric test, assumes that the distribution of the difference scores in the population is *normal*. Wilcoxon's signed-ranks test, a nonparametric test, assumes that the distribution of the difference scores in the population is *symmetrical*. The latter assumption is less restrictive. A rectangular distribution, an inverted-U distribution, and other distributions could fulfill the criterion of symmetry.

You have devoted a lot of time and effort to learn a variety of parametric statistical techniques. These tools may be applied to an array of problems to achieve the goal of applied statistics—making sense of numbers. Why should nonparametric statistics be added to the tool box? First, nonparametric techniques have been specifically designed to help make sense of ranks, signs, and frequencies. Research does not always end with numbers. Second, the logic underlying nonparametric tests is simpler. Third, the assumptions for nonparametric tests are not as restrictive. Fourth, many times nonparametric tests are easier to do. Fifth, with small batches the probabilities for some nonparametric statistics are exact. (With large batches known theoretical distributions can provide approximate probabilities.)

When Lucky Eddie read this list, he asked Hagar The Horrible a thoughtful question: "Why should we do parametric tests?" Hagar then did a horrible thing. He could not answer Lucky Eddie's question, so he hit him. The answers are (a) the application

---

1. Actually, you have already unknowingly encountered some nonparametric tests—the binomial test in Chapter 5 and Spearman's r_s in Chapter 6.

of a nonparametric test to numbers that meet the assumptions for a parametric test is "wasteful of data" (Siegel, 1956, p. 33) and (b) there are problems for which no nonparametric tests are available. Let us discuss these two disadvantages of nonparametric tests. The being-wasteful-of-data disadvantage is related to *statistical power,* the probability of rejecting the null hypothesis when it is false. When the assumptions for a parametric test are fulfilled, it is optimally powerful. The *power efficiency* of a nonparametric test under these conditions can be determined. Imagine that a parametric and a nonparametric test have equal power. Then, the power efficiency of the nonparametric test is $100N_p/N_{np}$, where N_p and N_{np} are the Ns for the parametric and nonparametric test, respectively, needed to achieve equal power. For example, if $N_p = 80$ and $N_{np} = 100$, then the power efficiency of the nonparametric test would be $100(80)/100 = 80\%$. Application of the nonparametric test requires 20 more cases than the parametric test. Two comments are in order, however. First, the power efficiencies for nonparametric tests are not always this low. Second, when the assumptions for the *parametric test* are *violated,* a nonparametric test may have greater power. Regarding the lack of available nonparametric tests, it should be pointed out that while there are many nonparametric tests in existence (Bradley, 1968; Conover, 1971; Hollander & Wolfe, 1973; Daniel, 1978), nonparametric tests are lacking for certain commonly occurring research paradigms, such as complicated factorial designs. From the host of nonparametric tests we will include only a few of those used most frequently. We will examine these tests in the same framework of a single batch, two separate batches, and so on that was followed previously for parametric tests.

TESTING THE CENTER OF A SINGLE BATCH

12.2

Recently I read a newspaper story with great interest. A town next to a large metropolitan center was described as "a giant radar trap." Irate commuters reported that they bypassed the town to avoid speeding tickets. Other angry citizens alleged that the chief of police had his traffic officers on a "quota system." If the officers did not issue a certain number of speeding citations per month, their job performance would be rated as unsatisfactory and three unsatisfactory ratings in a year meant dismissal. (I resonated to this story because I had helped two officers' performance ratings by being ticketed twice in one month.) First, the chief of police denied the charges. Then, the town council forced him to reveal the number of tickets that 10 of his traffic officers (randomly selected) had issued in the previous month. A resourceful council member also learned that the *median* number of citations issued by the metropolitan traffic officers in the same month was 135. The numbers of tickets issued by the 10 officers in the town were 153, 197, 147, 122, 165, 172, 186, 135, 186, 148. Are the police in Radar Trap Town issuing *more* tickets than the officers in the metropolitan area (suggesting, perhaps, that the chief does impose a quota system)?

In Chapter 8 we did a t test for a single batch on such numbers. Since this is clearly a one-tailed test problem, the alternative hypothesis for a t test would be $H_1: \mu > \mu_{MA}$. The null hypothesis would be $H_0: \mu \leq \mu_{MA}$, where μ_{MA} is the metro area mean. But we do not know μ_{MA}, and we want to demonstrate a nonparametric test. Accordingly, the $H_1: m > 135$, where m is the *population median*, $m_{MA} = 135$ is the metro area median, and $H_0: m \leq 135$. An appropriate nonparametric test for this situation is *Wilcoxon's signed-ranks test*. We will call Wilcoxon's test the Big T. Why? Because (a) the test statistic is T, and (b) T may remind you that Wilcoxon's test is the nonparametric analogue to small t, the t test for a single batch. How is the Big T test done? Reduce the numbers, Xs, to Ds, difference scores from the metro median, m_{MA}. That is, $D = X - m_{MA}$. Rank the *absolute* values of the Ds, $|D|$, from 1 ... N. Then, find the test statistic, Big T. T is defined as the rank sum for the Ds with the less-frequently occurring sign. When the Ds are calculated, some will usually be positive (+) and others (−). Big T is the rank sum of the Ds for whichever class of Ds is less frequent. If any Ds have the same values (are tied), their ranks are assigned by the midrank method (see Chapter 6). If a $D = 0$, it is deleted from the batch, and N, the number of Ds, is reduced by one. Let us set $\alpha = .01$ and analyze the ticket data in Table 12.1.

In column 2 the residuals have been calculated (i.e., $D = X - m_{MA} = 153 - 135 = 18$, etc). Disregarding the signs of the Ds, they have been ranked in column 3. Since case 8 had a $D = 0$, the case is dropped from the batch, and N becomes $10 - 1 = 9$. There are eight $+D$s and one $-D$. Therefore, the test statistic, T, is the rank of the single negative difference or 2.5. From Table G the one-tailed $P(T = 2.5)$ when $N = 9$ is $.01 > P > .005$. Accordingly, H_0 would be rejected and H_1 accepted. The officers from Radar Trap Town are handing out significantly more tickets than the median number in the adjacent metropolitan area. I want my $40 in fines back. I will add it to the buy-a-radar-detector fund.

TABLE 12.1
Numbers of tickets (X), difference scores ($D = X - m_{MA}$), and ranks for the radar trap example

X	D	Rank for |D|	Rank for +D	Rank for −D
153	18	4	4	
197	62	9	9	
147	12	1	1	
122	−13	2.5		2.5
165	30	5	5	
172	37	6	6	
186	51	7.5	7.5	
135	0			
186	51	7.5	7.5	
148	13	2.5	2.5	
Sums			42.5	2.5 = T

In the beginning of the chapter it was stated that for small batches the probability values for nonparametric tests are often exact. For large Ns the probabilities may be approximated by some known theoretical distribution. The Big T statistic conforms to these generalizations. If we have access to Wilcoxon's Table G, the probabilities for T can be obtained up to $N = 25$. For large batches the sampling distribution of T becomes approximately normal. Thus, a Z statistic for the Wilcoxon test can be computed and its probability determined in Table D, the unit-normal distribution. The Z formula is

$$Z = \frac{T - \frac{N(N+1)}{4}}{\sqrt{\frac{N(N+1)(2N+1)}{24}}}$$

What is this strange formula? It is nothing very esoteric. It is

$$\text{a test statistic} = \frac{\text{a statistic} - \text{a parameter}}{\text{random variability}}$$

T is the statistic. $T_E = N(N + 1)/4$ is the parameter, the T expected by chance. The rank sum for *any* set of N ranks is $\Sigma R = N(N + 1)/2$. By chance, half of the ranks should be positive and half negative. The T_E for half the Ds would be

$$\frac{\Sigma R}{2} = \frac{\frac{N(N+1)}{2}}{2} = \frac{N(N+1)}{4}$$

What is the denominator? It is the standard error of T_E or σ_{T_E}. The formula is not strange.

We have a small sample. If we did not have Table G, we might try the Z formula. Would it work with an $N = 9$? Let us see how the Z formula performs. $T = 2.5$ and it is calculated as above. Then,

$$Z = \frac{2.5 - \frac{9(9+1)}{4}}{\sqrt{\frac{9(9+1)(18+1)}{24}}}$$

$$= \frac{-20}{\sqrt{71.25}} = -2.37$$

The one-tailed $P(Z = -2.37)$ in Table D is .0089. From Wilcoxon's Table G the $P(T = 2.5)$ was $.01 > P > .005$. Therefore, for this problem with $N = 9$ the Z formula works very well. But don't calculate Z when you can read the answer from a T table.

Finding T is a lot easier than finding t. How about an even simpler test? Today, this test is sometimes referred to as *Fisher's sign test* because Fisher used it in the 1920s. Daniel (1978), however, gives credit to Arbuthnott (1710) for the test. What is the rationale for the sign test? For the radar trap example we said that by chance half of the Ds should be positive and half negative. In a *t*-for-twins test the *actual values* of the Ds (from the mean) are analyzed. In the Big T test, the *ranks* of the Ds were analyzed. In the sign test, only the *signs* of the Ds are counted. We observed eight +Ds and one −D. By chance, we would expect four and one-half +Ds and four and one-half −Ds. What is the one-tailed probability of obtaining eight or more +Ds from nine Ds? Are you having a flashback experience of a friendly cantina where some coins were flipped? In brief, we can answer the sign test question by the binomial distribution as we did for coin tossing in Chapter 5. We can calculate the $P(8H, 1T$ or $9H)$, which is like observing eight or more +Ds, by expanding our former Pascal's triangle from $N_T = 6$ to $N_T = 9$. Or, we can read the probability from a table for the binomial distribution. The coefficients for the expanded triangle are presented in Table 12.2.

Note that T in Table 12.2 is the total number (512) of possible outcomes, not Big T. The coefficient for 9H is 1, and the coefficient for 8H,1T is 9. That is, 9H can occur by chance in only one way; 8H can occur in nine ways. Since $p = q = .5$ the $P(8H,1T$ or $9H) = W/T = (1 + 9)/512 = .0195$. Or, the $P(8H,1T$ or $9H) = p^9 +$

SEC. 12.3 □ COMPARING CENTERS FOR TWO SEPARATE BATCHES—BIG T FOR TWO

TABLE 12.2
Coefficients for the binomial expansion for $N_T = 9$.

N_T	Coefficients	$\Sigma = T$
6	1 6 15 20 15 6 1	
7	1 7 21 35 35 21 7 1	
8	1 8 28 56 70 56 28 8 1	
9	1 9 36 84 126 126 84 36 9 1	512

$9p^8q = (.5)^9 + 9(.5)^8(.5) = .0195$, as before. Or, from the binomial table in Hollander and Wolfe (1973) P is also .0195. For this problem with an $\alpha = .01$ the two nonparametric tests result in *different* decisions with respect to H_0. The Big T test leads to rejection of H_0; with the sign test H_0 cannot be rejected. Do these conflicting decisions trouble you? Do not feel that way. Pick the test statistic that is appropriate for your data, and abide by the decision it leads you to make.

What are we assuming for these two nonparametric tests? The parametric t test for a single batch, Wilcoxon's signed-ranks test, and the sign test *all* assume that the Xs or Ds (a) have been randomly sampled, (b) are independent, and (c) were continuously measured. (Continuous measurement means that the probability of ties in the Xs or Ds is 0. In the real world that means that the probability of ties is near 0.) The three tests differ in their assumption about the shape of the Xs or Ds in the population. For the t test the distribution must be *normal*, for the Wilcoxon test it must be *symmetrical*, and for the sign test *no* assumption is made regarding the shape of the distribution. Thus, the sign test is truly a distribution-free test. A final comment: The Big T and sign test can, of course, be employed to test hypotheses that $m = 0$. Then, the Ds = Xs.

COMPARING CENTERS FOR TWO SEPARATE BATCHES—BIG T FOR TWO

12.3

The resourceful council member in Radar Trap Town also discovered the numbers of citations in the same month issued by 10 randomly selected traffic officers in another town bordering on the metro area where the police chief was highly regarded. The council member reasoned that if the chief in Radar Trap Town had his officers on a quota system, then his officers would have issued more tickets on the average than the officers in the other town. Now we have two separate batches of numbers. The H_1: $m_1 > m_2$, where m_1 is the population median for Radar Trap

Town and m_2 is the population median for the other town; and $H_0: m_1 \leq m_2$. We will set $\alpha = .01$. The nonparametric test for this situation is a different Wilcoxon test—*Wilcoxon's rank-sum test*. We will refer to this test as *Big T for two*. To calculate T, the N_1 Xs and N_2 Xs are tossed together into a pile. The N Xs ($N = N_1 + N_2$) are then ranked from 1 ... N. Tied Xs are assigned midranks. The ranks are then attached to the Xs for the separate groups and summed.[2] T is the *smaller* rank sum, and T' is the *larger* rank sum. The numbers of tickets (Xs) issued in the two towns are shown in Table 12.3.

TABLE 12.3
Numbers of tickets (Xs) and their ranks (R) for the two towns

Town 1		Town 2	
X	(R)	X	(R)
153	(15)	104	(2)
197	(20)	137	(9)
147	(12.5)	128	(6)
122	(4)	102	(1)
165	(16)	117	(3)
172	(17)	133	(7)
186	(18.5)	146	(11)
135	(8)	124	(5)
186	(18.5)	138	(10)
148	(14)	147	(12.5)
Sums	143.5 = T'		66.5 = T

The smallest X in the combined batch is 102; its rank is 1. Continuing, we reach $X = 197$ whose rank is 20. The ranks for the Xs are listed in columns two and four and summed to get T and T'. Since

$$\Sigma R = \frac{N(N+1)}{2} = \frac{20(20+1)}{2} = 210$$

a check on the ranking is possible:

$$\Sigma R = T + T' = 66.5 + 143.5 = 210$$

Note that if chance alone were operating, we would expect $T = T' = \Sigma R/2 = 105$. To test the hypotheses, the smaller rank sum T is evaluated in Table H for N cases per group ($N/2 = 10$).

2. It is, of course, possible that the numbers from a study could be in the form of ranks from 1 ... N for the combined batches.

The one-tailed $P(T = 66.5)$ is less than .005. Therefore, H_0 would be rejected and H_1 accepted. We conclude that the traffic officers in Radar Trap Town are issuing more tickets than the officers in the next town.

When $N_1 = N_2$ the Big T-for-two test is a dream come true for people who don't like to calculate. But what can be done when $N_1 \neq N_2$? You could search for an extended set of T tables for equal and unequal Ns (Wilcoxon, Katti, & Wilcox, 1963). Or, you could run an equivalent test, the Mann-Whitney U test, on a computer (Ryan et al., 1981). Finally, you could calculate a Z test version of the Big T for two. Again, when N is large, T is approximately normally distributed. And, once more, the general form of the test is

$$Z = \frac{T - T_E}{\sigma_{T_E}}$$

Now T_1 is the rank sum for batch 1 and T_2 is the rank sum for batch 2. Then,

$$Z = \frac{T_1 - \frac{N_1(N + 1)}{2}}{\sqrt{\frac{N_1(N_2)(N + 1)}{12}}}$$

or

$$Z = \frac{T_2 - \frac{N_2(N + 1)}{2}}{\sqrt{\frac{N_1(N_2)(N + 1)}{12}}}$$

Either formula can be applied. The Zs from the two formulas will be identical in value but opposite in sign. Let us demonstrate the first formula for the Z test with the traffic ticket data:

$$Z = \frac{143.5 - \frac{10(20 + 1)}{2}}{\sqrt{\frac{10(10)(20 + 1)}{12}}}$$

$$= \frac{38.5}{\sqrt{175}}$$

$$= \frac{38.5}{13.23} = 2.91$$

The one-tailed $P(Z = 2.91)$ from Table D is .0018. Thus, even with these small Ns the Z test version of the Big T for two performs well.

The assumptions for the Big T-for-two test need to be described. You will recall that the assumptions for the t-for-two test were: random sampling from normal populations, homogeneity of variance in the populations, and independence of observations in the populations. The Big T for two also assumes independence, but makes a less restrictive assumption of random sampling from the *same* population.[3] This implies that the populations have the same shape—not necessarily normal—and that, in turn, implies equal spreads. In statistics, as in life, the chances of getting something for nothing are not great.

COMPARING CENTERS FOR TWO RELATED BATCHES—A TIRED EXAMPLE

12.4

This type of problem can be solved by *Wilcoxon's signed-ranks test* and *Fisher's sign test*. Previously, in the one-batch case D was defined as $X - m$. With related batches the Ds are the difference scores (e.g., X for trial 1 $-$ X for trial 2) for each case. Then the Ds can be analyzed with the Big T for a single batch or with the sign test. Since the Ds are obtained from related batches, we will term this Wilcoxon T procedure the *Big T for twins*. Generating examples and numbers can turn your brain into yogurt. How about a tired example? Let us recycle Goodie's tire data from Chapter 8 where a t-for-twins test was demonstrated. To save space, only the Ds are displayed in Table 12.4. Remember, the Ds were the miles per gallon for cars equipped with steel-belted tires (X_1) minus the miles per gallon for the same models of cars equipped with polyester tires (X_2). The single $D = 0$ is deleted to make $N = 19$. There are quite a few ties among the Ds, but we won't apply possible corrections for the ties. Since

$$\Sigma R = \frac{N(N + 1)}{2} = \frac{19(20)}{2} = 190$$

and since the rank sums are $172 + 18 = 190$, the ranking procedure appears to be correct. The H_1: $m_1 > m_2$ and H_0: $m_1 \leq m_2$, as

3. This assumption holds for H_0. Under H_1 the numbers in one batch are shifted by $\Delta = m_1 - m_2$ (Hollander & Wolfe, 1976, p. 10).

SEC. 12.4 □ COMPARING CENTERS FOR TWO RELATED BATCHES—A TIRED EXAMPLE

TABLE 12.4
Difference scores (D = $X_1 - X_2$) and ranks for 20 pairs of cars in the tire example

Pair	D	Rank for |D|	Rank for +D	Rank for −D	Pair	D	Rank for |D|	Rank for +D	Rank for −D
1	0.7	13	13		11	0.7	13	13	
2	−0.1	1		1	12	−0.5	8.5		8.5
3	1.1	18.5	18.5		13	1.1	18.5	18.5	
4	−0.5	8.5		8.5	14	0.2	2.5	2.5	
5	0.8	15	15		15	0.3	4.5	4.5	
6	0.5	8.5	8.5		16	1.0	17	17	
7	0.6	11	11		17	0.9	16	16	
8	0.5	8.5	8.5		18	0.2	2.5	2.5	
9	0.7	13	13		19	0.0			
10	0.4	6	6		20	0.3	4.5	4.5	
								Sums 172	18 = T

in the original example, and α = .01. T is the rank sum for the −Ds. The one-tailed P(T = 18) for N = 19 from Table G is less than .005. Both a *t*-for-twins test and Big *T*-for-twins test confirm Goodie's hypothesis that steel-belted radials are superior.

The signs of these Ds can also be tested with the *sign test*. N = 19 and there are 16 +Ds and 3 −Ds. We could expand the binomial to $N_T = 19$, but that would be considerable work. We can also obtain an approximate P from the normal distribution as described in Chapter 5. Here, assuming p = q = .5, the formula is

$$Z = \frac{\left| X - \frac{N_T}{2} \right| - .5}{\sqrt{\frac{N_T}{4}}}$$

where X = 16 (or 3), and $N_T = 19$. Then Z is

$$Z = \frac{\left| 16 - \frac{19}{2} \right| - .5}{\sqrt{\frac{19}{4}}}$$

$$= \frac{6}{2.18} = 2.75$$

From Table D the one-tailed $P(Z = 2.75)$ is .003. Another solution is to look up the $P(16+, 3-$ or more) in a binomial table (e.g., Hollander & Wolfe, 1973) or in a binomial table generated by MINITAB. The exact P is .0022. Thus, the sign test also results in the rejection of H_0. Finally, the assumptions for Wilcoxon's signed-ranks test and the sign test were presented in section 12.2.

COMPARING THE CENTERS OF k SEPARATE BATCHES OF RANKS—THE SINGAPORE STORY ONE MORE TIME

12.5

The Big T-for-two test, which is applicable to the ranks for two separate batches, has been generalized to k separate batches by Kruskal and Wallis (1952). The test statistic is H, and the procedure is an ANOVA for k sets of ranks. Exact probabilities for H when $k = 3$ and $N_G \leq 5$ are given in Table I. For designs larger than $k = 3$ or $N_G > 5$, H is approximately distributed as a χ^2 (chi square) with $df = k - 1$. To repeat: χ^2 distributions are a family of asymmetrical distributions that vary with degrees of freedom (Table J). The computational formula for the *Kruskal-Wallis H* statistic is

$$H = \frac{12}{N(N+1)} \left[\frac{(R_1)^2}{N_1} + \cdots + \frac{(R_k)^2}{N_k} \right] - 3(N+1)$$

where N is the total number of cases, R_1 is the rank sum for batch 1, N_1 is the number of cases for batch 1, R_k is rank sum for the kth batch, and N_k is the number of cases for the kth batch. When N_Gs for the batches are equal, the formula becomes

$$H = \frac{12}{N(N+1)} \left[\frac{(R_1)^2 + \cdots + (R_k)^2}{N_G} \right] - 3(N+1)$$

where N_G is the batch size.

To compute H the Xs from all batches are combined as in the Wilcoxon rank-sum test and ranked from 1 . . . N. Tied Xs are assigned midranks. For an example we will go back to the Singapore knife problem in Chapter 10. The *ranks* for the knife outputs for the four music conditions are shown in Table 12.5.
(If you will repeat this ranking procedure, you will become aware that with larger batches nonparametric tests can become tedious.) The sum of the rank sums for the four batches is

$$\Sigma R = 163 + 272.5 + 167.5 + 63 = 666$$

None	Elevator	Chinese	Rock
27	23.5	23.5	3
18.5	34	13	1
29.5	32	27	5.5
18.5	32	13	2
5.5	18.5	9	5.5
13	36	18.5	5.5
18.5	35	23.5	13
23.5	32	27	18.5
9	29.5	13	9
Sums 163	272.5	167.5	63

TABLE 12.5
The ranks for knives assembled under the four music conditions

These rank sums should equal

$$\Sigma R = \frac{N(N+1)}{2} = \frac{36(36+1)}{2} = 666$$

If H_0 were true, all four ranks sums should be close to $\Sigma R/4 = 666/4 = 166.5$. The rank sums clearly are not equal.

The H_0: $m_1 = m_2 = m_3 = m_4$ and H_1: $m_i \neq m_j$. As in the original ANOVA, α will be set at .01. With the formula for equal N_Gs the test statistic is

$$H = \frac{12}{36(36+1)} \left[\frac{(163)^2 + (272.5)^2 + (167.5)^2 + (63)^2}{9} \right] - 3(36+1)$$

$$= .009009(14761.167) - 111$$

$$= 132.98 - 111 = 21.98$$

It is very important to carry out the $12/N(N+1)$ portion of the formula to *many decimal places*. If it is rounded to .01, then $H = 36.61$! Another way to obtain the correct H is $1/111(14761.167) - 111 = 132.98 - 111 = 21.98$. Since $k = 4$ and $N_G = 9$, the problem exceeds the bounds of Table I. Therefore, H is evaluated as a χ^2. From Table J the $P(\chi^2 = 21.98)$ with $df = k - 1 = 3$ is $< .001$. Using the H test instead of ANOVA does not change the decisions; H_0 would be rejected and H_1 accepted. There are a number of tied Xs in Reggie's data. H can be corrected for ties (Daniel, 1978, p. 203). But since the correction only increases the size of H, it would be pointless to apply it in this case.

Rejection of H_0 with the H test leaves us in the same situation as when the F test led to the rejection of H_0. The question now becomes: What population medians for the music conditions differ from another? One Monte Carlo study (Wike & Church, 1978) recommended comparing the music medians by *protected, pairwise Wilcoxon rank-sum tests* or *Big T-for-two-tests*. In the Monte Carlo study the Big T test had better Type 1 error rates than three other procedures when it was applied *only* after a significant H test. The bad news is this: $k(k-1)/2 = 6$ Big Ts must be carried out, and for each test the Xs in the two batches being compared must be ranked *anew*. For example, if the none and elevator music conditions are reranked, then $T = 52.5$ and $T' = 118.5$. From Table H the $P(T = 52.5)$ for $N = 9$ is less than .01. Thus, production was higher in the elevator condition. The reranking can be laborious, so seeking the help of a tireless slave—a computer—is a wise strategy. We will demonstrate the slave's efforts in the addendum to this chapter. While nonparametric tests are often quick and easy, doing protected t tests is easier than doing protected Big T tests.

Finally, the assumptions for the Kruskal-Wallis H test are worthy of mention. The assumptions are the same as those for the Wilcoxon rank-sum test described in Section 12.3. With the H test the population distributions do not have to be normal as required for the F test, but they must have the same shape.

COMPARING CENTERS FOR k RELATED BATCHES—A REANALYSIS OF THE SECRET MISSION NUMBERS

12.6

Surely you haven't forgotten Mr. G and Mr. N from Chapter 10. Actually, the statistician also did a nonparametric analysis of the secret data. The nonparametric test for k related batches is one of the earliest nonparametric tests—*Friedman's χ_r^2 test* (1937)—which was devised by the well-known economist. The r subscript on the χ^2 stands for ranks. However, in Friedman's test the Xs are ranked in a new way: from $1 \ldots k$ *across the tests for each case*. The results of this way of ranking (in parentheses) and the original numbers are displayed in Table 12.6

For a single case, the rank sum is

$$\frac{k(k+1)}{2} = \frac{5(5+1)}{2} = 15$$

For all cases,

$$\Sigma R = \frac{N_G k(k+1)}{2} = 10(15) = 150$$

SEC. 12.5 □ COMPARING THE CENTERS OF k SEPARATE BATCHES OF RANKS

TABLE 12.6
Test scores and ranks for 10 cases in the secret mission study

Cases	1	(R)	2	(R)	3	(R)	4	(R)	5	(R)
1	−0.5	(1)	0.5	(2)	2.7	(5)	1.5	(3)	1.6	(4)
2	1.7	(3)	2.6	(4)	0.6	(1)	1.4	(2)	3.4	(5)
3	1.9	(3)	−0.6	(1)	2.8	(4)	3.6	(5)	1.5	(2)
4	−1.6	(1)	2.9	(3.5)	−0.5	(2)	3.8	(5)	2.9	(3.5)
5	1.1	(2)	2.1	(3)	−0.7	(1)	3.2	(4)	3.3	(5)
6	−1.9	(1)	−0.6	(2)	2.2	(5)	1.5	(3)	2.1	(4)
7	−1.6	(1)	2.2	(3)	−0.6	(2)	3.3	(5)	3.0	(4)
8	1.2	(2)	−0.6	(1)	2.0	(3)	2.9	(4)	3.2	(5)
9	1.0	(2)	−0.3	(1)	2.0	(5)	1.3	(3)	1.7	(4)
10	−1.3	(1)	2.1	(5)	−0.4	(2)	1.1	(3)	1.3	(4)
Sums		17		25.5		30		37		40.5

The sums of ranks for the tests should again equal ΣR:

$$\Sigma R = 17 + 25.5 + 30 + 37 + 40.5 = 150$$

If H_0 were true, the rank sums for the tests should be close to $\Sigma R/k = 150/5 = 30$. Note that the rank sums in Table 12.6 diverge greatly from 30. The formula for the Friedman statistic is

$$\chi_r^2 = \frac{12}{N_G k(k+1)}[(R_1)^2 + (R_2)^2 + \cdots + (R_k)^2] - 3N_G(k+1)$$

where N_G is the number of cases, k is the number of tests, R_1 is the rank sum for batch 1, and R_k is the rank sum for the kth batch.

The $H_0: m_1 = m_2 = m_3 = m_4 = m_5$, $H_1: m_i \neq m_j$, and $\alpha = .01$. The test statistic is

$$\chi_r^2 = \frac{12}{10(5)(6)}[(17)^2 + (25.5)^2 + (30)^2 + (37)^2 + (40.5)^2] - 3(10)(6)$$

$$= 0.04(4848.5) - 180 = 13.94$$

(Once more, do not round $12/N_G k(k+1)$. Here, 0.04 is the exact value. Or employ 1/25 as the multiplier.) Critical values for χ_r^2 are provided in Table K. When k and N_G exceed the bounds of Table K, χ_r^2 is evaluated as a χ^2 with $df = k - 1$ in Table J. From Table J the $P(\chi^2 = 13.94)$ with $df = 5 - 1 = 4$ is $.01 > P > .005$. As in the ANOVA the H_0 would be rejected—the population test medians for the secret data differ significantly.

Because the χ_r^2 was significant we need to compare the test medians. Church and Wike (1979) did a Monte Carlo study on this problem. Six procedures for comparing medians were investigated in 9000 computer-simulated experiments. The Type 1 error rates were the best for the *Wilcoxon signed-ranks test* when it was applied for pairwise tests *after* a significant overall χ_r^2 test.[4] This procedure is laborious because the Ds for all pairwise comparisons must be obtained, ranked, and so on.

Let us demonstrate the Wilcoxon T technique for one comparison: Test 1 versus Test 5 of Mr. G and Mr. N's data. The Ds $(X_5 - X_1)$ for cases 1 ... 10 are $1.6 - (-0.5) = 2.1$; $3.4 - 1.7 = 1.7$; $1.5 - 1.9 = -0.4$; $2.9 - (-1.6) = 4.5$; $3.3 - 1.1 = 2.2$; $2.1 - (-1.9) = 4.0$; $3.0 - (-1.6) = 4.6$; $3.2 - 1.2 = 2.0$; $1.7 - 1.0 = 0.7$; and $1.3 - (-1.3) = 2.6$. The absolute Ds and ranks (in parentheses) are 0.4 (1), 0.7 (2), 1.7 (3), 2.0 (4), 2.1 (5), 2.2 (6), 2.6 (7), 4.0 (8), 4.5 (9), 4.6 (10). The single $-D$ of -0.4 receives a rank = 1. Therefore, $T = 1$. The $P(T = 1)$ from Table G for $N = 10$ is $< .01$. The population median, m, for Test 5 is larger than the m for Test 1. Since $k(k - 1)/2 = 5(5 - 1)/2 = 10$, nine more Wilcoxon signed-ranks tests would have to be done. As ks and Ns increase, doing nonparametric multiple-comparison tests becomes more awkward and time-consuming than parametric *LSD*s.

The assumptions for Friedman's test are somewhat unclear. Two commonly proposed assumptions (Conover, 1971) are independence of the data for different cases and that the Xs for the treatments can be ranked with only a few ties. Independence could be violated if the cases communicated about the experiment to one another. It is the experimenter's job, not the statistician's, to handle this problem.

You may have noticed two omissions in this chapter. First, nothing has been said about nonparametric tests for spreads. Second, we have not mentioned percentage of accounted-for variance for nonparametric statistics. Since this is a primer of data analysis and not a handbook, these topics were left out. If you should want to know about nonparametric tests for spreads, they are discussed elsewhere (see Bradley, 1968; Hollander & Wolfe, 1973; Daniel, 1978). The problem of accounted-for variance and non-

---

4. Another test that performed nicely was a modification of the sign test (the stepped-down sign test) that is described in Church and Wike (1979).

parametric statistics was addressed by Welkowitz et al. (1982). The exclusion of these two topics should not be interpreted to mean that they are unimportant. But as the late Pablo Picasso said after creating his remarkable three-line nude: "You've got to draw the line somewhere."

SUMMARY

Nonparametric or distribution-free statistics are applicable to data in the form of ranks, signs, and frequencies. The assumptions underlying these techniques are less restrictive than those for parametric tests. Advantages of nonparametric statistics include: They were designed specifically for ranks, signs, and frequencies; their logic is simpler and the assumptions are less demanding; they are often easier to do; and with small Ns the probabilities for these statistics are exact. Two disadvantages are that when parametric assumptions are fulfilled, a nonparametric test requires more cases than a parametric test to achieve equal power; and nonparametric tests are unavailable for some research paradigms like complex factorial designs.

Hypotheses regarding the median of a single batch of numbers can be tested by Wilcoxon's signed-ranks test (Big T) or the Fisher sign test. The Big T test is performed upon the Xs or the difference scores (Ds) from the population median; the sign test is applied to the signs of the Xs or Ds. T is the rank sum for the Ds with the less-frequently occurring sign. The $P(T)$ is found in Wilcoxon's table (Table G) or by a Z statistic version of the Big T test, which can be evaluated in Table D. The exact P for the frequencies of + and − signs is obtained from the binomial distribution or an approximate P by a Z test. The Big T test assumes that the Xs or Ds are symmetrically distributed in the population; the sign test makes no assumption about distribution shape.

The medians of two separate batches can be tested by Wilcoxon's rank-sum test (Big T for two). T is the smaller of the rank sums for the two batches. A Z test version of the Big T-for-two test is available when Ns are large. To perform the Big T-for-two test, the two batches are combined and the numbers are ranked from 1 . . . N. The Big T-for-two test assumes random sampling from the same population. Thus, population normality is not necessary, but equality of shapes and spreads is assumed.

To compare the medians of two related batches, the single-batch tests (Wilcoxon's signed-ranks test and the sign test) are applied to the difference scores, Ds, between the N matched pairs or the two repeated measures for N cases. The assumptions for Wilcoxon's test (Big T for twins) and the sign test are the same as those for a single batch.

The Kruskal-Wallis H test is a generalization of Wilcoxon's rank-sum test to k separate batches. Exact Ps for H for small k and N are provided in Table I; for larger arrays $H \cong \chi^2$ with $df = k - 1$. Ranking is done in the same manner as for the Wilcoxon rank-sum test. After observing a significant H, it was recommended that Wilcoxon rank-sum tests be done in a pairwise manner on the k treatments. The assumptions for the Kruskal-Wallis test are the same as for the Wilcoxon rank-sum test.

Ranks for k related batches can be analyzed by Friedman's χ_r^2 test. Instead of ranking the combined batch from $1 \ldots N$ as in the Wilcoxon rank-sum test and Kruskal-Wallis H test, the test values for each case are ranked from $1 \ldots k$. $P(\chi_r^2)$ can be found in Table K for small k and N. Beyond the table's bounds, $\chi_r^2 \cong \chi^2$ with $df = k - 1$. After a significant χ_r^2, pairwise tests of the k treatment medians can be carried out by Wilcoxon's signed-ranks test. The χ_r^2 test assumes independence of the treatment scores between cases and the ranking of treatments within cases with a minimum of ties.

Two topics not included in the discussion of nonparametric tests are nonparametric tests for spreads and percentage of accounted-for variance and nonparametric tests. References on these topics were provided.

PROBLEMS

1. A noted scientist once suggested that courses and lectures that are too organized may lull students to sleep and be less effective than disorganized courses. In a pilot study an instructor randomly split his math class into two sections and tried to achieve a high degree of organization in one section and little or no organization in the other section. Given the scores on the common final examination below, use a nonparametric test to evaluate the scientist's teaching hypothesis ($\alpha = .05$).

Organized:	74,	67,	83,	47,	51,	63,	68,	72,	59
Not :	88,	87,	66,	69,	94,	92,	91,	79,	89

2. The students in Problem 1 were asked to rate the quality of the course on a 10-point scale (with 10 being "one of the best courses I have taken"). Do a two-tailed nonparametric test at $\alpha = .05$ on the ms of the course evaluations.

Organized:	7.5,	8.0,	8.5,	9.0,	6.5,	5.5,	7.0,	7.5,	9.5
Not :	3.0,	5.5,	6.0,	3.0,	3.5,	4.0,	4.5,	8.0,	7.0

(Note these scores are ratings, not ranks.)

3. The investigator in Problem 1 repeated the study using eight pairs of subjects who were carefully matched on their GRE quantitative scores. Test the same hypothesis at $\alpha = .05$ with their final examination scores. Apply a nonparametric test.

	Pairs							
	1	2	3	4	5	6	7	8
Organized:	72,	61,	71,	67,	63,	50,	47,	82
Not :	81,	69,	74,	72,	62,	49,	54,	88

4. Test the same hypothesis in Problem 3 using the signs of the difference scores and set $\alpha = .05$.
5. Using the persistence data (Problem 1, Chapter 10) test the H_0: $m_E = m_A = m_H$ at $\alpha = .05$.
6. If the H_0 is rejected in Problem 5, what should be done next?
7. Using the data from the verbal discrimination experiment (Problem 8, Chapter 10) test the H_0: $m_0 = m_4 = m_{16}$ at $\alpha = .05$.
8. If H_0 is rejected, perform pairwise tests upon the medians for the three treatments ($\alpha = .05$).
9. An investigator measures the heart rate of 10 subjects twice: after smoking two high-dosage marijuana cigarettes and after smoking two low-dosage marijuana cigarettes. Eight subjects' heart rates were higher after high dosage, one subject's heart rate was higher after low dosage, and one subject's heart rate was the same after both dosages. Test the researcher's hypothesis that higher dosage increases heart rate. Use $\alpha = .05$.

ADDENDUM: USING COMPUTERS IN STATISTICS

Instead of Wilcoxon's rank-sum test MINITAB includes the equivalent MANN-WHITNEY test. In the first example this test is applied to the two-batch problem in the chapter (1). The MANN-WHITNEY program is especially useful for computing pairwise tests of medians after a significant Kruskal-Wallis H test. In the second example this procedure has been carried out to compute the six pairwise comparisons (2) in the Kruskal-Wallis example. Note that the reranking of the pairs of batches was accomplished by reranking the original ranks.

The Wilcoxon signed-ranks test has not yet been implemented in MINITAB. Sign tests are possible for matched batches, and the resulting frequencies of + and − signs can be evaluated by outputting a binomial table for $p = q = .5$ and the desired N_T.

```
--MANN-WHITNEY ON C1 C2
  C1   N = 10    MEDIAN = 159.00
  C2   N = 10    MEDIAN = 130.50                          (1)

  A POINT ESTIMATE FOR ETA1-ETA2 IS   33.5
  A 95.5 PERCENT C.I. FOR ETA1-ETA2 IS ( 11.0, 53.1)

  W = 143.5
  TEST OF ETA1 = ETA2 VS. ETA1 N.E. ETA2
  THE TEST IS SIGNIFICANT AT 0.0041

--SET IN C3
--27 18.5 29.5 18.5 5.5 13 18.5 23.5 9
--SET IN C4
--23.5 34 32 32 18.5 36 35 32 29.5
--SET IN C5
--23.5 13 27 13 9 18.5 23.5 27 13
--SET IN C6
--3 1 5.5 2 5.5 5.5 13 18.5 9

--MANN C3 C6
  C3   N = 9    MEDIAN = 18.500
  C6   N = 9    MEDIAN =  5.500                           (2)

  A POINT ESTIMATE FOR ETA1-ETA2 IS  13.0
  A 95.8 PERCENT C.I. FOR ETA1-ETA2 IS ( 3.5, 18.0)

  W = 116.0
  TEST OF ETA1 = ETA2 VS. ETA1 N.E. ETA2
  THE TEST IS SIGNIFICANT AT 0.0081
--MANN C3 C5
  C3   N = 9    MEDIAN = 18.500
  C5   N = 9    MEDIAN = 18.500

  A POINT ESTIMATE FOR ETA1-ETA2 IS  0.0
  A 95.8 PERCENT C.I. FOR ETA1-ETA2 IS ( -8.5, 6.0)

  W = 84.5
  TEST OF ETA1 = ETA2 VS. ETA1 N.E. ETA2
  THE TEST IS SIGNIFICANT AT 0.9648
  CANNOT REJECT AT ALPHA = 0.05
```

```
--MANN C3 C4
  C3   N = 9    MEDIAN =  18.500
  C4   N = 9    MEDIAN =  32.000

  A POINT ESTIMATE FOR ETA1-ETA2 IS  -13.5
  A 95.8 PERCENT C.I. FOR ETA1-ETA2 IS ( -19.0,  -5.0)

  W =  52.5
  TEST OF ETA1 = ETA2 VS. ETA1 N.E. ETA2
  THE TEST IS SIGNIFICANT AT 0.0041
--MANN C4 C6
  C4   N = 9    MEDIAN =  32.000
  C6   N = 9    MEDIAN =   5.500

  A POINT ESTIMATE FOR ETA1-ETA2 IS  26.0
  A 95.8 PERCENT C.I. FOR ETA1-ETA2 IS ( 16.5,  30.0)

  W = 125.5
  TEST OF ETA1 = ETA2 VS. ETA1 N.E. ETA2
  THE TEST IS SIGNIFICANT AT 0.0005
--MANN C4 C5
  C4   N = 9    MEDIAN =  32.000
  C5   N = 9    MEDIAN =  18.500

  A POINT ESTIMATE FOR ETA1-ETA2 IS  10.5
  A 95.8 PERCENT C.I. FOR ETA1-ETA2 IS (  5.0,  19.0)

  W = 118.5
  TEST OF ETA1 = ETA2 VS. ETA1 N.E. ETA2
  THE TEST IS SIGNIFICANT AT 0.0041
--MANN C5 C6
  C5   N = 9    MEDIAN =  18.500
  C6   N = 9    MEDIAN =   5.500

  A POINT ESTIMATE FOR ETA1-ETA2 IS  11.0
  A 95.8 PERCENT C.I. FOR ETA1-ETA2 IS (  5.0,  20.5)

  W = 118.5
  TEST OF ETA1 = ETA2 VS. ETA1 N.E. ETA2
  THE TEST IS SIGNIFICANT AT 0.0041
```

CHAPTER OUTLINE

13.1 BRING ON THE FREAKS—IN A SINGLE BATCH
13.2 MORE FREAKS—IN TWO SEPARATE BATCHES
13.3 MORE FREAKS—IN k SEPARATE BATCHES
13.4 MORE FREAKS—IN TWO RELATED BATCHES
13.5 AN ENDING—FREAKS IN k RELATED BATCHES

CHAPTER THIRTEEN
MEETING FREAKS AT HIGH NOON

Some writers regard statistical procedures for *frequencies* as nonparametric statistics; other writers do not. Since we have employed the terms *nonparametric* and *distribution-free* to refer to statistical tests whose assumptions are less restrictive than those for parametric tests, we will include statistical tests for frequencies in the class of nonparametric tests.

Actually, we encountered tests for frequencies in Chapter 5 when binomial problems were analyzed and in Chapter 12 when signs were analyzed. In each instance the frequencies for two bins or classes (e.g., heads and tails or + and − signs) constituted the data to be evaluated. In the present chapter we will extend the coverage to frequencies for two and k batches. Tests of frequencies for both separate and related batches will be described as well as tests for some arrangements with more than two bins.

BRING ON THE FREAKS—IN A SINGLE BATCH

13.1

Now we will examine nonparametric tests for *frequencies*. The Secretary of Agriculture wanted to know what was on the minds of farmers. So the secretary commissioned a famous pollster, Elmo "Fast" Trotter, to find out. The pollster secured a carefully selected, representative sample of 2000 farmers from the seven leading agricultural states and sent his skilled interviewers out to poll them. The secretary was particularly concerned with the attitudes of the farmers toward the PIK (payment in kind) program. The secretary believed that the PIK program was viewed favorably by only *half* of the farmers. Elmo, therefore, included this question in the interview schedule: "Do you believe that the PIK program is a good program for the American farmer?" In one farm state when 120 farmers (a single batch) were asked the question, 72 replied "Yes" and 48 "No." A data sheet summarizing this outcome looked like Table 13.1.

TABLE 13.1
Data sheet for the PIK question

Case	Question 17
1	1
2	1
3	0
.	.
.	.
.	.
120	1

Here "Yes" has been coded 1 and "No" coded 0 by the interviewer.

The *dependent variable* is the respondent's answer to the question. The measure of the dependent variable is not fancy—it is Yes or No. If we toss all the 1s into one *bin* (class or category) and the 0s into another bin and count the number of cases in each bin, then 72 and 48, respectively, are the *frequencies*.

How could such frequencies be analyzed to evaluate the secretary's belief about the popularity of the PIK program in this state? One approach is to determine if the *proportion* of Yes responses deviates from .50, the proportion the secretary believes to exist. For a two-tailed test, H_0: $\pi = .50$ and H_1: $\pi \neq .50$, where π is the population proportion of Yes responses expected by the secretary. The test statistic is $Z = (p - \pi)/\sigma_p$, where p is the observed proportion of Yes and σ_p is the standard error of a proportion. The *observed proportion* is $p = f/N = 72/120 = .60$. Since $\pi = .50$, the *expected frequency* (f_e), if the secretary's belief is true, is $f_e = N\pi = 120(.50) = 60$. Or, $f_e = (72 + 48)/2 = 60$. The *observed frequencies* (f_o) for Yes and No are 72 and 48, respectively. Although we have discussed the problem in terms of proportions, we are not going to do the proportion test. Since we can readily convert proportions to frequencies and vice versa, and because there is an equivalent, but more widely applicable test for frequencies, we will test the frequencies.

The f_os and f_es for the bins are displayed in Table 13.2.

Response	f_o	f_e
Yes (1)	72	60
No (0)	48	60

TABLE 13.2
Observed (f_o) and expected (f_e) frequencies in response to the PIK question

We can assess the validity of the secretary's belief with a χ^2 test for goodness of fit. The formula is

$$\chi^2 = \sum_{i=1}^{r} \frac{(f_o - f_e)^2}{f_e}$$

where r is the number of bins (rows). H_0 is that frequencies of Yes and No in the population fit a specific distribution; H_1 is that they do not fit. The specific distribution in this problem is a rectangular one with $f_e = 60$ in each bin. To test the hypotheses, χ^2 is computed and evaluated for $df = r - 1$ in Table J. The criterion of significance is $\alpha = .05$. The test statistic is

$$\chi^2 = \frac{(72 - 60)^2}{60} + \frac{(48 - 60)^2}{60}$$
$$= 2.4 + 2.4 = 4.8$$

From Table J the $P(\chi^2 = 4.8)$ for $df = 1$ is $.05 > P > .025$. Thus, H_0 would be rejected and H_1 accepted. At least in this state, the secretary was wrong—more than half of the farmers liked the PIK program.

Here are a few odds and ends:

1. The χ^2 goodness of fit test is not limited to $r = 2$ bins. Several years ago a newspaper article reported a study on the frequencies of fatal heart attacks for the days of the week. Here, $r = 7$.
2. The population distribution in H_0 is not limited to a rectangular distribution—it could be a normal distribution, a binomial distribution, or whatever distribution a theory might dictate.
3. When $df = 1$, one-tailed tests of hypotheses may be assessed. For a *one-tailed test* χ^2_α should be located under $\chi^2_{2\alpha}$ in Table J, (i.e., .05 under .10, etc.). For example, when $df = 1$, $\chi^2_{.05} = 3.84$, *two-tailed*, and $\chi^2_{.05} = 2.71$, *one-tailed*. One way to understand this is to know that with $df = 1$, $\chi^2 = Z^2$. From the microtable $Z_{.05} = 1.96$, *two-tailed*, and $Z^2_{.05} = 3.84$; $Z_{.05} = 1.645$, *one-tailed*, and $Z^2_{.05} = 2.71$. However, when $df > 1$, the Ps in the χ^2 table are read "as is."
4. From the calculation of χ^2 the necessity for the limits on Σ should be clear. The frequencies in *both* bins are included.
5. Store the χ^2 formula in memory. As you will see, it can be applied to many other situations with frequencies.

As usual, we will finish by describing the *assumptions* for the χ^2 test for a single batch. They are that (a) the N responses are a random sample, (b) the responses are independent of one another, and (c) the responses fit into nonoverlapping, exhaustive bins. Regarding c, a case's response cannot fall into more than one bin, and "everybody has to be someplace"—all N responses must fit into the bins.

MORE FREAKS—IN TWO SEPARATE BATCHES

13.2

As we have said before, investigators generally do research involving more than one batch. If there are two separate batches and two bins, then we have a so-called 2×2 *contingency table*. A statistics teacher did the following study with a beginning statistics class. Half of the 74 students (randomly selected) had a short course in algebra before the statistics course, and the other half did not. The dependent variable was the score of the final statistics examination. The 74 test scores on the final were cut at the median. Students above the median were put in the + bin; those below the median in the − bin. When these cases were sorted into those who did and did not have the shorter course, a 2 × 2 contingency table (Table 13.3) resulted.

| | Short Course | | |
Final	No	Yes	Sums
+	14	23	37
−	24	13	37
Sums	38	36	74 = N

TABLE 13.3
Frequencies of students in relation to taking a short course and performance on the final examination

As in the PIK example, we could do a Z test on the *difference in the proportions* of students who scored above the median (+) in the two batches, (i.e., $f/N = 14/38 = .368$ vs. $23/36 = .639$). But why learn another formula, when it is possible to do a χ^2 test, with the same formula as above? Why carry around a knife and a screwdriver when a Swiss Army Knife will do the job?

The χ^2 test on a two-dimensional table is called a *test of independence*. The H_0 is that the two variables are independent, and H_1 is that they are *associated*. In the study the two variables are (a) taking or not taking the short course in algebra and (b) scoring high or low on the exam. The teacher predicted that the short course would be beneficial, and did a one-tailed test at $\alpha = .01$. The precise directional prediction was that the frequencies would pile up in lower left and upper right bins and would be low in the bins for the opposite diagonal. Indeed, they do. So H_1 then is an association in which the frequencies are distributed in the predicted manner in the bins.

To perform a χ^2 test, the f_es are needed. When an answer is unknown, it is sometimes possible to estimate it. The f_es can be estimated from the marginal sums in the table:

$$f_e = \frac{R(C)}{N}$$

where R is the marginal row sum for a bin, C is the marginal column sum for a bin, and N is the sum of all frequencies in the table. For the upper left cell,

$$f_e = \frac{37(38)}{74} = 19$$

For the upper right cell,

$$f_e = \frac{37(36)}{74} = 18$$

All f_es are summarized in Table 13.4.

TABLE 13.4
Expected frequencies in relation to taking the short course and performance on the final examination

Final	Short Course No	Short Course Yes	Sums
+	19	18	37
−	19	18	37
Sums	38	36	74 = N

Notice that the marginal sums for the f_es in Table 13.4 are the same as those for the f_os in Table 13.3.

The test statistic is

$$\chi^2 = \sum_{i=1}^{rc} \frac{(f_o - f_e)^2}{f_e}$$

where the $\sum_{i=1}^{rc}$ signifies a summation over all bins. The degrees of freedom for any two-dimensional table are like those for an interaction in ANOVA—that is, $df = (r - 1)(c - 1)$. Here, $df = (2 - 1)(2 - 1) = 1$. An "association" between two factors is an interaction. Then,

$$\chi^2 = \frac{(14 - 19)^2}{19} + \frac{(24 - 19)^2}{19} + \frac{(23 - 18)^2}{18} + \frac{(13 - 18)^2}{18}$$

$$= 1.32 + 1.32 + 1.39 + 1.39 = 5.42$$

From Table J the *one-tailed* $P (\chi^2 = 5.42)$ for $df = 1$ is $.0125 > P > .005$. What is χ^2 more exactly? When $df = 1$, $\chi^2 = Z^2$, so

$$\sqrt{\chi^2} = Z$$
$$\sqrt{5.42} = 2.33$$

The one-tailed $P(Z = 2.33)$ from Table D is .0099 or .01. Therefore, the teacher can reject H_0 and accept H_1—taking the short course is associated in the population with higher final scores and not taking it is associated with lower final scores.

When χ^2 is significant, another question can be asked: How strong is the association? This question can be answered by computing a Pearson product-moment r for the 2 × 2 table. For a 2 × 2 table r is termed ϕ (phi) and ϕ is readily computable:

$$\phi = \sqrt{\frac{\chi^2}{N}}$$

For this problem,

$$\phi = \sqrt{\frac{5.42}{74}} = .27$$

Is the $\phi = .27$ significant? Yes, because the χ^2 was significant. In summary, while the direct relationship between the short algebra course and exam grades was significant, the strength of the relationship was not overwhelming.

The assumptions for the χ^2 test for independence are the same as those for a χ^2 test for goodness of fit, except that the word *sample* becomes *samples*. For a long time a fourth assumption was prescribed for the 2×2 χ^2 test: that the f_es in all cells had to be equal to or greater than 5. When this assumption was violated, a different test was applied or the χ^2 test was corrected by *Yates' correction for continuity*.[1] Monte Carlo studies by Camilli and Hopkins (1978, 1979) demonstrated that this assumption regarding minimal size of f_e can be largely ignored and that correcting for continuity is inadvisable because it reduces the power of the χ^2 test.

In writing this section the distinction between *goodness of fit tests* and *independence tests* became unclear to me. The same formulas are employed for both types of tests and the assumptions are the same. Can't you view the 2×2 χ^2 example as a goodness-of-fit test to a distribution whose frequencies are the f_es of 19, 19, 18, and 18? Does the distinction make any difference?[2]

MORE FREAKS—IN k SEPARATE BATCHES

13.3

Now we will examine a problem with more than two bins and more than two separate batches. Do women's attitudes toward abortion differ as a function of the year in college? Imagine that 320 women were asked the question: "Do you believe that a woman has the right to have an abortion?" The hypothetical results are shown in Table 13.5.

1. Yates' correction consists of reducing the absolute differences between f_o and f_e by 0.5 prior to squaring the differences.
2. A colleague, Professor Edwin Martin, who has investigated χ^2 tests extensively, questions this distinction also. See, too, Hopkins and Glass (1978, p. 315, note 7).

TABLE 13.5
Frequencies in relation to responses to a question on abortion and class in college

		Class in College			
Responses	Fr.	So.	Jr.	Sr.	Sums
Yes	25	31	34	41	131
No	24	23	28	29	104
Undecided	31	26	18	10	85
Sums	80	80	80	80	320

This example includes three bins and four separate batches, forming a 3 × 4 table. The expected frequencies are computed as in a 2 × 2 table. For the first bin, $f_e = R(C)/N = 131(80)/320 = 32.75$. For the whole table the f_es are displayed in Table 13.6.

TABLE 13.6
Expected frequencies in relation to responses to a question on abortion and class in college

		Class in College			
Responses	Fr.	So.	Jr.	Sr.	Sums
Yes	32.75	32.75	32.75	32.75	131
No	26	26	26	26	104
Undecided	21.25	21.25	21.25	21.25	85
Sums	80	80	80	80	320 = N

We will use an α of .01 and the same formula for χ^2. When computing the 12 χ^2 values, it is convenient to assemble the calculations in a worksheet (Table 13.7).

TABLE 13.7
A worksheet for computing χ^2

f_o	f_e	$(f_o - f_e)$	$(f_o - f_e)^2$	$(f_o - f_e)^2/f_e$
25	32.75	−7.75	60.06	1.83
24	26	−2	4	0.15
31	21.25	9.75	95.06	4.47
31	32.75	−1.75	3.06	0.09
23	26	−3	9	0.35
26	21.25	4.75	22.56	1.06
34	32.75	1.25	1.56	0.05
28	26	2	4	0.15
18	21.25	−3.25	10.56	0.50
41	32.75	8.25	68.06	2.08
29	26	3	9	0.35
10	21.25	−11.25	126.56	5.96
Sums		0.0		17.04 = χ^2

The $P(\chi^2 = 17.04)$ for $df = (r - 1)(c - 1) = (3 - 1)(4 - 1) = 6$ is less than .01. Therefore, we may say either that the two variables are associated or that the observed frequencies do not fit the model (the distribution of f_es generated by the marginal totals).

As in the 2 × 2 table, it is possible to calculate an *index of association* between the two variables, the college class and response to the abortion question. One index is called *Cramér's V* coefficent and V ranges from 0 to 1 in size. The formula for Cramér's statistic is

$$V = \sqrt{\frac{\chi^2}{N(df_{min})}}$$

where df_{min} is the *smaller* of the two dfs for the variables. In this example the $df = r - 1 = 3 - 1 = 2$, for the response variable, is the smaller. Thus, Cramér's V is

$$V = \sqrt{\frac{17.04}{320(2)}}$$
$$= \sqrt{0.0266} = .16$$

Cramér's V is small, but significant. Like ϕ, Cramér's V is tested for significance by the χ^2 test for the 3 × 4 table.

Most elementary statistics books stop at this point in discussing χ^2 with $df > 1$. Is there anything to do after finding a significant χ^2 and calculating Cramér's V? Stopping at this point is like stopping after a significant F in ANOVA. The name of the game in science is *analysis*. Four other procedures are possible:

1. χ^2 tests of any specific research hypothesis by partitioning the 3 × 4 table into the relevant smaller table
2. χ^2 tests on orthogonal partitions of the 3 × 4 table
3. χ^2 comparisons on the rows, columns, or both for the 3 × 4 table
4. χ^2 tests on partitions based upon eyeballing the data

Regarding 2, it is possible to partition the 3 × 4 table with $df = 6$ into six 2 × 2 tables (each with df = 1) whose separate χ^2s sum to the original χ^2 value of 17.04. Some statisticians contend that such *orthogonal comparisons* are the only legitimate way to partition a large table.[3] (Orthogonal comparisons can also be per-

---

3. Orthogonal comparisons are independent of one another. Pairwise comparisons of say four means are not independent because a given mean is compared more than once, e.g., 1 vs. 4, 1 vs. 3, 1 vs. 2, etc. Likewise, the tests of all possible 2 × 2 tables from a large contingency table would not be independent.

formed on means after an ANOVA.) Orthogonal comparisons of contingency tables have been discussed by Castellan (1965), Bresnahan and Shapiro (1966), and others. In the case of partitioning with the aid of the eyeball (4), the χ^2s for the cells can be inspected. Since the subsequent tests would be based upon eyeballing the data, the findings should be regarded with caution, and it has been suggested that the *df*s for such tests be increased (Miller, 1966; Fleiss, 1973, p. 95). Delucchi (1983) has recently reviewed partitioning problems in χ^2. I don't know which partitioning procedure to apply after observing a significant χ^2 when $df > 1$. But one should not stop the analysis at that point.

MORE FREAKS—IN TWO RELATED BATCHES

13.4

When a study produces frequencies from two or k related batches, the χ^2 tests described above are inapplicable. Why? Because they assume independence of the frequencies. Related, as before, denotes matching or repeated measurements. Therefore, the frequencies from related batches violate the assumption of independence.

First, let us consider the simple case of two related batches by taking a trip to the dentist's office. A sample of 140 patients had their teeth cleaned. For every patient two different ultrasonic scalers, Brand X and the best selling Brand M, removed the accumulated plaque. The patient had a button to push whenever pain was experienced. The research question was: Did one scaling device produce more pain than the other? The data sheet looks like Table 13.8.

TABLE 13.8
Data sheet for frequencies of pain responses to two ultrasonic scalers

	Scaler	
Pt.	M	X
1	1	0
2	0	1
3	0	0
.	.	.
.	.	.
.	.	.
140	1	0

Pain has been coded as 1 and no pain as 0. The patients' responses can be tallied in the four bins of a 2×2 table and the bin frequencies are summarized in Table 13.9.

	Scaler	X		
M	Pain	0	1	Sums
	1	18 (A)	3 (B)	21
	0	112 (C)	7 (D)	119
	Sums	130	10	140 = N

TABLE 13.9
Frequencies of pain responses in relation to ultrasonic scalers

Notice that $N = 140$, the *number of cases,* not the number of 1 and 0 codes on the data sheet. We could again think about the *difference in the proportions* of subjects experiencing pain with the two scalers. For scaler M, $p_1 = (A + B)/N = 21/140 = .15$; for scaler X, $p_2 = (B + D)/N = 10/140 = .07$. The difference is $p_1 - p_2 = [(A + B)/N] - [(B + D)/N] = (A - D)/N$. Then the standard error of $p_1 - p_2$ could be found, and so on. McNemar (1947) derived a χ^2 test in terms of the table frequencies. For a two-tailed test the H_0 is that the frequencies in the population are equal for the A and D bins; H_1 is that they are not. Bins C (the feeling-no-pain patients) and B (the I can't-stop-hurting patients) are not included in the analysis. *McNemar's χ^2 test for correlated frequencies* is

$$\chi^2 = \frac{(A - D)^2}{A + D}$$

As we now have only two bins, A and D, $df = 2 - 1 = 1$, and $N = A + D$. We will set $\alpha = .05$. The computed χ^2 is

$$\chi^2 = \frac{(18 - 7)^2}{18 + 7}$$

$$\frac{121}{25} = 4.84$$

The two-tailed $P(\chi^2 = 4.84)$ from Table J is $.05 > P > .025$. Thus, H_0 would be rejected and H_1 accepted—fewer patients pushed the pain button when the Brand X scaler was applied. (The good news for us and for both manufacturers is in bin C: 112 patients out of 140 reported no pain.)

Another analysis can be done instead of McNemar's test. Conover (1971) considers the McNemar test to be a variant of the *sign test.* It follows that the frequencies in the A and D bins can be evaluated by the binomial distribution. From Hollander and Wolfe's binomial table (1973) the two-tailed probability of an 18+ and 7− outcome is .0432. The sign test should be applied when N, the sum of the frequencies in A and D, is less than 10. But if a binomial table is handy, why calculate χ^2? Lastly, the assumptions for testing frequencies in related batches are that (a) the observations *between* the N cases tested twice or N matched pairs are independent and (b) each of the N cases or N pairs can be put in one and only one of the four bins.

AN ENDING—FREAKS IN k RELATED BATCHES

13.5

Now we consider the second case in which k related batches yield frequencies. As in the dental example, the dependent variable is dichotomous (there are only two bins). A longitudinal study was carried out on attitudes towards abortion. Fifty women students were randomly selected in their first year of college and individual, in-depth interviews were conducted with them every spring thereafter on a variety of topics, including the issue of abortion. Transcripts of the sections of the interviews on abortion were submitted to a panel of expert judges who classified the women as either for (1) or against (0) abortion. Fortunately, only two women dropped out of college. Two other cases were deleted from the sample because the judges failed to agree on the women's attitudes toward abortion. The data sheet for the $N = 46$ cases looks like Table 13.10.

TABLE 13.10

Data sheet for attitudes toward abortion in four interviews

| | \multicolumn{4}{c}{Interview} | | | |
Case	1	2	3	4
1	1	1	1	1
2	1	1	0	0
3	0	0	1	1
.	.	.	.	.
.	.	.	.	.
.	.	.	.	.
46	0	0	0	0

The data sheet can be converted to a smaller table (Table 13.11) by *aggregating cases with the same response patterns over the four interviews.* The first four women in Table 13.11 were proabortion but changed after their second interview. The next five women changed from for to against abortion after their first interview, and so on. The research question is: Did a change in attitude toward abortion occur in these college students over the four-year

TABLE 13.11 Frequencies of different response patterns in four interviews

Number of Cases	Interview 1	2	3	4	L_i	L_i^2
4	1	1	0	0	8	16
5	1	0	0	0	5	5
5	0	0	0	1	5	5
4	0	0	1	1	8	16
8	0	1	1	1	24	72
$N = 26$	G_j 9	12	12	17	50	$114 = \Sigma L_i^2$

period? The H_0 is $\pi_1 = \pi_2 = \pi_3 = \pi_4$, where π is the proportion of cases in the population favoring abortion. The H_1 is $\pi_i \neq \pi_j$. The proper test is *Cochran's Q test for correlated frequencies*. The Q statistic is approximately distributed as a χ^2 with $df = k - 1 = 4 - 1 = 3$. The formula is

$$Q = \frac{k(k-1)[\Sigma(G_j - \overline{G})^2]}{k(\Sigma L_i) - \Sigma L_i^2}$$

where k is the number of interviews, G_j is the sum of the 1s for a column, $\overline{G}$ is the mean of the G_js, and L_i is the sum of the 1s for a row.

Calculating Q is a little tricky, so be careful. Before beginning, we will answer the obvious question: What happened to the other 20 women? We had a sample of $N = 46$. Table 13.11 shows an $N = 26$. The 20 "lost cases" were dropped from the analysis: 11 had a response pattern of all 0s and 9 had all 1s. The Q statistic is identical with and without the all 1s and all 0s (Fleiss, 1973, p. 87). Let us obtain some pieces of the formula to simplify the calculations. $\overline{G} = G_j/k = 50/4 = 12.5$. $\Sigma L_i^2 = 114$. Why? Consider the four cases in the first subbatch in Table 13.12.

TABLE 13.12 An example showing the calculation of L_i and L_i^2 for the first four cases

Case	Interview 1	2	3	4	L_i	L_i^2
1	1	1	0	0	2	4
2	1	1	0	0	2	4
3	1	1	0	0	2	4
4	1	1	0	0	2	4
Sums					8	16

If ΣL_i^2 still isn't clear, list the other subbatches. We will set $\alpha = .05$ and calculate Q:

$$Q = 4(3)\frac{(9 - 12.5)^2 + (12 - 12.5)^2 + (12 - 12.5)^2 + (17 - 12.5)^2}{4(50) - 114}$$

$$= \frac{12(33)}{86} = 4.60$$

The $P(\chi^2 = 4.60)$ for $df = k - 1 = 3$ is $.25 > P > .10$. Therefore, H_0 cannot be rejected. There is no evidence that the changes in attitude toward abortion over the four years differed from random fluctuations. As a check on the calculations, the nine cases with all 1s were restored to the sample and Q was recalculated. Q was again 4.60.

What next? After watching four years of work go down the tube, what should the investigator do? Can the investigator conclude that the study *proves* that four years in college did not change the women's attitudes toward this highly charged issue? No, unfortunately you cannot prove the null hypothesis. Why not? Failure to reject the H_0 can occur when H_0 is true or when H_0 is false. In the latter case, the particular batches included in the study may not permit rejecting H_0 or the study may lack sufficient power to detect an effect. (Power will be discussed in Chapter 14.) Thus, an investigator cannot conclude that a failure to reject H_0 means that H_0 is true.

If Q had been significant, then there is a what-next problem. It is comparing the frequencies or proportions of the k interviews. How might this be done? One could extrapolate from the Monte Carlo studies cited previously, and do protected, pairwise *sign tests* among the k interviews. As an example of this procedure we will compare frequencies for the first and last interviews. Subtracting the 1s for the first interview from the 0s for the last interview for the first four cases in Table 13.11 gives 4 − signs. The next five cases yield 5 − signs. The next 17 cases give 17 + signs. (The all 0 and all 1 cases are excluded as 0 differences.) So there are 17 + and 9 − signs. This outcome cannot be evaluated by the binomial table (Hollander & Wolfe, 1973), since it stops at $N_T = 25$. What could be done to determine the probability of this outcome?

1. Generate a binomial table with a computer.
2. Expand the binomial.
3. Use the normal distribution to secure an approximate P value.
4. Do a χ^2 test like the one in the first part of Section 13.1.

What are the assumptions for Cochran's Q test? They are the same as those described above for McNemar's χ^2 test.

SUMMARY

When the dependent variable is of the yes/no type or falls into two bins, the data for a single batch consist of the two bin counts or frequencies. A χ^2 goodness-of-fit test can be applied to determine whether the observed frequencies (f_os) in the bins deviate from the expected frequencies (f_es) or if the observed frequencies deviate from a model, a specific distribution. For a single batch with two bins, $df = r - 1$, where r is the number of bins or classes. When $df = 1$, the $P(\chi^2)$ in Table J is halved for a one-tailed test and used "as is" for a two-tailed test. The χ^2 test assumes random sampling, independence of observations, and that the responses can be put into exhaustive and nonoverlapping bins.

Frequency data from two or k separate batches are analyzed by a χ^2 test of independence. (It was argued that the distinction between goodness of fit and independence is questionable.) A significant χ^2 means that variables are associated. The df for two-dimensional table of frequencies are like those for an interaction: $df = (r-1)(c-1)$, where r is the number of rows and c is the number of columns. With a 2 × 2 table, ϕ, a Pearson product-moment r, may be calculated to show the strength of association. For tables with $df > 1$, one index of association is Cramér's V, which ranges from 0 to 1. Both ϕ and V are tested for significance by the χ^2 from their parent tables. The assumptions for χ^2 tests for two or more separate batches are the same as for a χ^2 test on a single batch. Recent research has questioned the older assumption that the f_es in a 2 × 2 table must be ≥ 5, and that χ^2 should be corrected for continuity—the correction lowers the power of the χ^2 test.

With $df > 1$ a significant χ^2 should be followed by partitioning of the table. Four possible ways to do this were suggested: (a) partitions and χ^2 tests to evaluate specific research hypotheses; (b) χ^2 tests of orthogonal partitions; (c) χ^2 tests of rows, columns, or both; and (d) partitions and χ^2 tests based on eyeballing the data.

When frequencies come from two or k related batches, the usual χ^2 tests are inappropriate because the frequencies are not independent. In the case of frequencies from two related batches and two bins, McNemar's test of correlated frequencies is applicable. Only the frequencies for two bins from the 2 × 2 table are evaluated by McNemar's χ^2 test or by the sign test. The McNemar test assumes independence of observations between pairs and exhaustive, nonoverlapping classification of responses.

The frequencies for k related batches and two bins can be evaluated by Cochran's Q test. $Q \cong \chi^2$ with $df = k - 1$. After observing a significant Q, pairwise sign tests may be performed upon the k treatments. The assumptions for Cochran's Q test are the same as those for McNemar's χ^2 test.

PROBLEMS

1. In an experiment on drinking and driving, 20 "inexperienced" and 20 "moderate" drinkers were tested after consuming four martinis. The numbers of pylons knocked down in a driving test were recorded for all drivers and each driver was classified into + (above the median) and − (below). Is drinking experience related ($\alpha = .05$, two-tailed) to driving performance?

	Driving +	Driving −	Sum
Inexperienced	16	4	20
Moderate	4	16	20
Sum	20	20	40

2. Calculate an index of relationship between drinking and driving in Problem 1.

3. At an international conference on the nuclear crisis, 200 delegates voted on the solution they should work towards. There were 80 "young" delegates (less than 40 years of age) and 120 "old" delegates (more than 40). Is there an association ($\alpha = .05$) between the age of the delegate and the type of solution?

Age	World Government	Disarmament	Determent	Sum
Old	28	25	67	120
Young	32	40	8	80
Sum	60	65	75	200

4. Compute an index of relationship between the age of the delegates and solution type.

5. Twenty-four subjects participated in two problem-solving tasks—I and II. The performance in terms of + (success) and − (failure) on

the two tasks is given below. Is there a relationship ($\alpha = .05$) in the performance on the two tasks? What would you conclude?

Task		Response	II −	+	Sum
I		+	1	9	10
		−	7	7	14
		Sum	8	16	24

6. In another study on problem solving, 54 subjects did five similar tasks. In the data below 1 = success and 0 = failure. Test the H_0: $\pi_1 = \pi_2 = \ldots = \pi_5$ at $\alpha = .05$.

No. of subjects	Tasks				
	1	2	3	4	5
5	0	0	0	0	0
7	1	1	1	1	1
2	1	1	0	0	0
15	0	1	1	1	1
12	0	0	1	1	1
13	0	0	0	1	1

7. If H_0 is rejected in Problem 6, what should be done next?
8. Forty student pilots are tested for their flying proficiency by two examiners. One examiner passes 50% and fails 50%; the other passes 60% and fails 40%. Is the difference in the percentages of those passing significant at $\alpha = .05$? Why is this problem insoluable?
9. Twenty scientists are asked to select the *best* research proposal from three proposals. Eight judge proposal A to be the best, five judge B to be the best, and seven judge C to be the best. Do these judgments differ from chance? Use $\alpha = .05$.

ADDENDUM: USING COMPUTERS IN STATISTICS

MINITAB's CHISQUARE program has been applied to analyze the 2 × 2 and 3 × 4 contingency tables in the chapter (1), (2). The expected frequencies are listed below the observed frequencies in all cells. The programs outputs the χ^2 values for each cell. Inspection of the χ^2 values discloses which cells contribute the most to the χ^2 for the total table.

```
--READ C7 C8
--14 23
--24 13
--CHISQUARE C7 C8                                                           (1)

EXPECTED FREQUENCIES ARE PRINTED BELOW OBSERVED FREQUENCIES
         I   C7   I   C8   I  TOTALS
-------I--------I--------I-------
     1 I    14  I    23  I    37
       I   19.0I   18.0I
-------I--------I--------I-------
     2 I    24  I    13  I    37
       I   19.0I   18.0I
-------I--------I--------I-------
TOTALS I    38  I    36  I    74

TOTAL CHI SQUARE =
          1.32 + 1.39 +
          1.32 + 1.39 +

                = 5.41

DEGREES OF FREEDOM = ( 2-1) x ( 2-1) = 1

--READ C30-C33
--25 31 34 41
--24 23 28 29
--31 26 18 10
--CHISQUARE C30-C33

EXPECTED FREQUENCIES ARE PRINTED BELOW OBSERVED FREQUENCIES
         I  C30  I  C31  I  C32  I  C33  I  TOTALS
-------I--------I--------I--------I--------I-------
     1 I    25  I    31  I    34  I    41  I   131
       I   32.8I   32.8I   32.8I   32.8I
-------I--------I--------I--------I--------I-------                         (2)
     2 I    24  I    23  I    28  I    29  I   104
       I   26.0I   26.0I   26.0I   26.0I
-------I--------I--------I--------I--------I-------
     3 I    31  I    26  I    18  I    10  I    85
       I   21.3I   21.3I   21.3I   21.3I
-------I--------I--------I--------I--------I-------
TOTALS I    80  I    80  I    80  I    80  I   320

TOTAL CHI SQUARE =
          1.83 + 0.09 + 0.05 + 2.08 +
          0.15 + 0.35 + 0.15 + 0.35 +
          4.47 + 1.06 + 0.50 + 5.96 +

                = 17.04

DEGREES OF FREEDOM = ( 3-1) x ( 4-1) = 6
```

CHAPTER OUTLINE

14.1 THE RETURN OF THE OUTLIERS

14.2 SPREADING IT AROUND SOME MORE

14.3 RE-EXPRESSION SOMETIMES HELPS

14.4 POWER—A CRITICAL BUT OVERLOOKED PROBLEM

14.5 DOING DATA ANALYSIS ON A COMPUTER

14.6 THE DATA-ANALYTIC ORIENTATION REVISITED

14.7 SOMETHING BLUE

CHAPTER FOURTEEN
NONTRIVIAL ODDS AND ENDS

During the big band days, the late Glenn Miller did a nightly air shot on a pictureless box called radio. A regular feature of the program was a medley of four numbers that followed the bridal tradition of "something old, something new, something borrowed, and something blue." That is the script for this chapter. We will consider some old, some new, and some borrowed ideas, and then we will finish with something blue. The ideas will be dealt with more succinctly and there will be somewhat less emphasis on "doing ciphers."

RETURN OF THE OUTLIERS

14.1

The other day a graduate student showed me some actual numbers. They were the reaction times for three human subjects in a memory experiment. Subject 1: 1465, 971, 1041, 921, 836, 839, 856, 889, 822, 1008, 736, 875, 1119, 1106, 855, 1108, 775, 992, 919, 905, 849, 887, 865, 904. Subject 2: 712, 761, 814, 652, 866, 1002, 790, 832, 938, 872, 775, 813, 1081, 991, 835, 850, 870, 856, 1423, 786, 870, 1564. Subject 3: 1061, 1031, 819, 1087, 779, 1056, 996, 1103, 820, 715, 2041, 1240, 869, 1059, 815, 803, 1046, 959, 843, 902, 861, 995. Are there any outliers among these numbers? Previously, the impact of outliers on data analysis was stressed. Outliers are numbers that differ greatly from the bulk of the batch. Barnett offered this definition: "An *outlier* is an observation (or subset of observations) that *appears surprising or discrepant to the investigator!*" (1983, p. 150).

What do outliers mean? They may represent different things. They may be *errors* in observing, recording, or calculating. They may be *instrument errors*—dirty test tubes, instrument warm-up, and so on. Or they may indicate *genuine phenomena* and even phenomena of singular importance. If outliers are "bad numbers," then they do not belong in the batch. We have already pointed out that one or more extreme numbers can have disastrous effects upon means and variances. But we cannot evoke a simple rule, such as *always* discard extreme values. The outlier-analysis game is to separate "the wheat from the chaff." It is always possible, however, that the chaff is the wheat.

To decide what an outlier might mean and what to do with it, the first task is to identify the outlier. Beckman and Cook (1983) report that researchers have been battling outliers for 200 years. Barnett (1983) estimated that the statistical literature on outlier problems includes over 600 sources! Faced with a plethora of tests and a variety of approaches to outliers, how can we determine whether the graduate student's data contain any outliers? Given the present state of outlier research, I propose that our approach should be simple, easy, and unpretentious. At least with such an approach we and others will know and comprehend what we did. Instead of a "significance test" for outliers, what we want is a simple procedure for tagging some numbers as suspiciously large (or small) so that they can be set aside for careful scrutiny. It is the *investigator's task* to find out what these suspicious numbers might represent and, in light of such suppositions, to decide how to cope with them.

Guess who is going to help us? Tukey (1977) is again coming to our rescue with a "rule of thumb." I dearly love rules of thumb—they are often simple and can serve as road maps (or gas stations) when one is lost. Tukey's rule of thumb has two parts: (a) if a number exceeds a limit (*a fence*), it is classified as *outside* and (b) if a number exceeds a second limit, it is classified as *far out*. In defining the two fences, Tukey incorporates a spread, the interquartile range (IR), that is uncontaminated by outliers. You will recall that $IR = Q_3 - Q_1$, where Q_3 is the 75th percentile and Q_1 is the 25th. The near fence is $1.5(IR)$ from the nearest Q and the far fence is $3(IR)$ from the nearest Q. Take an X, a suspicious number. If X is farther than $1.5(IR)$ from the nearest quartile (Q_1 or Q_3), X is outside. If X is farther than $3(IR)$ from the nearest quartile, X is far out.

If you eyeball Subject 1's reaction times, one number definitely looks suspicious. But we will sort the $N = 24$ numbers and apply the rule of thumb: 736, 775, 822, 836, 839, 849, 855, 856, 865, 875, 887, 889, 904, 905, 919, 921, 971, 992, 1008, 1041, 1106, 1108, 1119, 1465. $(N + 1)/4 = 6.25$ and $3(N + 1)/4 = 18.75$. Therefore, the 6.25th number in the ordered batch is Q_1, and the 18.75th number is Q_3. By linear interpolation, $Q_1 = 849 + (.25)(855 - 849) = 850.5$; and $Q_3 = 992 + (.75)(1008 - 992) = 1004$. Since $IR = Q_3 - Q_1 = 1004 - 850.5 = 153.5$, then the near fence is $1.5(IR) = 1.5(153.5) = 230.25$; and the far fence is $3(IR) = 3(153.5) = 460.5$. How far is the suspicious number (1465) from the nearest Q? It is $X - Q_3 = 1465 - 1004 = 461$. As $X - Q_3$ is greater than the far fence ($461 > 460.5$), the number 1465, is far out.

Notice that while 1465 appears, indeed, to be far out for Subject 1, it is not so far out when compared with the reaction times of Subjects 2 and 3. For k batches Mosteller, Fienberg, and Rourke (1983) suggested a modification of the rule-of-thumb procedure. Compute the residuals ($X - \overline{X}_j$) for each batch, then apply the rule of thumb to the combined residuals for all batches. Are you ready for some good news? Velleman and Hoaglin (1981) have a computer program, which has been incorporated into MINITAB (Ryan et al., 1981), to do the fence work. Boxplots of the numbers in a batch or the combined residuals from the batches designate an *outside number* by an * and a *far out number* by a 0. In the addendum to this chapter we will demonstrate this convenient method of detecting extremes. For the reaction time data above the investigator has one * and two 0s to ponder over and to decide upon their fate.[1]

Here are a few final points on outliers. If outliers are revealed by the rule-of-thumb method, do the outliers make a difference in the conclusions for an experiment? Why not analyze the data with and without the outliers and see? Or analyze the data in its entirety and contrast this analysis with a trimmed-means analysis (Dixon et al., 1983) in which, for example, 15% of the cases at each end of a distribution are omitted. If the outliers don't matter, then the investigator does not have to worry about them. If the outliers affect the conclusions, then report both sets of results. Can the outliers be cleaned up by re-expression of the data? Instead of analyzing the Xs, try converting them to

$$\sqrt{X}, \log X, \text{ or } \frac{1}{X}$$

when the distributions are positively skewed. Test these transformed scores or their residuals by MINITAB boxplots to learn if the re-expression helped. Finally, whatever happened to *Winsorizing?* Winsorizing consists of replacing the high and low extremes by the next-to-the-highest and next-to-the-lowest numbers. Windsorizing strikes me as a clever and simple procedure. It deserves to be tried more frequently to see how it works in practice.

1. This exposition of the rule-of-thumb procedure follows Mosteller et al. (1983). Tukey uses *hinges* rather than Q_1 and Q_3 in five-number summaries and the difference between hinges instead of IR. A hinge is "crudely, a quartile." A low hinge is the median of the lower half of the numbers when N is *even,* and it is the median of the lower half of the numbers *plus the median* when N is *odd.* A high hinge includes the same rules for the upper half of the numbers. Thus, a low hinge $\cong Q_1$, a high hinge $\cong Q_3$, and a high hinge − a low hinge $\cong$ IR.

SPREADING IT AROUND SOME MORE

14.2

In Chapter 4 it was asserted that *spread* is *the fundamental concept* in statistics. Why? It is essential in descriptive statistics and it is the foundation of inference. Remember that the general formula for a test statistic is how far a statistic *varies* from a parameter when compared to *random variability*. Spread was involved also when we adopted Cohen's recommendation that a test statistic and its probability should be accompanied by a statement of accounted-for variance. Lastly, a major tool in our kit, ANOVA, is a method of partitioning variance into its components.

One of my professors, who was truly a dedicated researcher, once remarked, "Research is like fishing, most of the time you don't catch anything." In nearly every problem in this book we experienced the "thrill of victory" by rejecting H_0. But how about the "agony of defeat" when H_0 is not rejected? This outcome occurs often in the jungle of real research. How are spreads implicated in these victories and defeats? The denominator in the general formula for a test statistic is *random variability*. If the random variability is too large, our research becomes like the parrot in the marvelous Monty Python sketch—it is an "expired parrot." On the other hand, another spread—the spread between the statistic and parameter or the variability of treatment means—must be large for us to enjoy the thrill of victory. In other words, if we are going to score in research, we must attend to both these spreads.

How can an investigator influence these two critical spreads? To increase the spread of the treatment means, we must wisely select the levels of our independent variables. Often, spacing the levels farther apart and imposing more intense treatments will help spread out the means. How can we reduce the random variability "downstairs?" An examination of the formula for the nonpooled error term in the t test suggests some answers:

$$s_{\bar{x}_1 - \bar{x}_2} = \sqrt{\frac{s_1^2}{N_1} + \frac{s_2^2}{N_2}}$$

To reduce $s_{\bar{x}_1 - \bar{x}_2}$, we can increase the Ns, decrease the s^2s, or both. The s^2s are the estimated variances of the numbers within the two separate batches. Three ways to reduce these s^2s are

1. Exercise control over the experimental situation to diminish the effects of "noise."

2. Increase the reliability of the X measures (e.g., use pointer-readings rather than ratings, trained observers rather than untrained ones, etc.).
3. Impose within designs instead of between designs (this tactic unfortunately brings with it the specter of carry-over effects).

While we would argue that experimental design is *the* basic problem in research, a knowledge of statistics (and particularly an awareness of the central role of spread) is valuable in design.

Our last topic regarding spreads is testing variances for equality. In Chapter 9 we advocated testing variances as a way of evaluating the assumption of homogeneity of variance and as a legitimate undertaking in its own right. The recommended procedures were

1. Test two variances from separate batches with the F_{max} test (equal Ns) and the F test (unequal Ns).
2. Test k variances with the F_{max} test (equal Ns) and Bartlett's test (unequal Ns).
3. If H_0 is not rejected with these tests, assume homogeneity of variance.
4. If H_0 is rejected, test the batches for normality by plots of residuals or Lilliefors tests.
5. If the batches are normally distributed, assume heterogeneity of variance.
6. If the batches are not normal and cannot be normalized by re-expression, perform a more form-robust variance test.

We are now at step 6 and it is show-and-tell time. The suggested form-robust test was developed by Box (1953). The logic of *Box's test* is both circular and intriguing. The logic is circular because it employs ANOVA to test the ANOVA assumption of equal variances! The logic is intriguing because it is understandable and intuitively appealing. The Box test procedure is (a) randomly break each batch of numbers into C_k subbatches of $n \geq 2$ numbers, (b) calculate s^2s for each subbatch, (c) re-express the s^2s into natural logarithms (ln) to help normalize the s^2s, (d) perform an ANOVA on the $ln\ s^2$s as numbers, and (e) if F is significant, conclude that heterogeneity of variance is present. What is being done in the Box test? The means of the subbatch logged variances in the k batches are being compared by an ANOVA test.

We will go through the complete variance-testing routine by

an example with $k = 3$ and $N_G = 12$ (Table 14.1). The numbers within each batch have been divided by a table of random numbers into $C_k = 4$ subbatches each with $n = 3$.[2]

TABLE 14.1 Three batches of hypothetical numbers randomly divided into four subbatches

	Batch	
1	2	3
8	8	10
9	8	11
9	10	8
10	7	11
9	9	7
7	8	10
8	7	8
9	10	11
15	8	11
10	9	8
10	9	8
8	8	20
s^2 4.06	0.99	11.66

First, the s^2s for the *whole* batches will be tested by an F_{max} test. Then, the Box procedure will be applied to the subbatches. With our friendly calculator the s^2s for the $k = 3$ *whole* batches have been found. Then,

$$F_{max} = \frac{\text{largest } s^2}{\text{smallest } s^2}$$

$$= \frac{11.66}{0.99} = 11.78$$

The $P(F_{max} = 11.78)$ for $k = 3$ and $df = N_G - 1 = 11$ is less than .01. The variances of the original numbers in the three complete batches are not equal. To check on normality, we examined the numbers and did plots of the residuals $(X - \overline{X}_j)$. They do not appear too normal and two batches have extreme scores. Normally, at this juncture re-expression would be attempted, but instead we will proceed with the Box test.

2. How many C_ks is a problem. If there are too many, then the s^2s are likely to be unreliable. If there are too few, then the df_W will be too small for the F test.

Table 14.2 shows the variances (s^2s) and the logged variances ($\ln s^2$s) in parentheses for the four subbatches within each batch.

TABLE 14.2
Variances and ln variances for the subbatches within three batches

	Batch 1	Batch 2	Batch 3
	0.33 (−1.10)	1.33 (0.29)	2.33 (0.85)
	2.33 (0.85)	1.00 (0.00)	4.33 (1.47)
	14.33 (2.66)	2.33 (0.85)	3.00 (1.10)
	1.33 (0.29)	0.33 (−1.10)	48.00 (3.87)
Sums	2.70	0.04	7.29

The $\ln s^2$s are then subjected to an ANOVA for k separate batches.

$$SS_T = \Sigma X^2 - \frac{(\Sigma X)^2}{N}$$

$$= 30.18 - \frac{(10.03)^2}{12} = 21.80$$

$$SS_G = \frac{(\Sigma X_1)^2 + (\Sigma X_2)^2 + (\Sigma X_3)^2}{N_G} - \frac{(\Sigma X)^2}{N}$$

$$= \frac{(2.70)^2 + (0.04)^2 + (7.29)^2}{4} - 8.38 = 6.73$$

$$SS_W = SS_T - SS_G$$

$$= 21.80 - 6.73 = 15.07$$

Then,

$$df_G = k - 1 = 3 - 1 = 2 \quad \text{and} \quad df_W = N - k = 12 - 3 = 9$$

$$MS_G = \frac{SS_G}{df_G} = \frac{6.73}{2} = 3.36$$

$$MS_W = \frac{SS_W}{df_W} = \frac{15.07}{9} = 1.67$$

Then F is

$$F = \frac{MS_G}{MS_W}$$

$$= \frac{3.36}{1.67} = 2.01$$

The $P(F = 2.01)$ for $df_G = 2$ and $df_W = 9$ is greater than .10. Accordingly, by Box's test the variances do not differ. Since Box's test is more form-robust than the $F_{\max}$ test and since the original numbers did not appear normal, the final decision regarding equality of the variances has been based on Box's test. The risk with Box's test is that it lacks power. Its Type 1 error rates are good, but it is not a sterling performer for detecting differences in variances when they, in fact, exist.[3] Life and statistics are often a series of tradeoffs. Here we are gaining form-robustness with the Box test, but losing power. Another possible variance test is the jackknife test. Church and Wike (1976) observed that this test was quite form-robust, but it also lacked power. Haertel and Lane (1979) constructed a computer program to ease the computational labor for the jackknife test. Recently, O'Brien (1981) proposed a transformation that results in a variance test. O'Brien alleges that his test is both robust and relatively powerful. These claims warrant further investigation. The novel idea in O'Brien's paper is to test the variances for factorial designs rather than the usual procedure of testing k variances.

RE-EXPRESSION SOMETIMES HELPS

14.3

In various places we have talked about re-expression—transforming Xs to another scale like reciprocals, logarithms, square roots, and the like. Why re-express? Because re-expression may produce some potential benefits: outliers might be eliminated or reduced, distributions might become more normal, and variances might be equalized. Notice all the "mights" in these potential benefits. Sometimes re-expression simply does not help. We tried re-expressing the student's reaction times in Section 13.1 into logs (see the addendum). The result was to reduce the outliers from one outside value and two far-out values to two outside values and one far-out value. Reciprocal re-expression of the data shifted the tail of the distribution from the right to the left, and reduced the extreme values to three outside. For positively skewed distributions, reciprocals ($1/X$) work best with extreme skewness, logarithms ($\log X$) with less extreme skewness, and square roots with

---

3. Some evidence on Box's test can be found in Layard (1973) and Tsai, Duran, and Lewis (1975). Keppel (1982, pp. 98–99) carefully reviewed the variance testing problem and supplied additional references.

slight skewness. When the numbers are proportions it is recommended that they be re-expressed as arcsins. A table for

$$2 \arcsin \sqrt{X}$$

where X is a proportion, is available in Winer (1971).

Often students and even some behavioral scientists manifest, what is to me, a curious attitude of nervousness regarding re-expression. They raise issues like: "I did an experiment and recorded the errors made by the subjects. If I re-expressed the errors into logs, what would the log errors mean?" Or: "I measured the times (Xs) it took my rats to traverse a runway. What do reciprocals ($1/X$) of the times mean?" Suppose that five subjects committed 12, 14, 16, 37, and 54 errors, respectively. Converted to $\log_{10} X$ we have 1.08, 1.15, 1.20, 1.57, and 1.73. What is so mysterious about the logged errors? Larger logs mean more errors and smaller logs mean fewer errors. If an article reports that the $\overline{X}$ log errors were 1.18 in one treatment and 1.46 in another, we can easily get back to errors by taking antilogs. In the rat example small reciprocals signify large times and large reciprocals signify small times. Furthermore, if an L/X re-expression is employed (where L is the length of the runway), then the re-expressed numbers become *speed of running*—a very meaningful measure. The comments of Mosteller and Tukey are relevant to the concern about re-expression: "Numbers are primarily recorded or reported in a form that reflects habit or convenience rather than suitability for analysis. As a result, we often need to re-express data before analyzing it" (1977, p. 89). When re-expression enables us to meet the assumptions for a statistical technique or facilitates understanding of our numbers, it should be used. Finally, the critics of re-expression should recognize that converting numbers to ranks, percentiles, and Z scores constitutes re-expression.

Years ago statisticians pushed re-expression as a potential solution to the problems of heterogeneity of variance and nonnormality. With the advent of the belief in the robustness of ANOVA, this push has subsided. Furthermore, some statisticians (e.g., Glass et al., 1971, p. 241) are less sanguine about the benefits of normalizing transformations in ANOVA. Nevertheless, I'm not ready to abandon re-expression. For one thing, the availability of calculators and computers offers the opportunity to easily re-express numbers and check upon the results of re-expressions. Re-expression was a very dreary undertaking when the tools were tables of logarithms and reciprocals and rotary calculators. Today it is far less laborious. Rapidly calculated statistics and displays of re-expressed numbers permit the data analyst to observe quickly whether or not a re-expression was of value. Before abandoning re-expression, it merits a fairer trial with an assist from today's more efficient tools.

One final comment on re-expression concerns interactions in ANOVA. Case one: Suppose that an ANOVA reveals significant main effects and an interaction. Suppose further that after re-expression the interaction *disappears,* leaving only the main effects. A major crusade in science is the quest for *simplicity.* Main effects are simpler than main effects *and* an interaction. Saying that a variable does something is simpler than saying that a variable does something depending upon another variable. For a concrete example of a transformation that removes an interaction, see Box et al. (1978, pp. 228–238). Imaginary case two: An interaction becomes more *interpretable* following re-expression. If that happened, wouldn't it be valuable? Fantasy case three: an observed interaction remains *invariant* under a variety of transformations. This outcome indicates that there is an interaction that is not an artifact of scale. If you accept the argument that re-expression deserves more serious testing and try it, check for the possible effects of re-expression upon interactions. Smith (1976) has written a thoughtful paper on various problems of re-expression.

POWER—A CRITICAL BUT OVERLOOKED PROBLEM

14.4

The 2 × 2 statistical decision-reality matrix was presented in Chapter 5. We talked about Type 1 errors and correct decisions *when H_0 is true.* Now we need to examine the column on the right side of the matrix—what transpires *when H_0 is false.* Then, if we do not reject H_0, a Type 2 error of inference has been committed. There was something there, but we failed to detect it. If we reject H_0, a correct decision has been made. There was something there, and we detected it. The P(Type 2 error) is β. The P(correct decision) is $1 - \beta$. The P(correct decision) is also *power.* It is the probability of finding something when that something is there to be found. And we must never lose sight of the fact that Type 1 and Type 2 errors are related inversely. If we increase Type 1 errors, we decrease Type 2 errors and vice versa. Because Type 2 errors and power are also linked, increasing Type 1 errors *increases* power and decreasing Type 1 errors *decreases* power. It is tradeoff time again.

What is *power analysis?* What factors are involved? How is it done? When should it be done? The factors involved are (a) α, *the criterion of significance;* (b) *N, batch size;* (c) γ (gamma), *the effect size in the population;* and (d) $1 - \beta$, *power*. The two major problems in power analysis are: first, finding power, $1 - \beta$, as a function of the remaining factors (α, N, γ); and second, finding batch size, N, as a function of the remaining factors (α, $1 - \beta$, γ). A power analysis should be performed *before* an experiment to determine what power a researcher has and what batch size is needed to achieve a desired power. A power analysis can be performed *after* an experiment to learn what power the researcher had and what batch size might be needed to achieve a desired power in a future study. An investigator who had reasonable power but failed to reject H_0 might be more convinced that there was nothing there to be found. If the power was low, then it could be increased in a future study. Power analysis is critical because it is a grievous research error to do an experiment that has a low probability of finding something when it is there to be found.

Cohen (1969) made a major contribution to power analysis. When I first looked at Cohen's book, I decided it was not for me. To do power analysis you must know population parameters (e.g., μs) or be able to predict them from quantitative theory. While you can attempt to estimate such parameters from previous research or pilot studies, such estimation is somewhat hazardous. On a more careful reading, I discovered that Cohen had devised a clever solution to the parameter problem. I got back on board. Power analysis is possible for various experimental designs and different statistics.[4] Either power charts or power tables can be used—I prefer the latter, which are provided in Cohen's book.

Let us demonstrate power analysis for the simple case of comparing two means from separate batches. From a quantitative theory an investigator expects a *difference in the population means* for two separate batches *in either direction* of 3.5. The investigator also anticipates that $\sigma = 7$, and we asume that $\sigma_1 = \sigma_2 = \sigma$. The investigator plans to do a two-tailed test with $\alpha = .05$. We need to describe the new concept of γ, *effect size*. Effect size is how large a difference exists between two population means when the difference is standardized. It is the difference between two population means in σ units. For the means of two separate batches

$$\gamma = \frac{\mu_1 - \mu_2}{\sigma}$$

---

4. See Welkowitz et al. (1982) for power analyses of simple statistics and Cohen (1969) for more complex problems.

Welkowitz et al. (1982) introduced a summarizing concept, δ (delta), to bridge γ and N. For two independent means δ is

$$\delta = \gamma\sqrt{\frac{N_G}{2}}$$

where $N_G = N_1 = N_2$. If the researcher planned to test 16 cases in each batch, what power would the researcher have? Effect size is

$$\gamma = \frac{3.5}{7} = 0.5$$

Then delta is

$$\delta = 0.5\sqrt{\frac{16}{2}} = 1.41$$

Entering the power table (Table L) for $\alpha = .05$, two-tailed, across the top and $\delta = 1.41$ down the left margin, the power or $1 - \beta = .29$. In other words, the investigator has roughly only one chance in three to find the predicted difference in the means.

Suppose that the investigator is a gambler, but not a wild one. The investigator wishes to have at least an even shot ($1 - \beta = .50$) at finding the same difference in the population means. What sample size (N_G) would be required to achieve this power value? Since

$$\delta = \gamma\sqrt{\frac{N_G}{2}}$$

$$\delta^2 = \gamma^2\left(\frac{N_G}{2}\right)$$

and

$$N_G = \frac{2\,\delta^2}{\gamma^2}$$

Entering table M for δ as a function of α (.05, two-tailed) across the top and γ (0.5, as determined above) down the left margin, $\delta = 1.96$. Then,

$$N_G = \frac{2(1.96)^2}{(0.5)^2}$$

$$\frac{7.68}{0.25} = 30.7$$

Or, 31 cases per group would be required to have the power, $1 - \beta = .50$, in the experiment.

In these determinations of power and sample size the parameters (μs and σ) had to be known or estimated. Suppose the investigator does not have such information. Is power analysis impossible? No, because Cohen has proposed an ingenious solution of *conventional effect sizes*. For two means he proposes that a *small effect size* (γ) is .20, a *medium effect size* is .50, and a *large effect size* is .80.[5] If an investigator expects a medium effect size, we don't know μ_1, μ_2, or σ, but we do know that $\gamma = (\mu_1 - \mu_2)/\sigma = .50$. Now we will apply the conventional effect size method to the original two-batch problem to obtain power and N_G. Assume the researcher anticipates a *large effect* ($\gamma = .80$). Then,

$$\delta = \gamma \sqrt{\frac{N_G}{2}}$$

$$= .80 \sqrt{\frac{16}{2}} = 2.26$$

From Table L for $\alpha = .05$, two-tailed, $.59 < 1 - \beta < .63$. The investigator's chances (roughly six of ten) of finding a large difference are much greater than in the original example.

Pretend that a power or $1 - \beta = .80$ (Cohen's goal in research) was desired. What N_G would be required? From Table M, $\delta = 2.80$. Then,

$$N_G = \frac{2\delta^2}{\gamma^2}$$

$$= \frac{2(2.80)^2}{(.80)^2}$$

$$= \frac{15.68}{0.64} = 24.5$$

Or 25 cases per group are required. Finally, when $N_1 \neq N_2$, the harmonic mean, $N_h = 2N_1N_2/(N_1 + N_2)$, replaces N_G in the power calculations for two means from separate batches.

A few details on power follow.

5. Cohen's conventional effect sizes have been criticized for being too small. While they are arbitrary values, he based them on his observations of research outcomes. I do not think they are unrealistic. They are not carved in stone and if they prove to be too small, they can be changed. In the meantime, they permit the important tasks of power analysis to be done.

1. The formulas for γ, δ, and N_G vary for different statistics.
2. Cohen's proposed conventional effect sizes vary for different statistics but are internally consistent across statistics.
3. It takes fewer cases to detect a large effect than a small one.
4. One-tailed tests are more powerful than two-tailed tests.
5. Relaxing α, such as from .01 to .05, increases power.
6. Besides increasing sample size, reducing variability will increase power, as described in Section 14.2.

If you do research with too litle power, it is a waste of effort, time, and money. If you do research with too many cases (a rare occurrence), it is also a waste. My advice on power analysis is "Don't leave home without it."

DOING DATA ANALYSIS ON A COMPUTER

14.5

Throughout the book the addenda have displayed output from a statistical package, MINITAB. MINITAB was designed especially for students in elementary statistics. In addition to the usual elementary statistics, MINITAB includes multiple regression, time series, and exploratory data analysis. It is an *interactive* package: The user types a command on a terminal for a statistical procedure to be executed upon a batch of numbers and immediately sees the results (or error messages) in the terminal.

More elaborate statistical packages, such as BMDP and SPSS[x], are usually run in the *batch* mode. In this mode the user submits a program and the data (or retrieves the data from a file), then executes the run. When the job is executed, which may be hours later, the user sees (and/or has printed) the results of the run. At this time also error messages are output, and if they are serious the run may be terminated. These large statistical packages can do incredibly complex tasks of data analysis.

There is now an international electronic war being waged to decide who can produce the biggest and best computer chips. Another war is going on to decide who is going to dominate the microcomputer market. Microcomputers are becoming more powerful and their software, including statistical packages, is multiplying at a rapid rate. These developments in microcomputers greatly increase the possibilities for computer analysis of data by a larger population of users.

What are the benefits of computer data analysis? King and Julstrom listed seven benefits that computers provide.

> a. mass storage of and immediate access to the data which will be analyzed, b. the set of instructions to perform the analytical procedures (the program) to be stored in the machine, c. the performance of more complex computations, d. the faster performance of computations, e. the calculations to be performed without constant human intervention, f. management (create, store, modify, rearrange, destroy) of data and sets of instructions, and g. the production of more than one result (1983, p. 10).

Despite their power and speed, computers are not without pitfalls. Analyzing a set of data on a computer does not guarantee that the analysis is correct. The data can be input incorrectly. The data being analyzed may not be the data the user wanted to analyze. While the compiler for a statistical package may be very picky about syntax and punctuation, it does not prevent the user from doing perfectly ridiculous analyses upon command. For example, SPSS^x has a χ^2 program (CROSSTABS) for testing frequencies in contingency tables. If the user applies CROSSTABS to variables that are in number form, it will generate gigantic, nonsensical contingency tables whose bins are filled largely with zeros. To repeat, the computer is an idiot. If the user issues absurd commands, the computer will generally do as it is told.

How can the user avoid producing nonsense? How can the user know when the pretty computer output is really garbage in disguise? It is essential for the user to understand statistics, statistical packages for computers, and how to apply the packages. If the user does not know, help should be sought. And whenever a computer analysis of data is performed, the output should be studied with great care to decide if it makes sense. Some means for the variables should be hand calculated on the raw data and compared with Big Brother's. If the two sets of means do not agree, then the data may not have been input correctly, the data are not being read properly, or both. Take a healthy paranoid view of your computer output. Always test it against your eyes, your calculator, and your knowledge.

THE DATA-ANALYTIC ORIENTATION REVISITED

14.6

Let us reiterate the stand on data analysis. Don't rush into inference. Explore your data first.[6] Sort it, picture it, re-express it, and try to look at it in novel ways. If it is spotty, try to clean it up.

There is an element of playfulness in science and a playfulness in data analysis. If you leap into inference without some time in the playpen, you may be overlooking something important.

A single study is just a single study. Don't become overly impressed by it. All you have are some *indications*. When your indications hold up in replications by you, by Harvey down the road, by Serena in Dallas, and some other folks, you may be onto something.

The end result of research is numbers. Your job as a data analyst is to try to make sense of the numbers. Statistics is a tool to help you make some sense of the numbers. But statistics is an imperfect tool. The beautiful normal curves that decorate statistics books are not always like the numbers obtained in research. The assumptions of the statistical test makers may not be realized out in the pits of research. Always work on your numbers hand-in-hand with your eyeball, your common sense, and your knowledge. Don't apply statistics in a vacuum. Making sense of numbers is an important endeavor. The numbers represent phenomena, and understanding phenomena is the name of the game in science.

SOMETHING BLUE 14.7

What's blue? Blue is that we have reached the end of the line. I have run out of number 2 pencils, Big Chief tablets, and my brain is yogurtized. As you deplane, I want to thank you for flying with us. Perhaps we will meet someday. Perhaps, if you are lost and go into a gas station to ask for directions, I might be there. I will be the one with a Genuine Swiss Army Knife on my belt, and I *won't* be asking how to get to Bad Axe, Michigan.

SUMMARY

Outliers are numbers that differ greatly from the bulk of a batch and are surprising or discrepant to a researcher. They may be errors in observing, recording, and computing, instrument errors, or indicants of genuine phenomena. Tukey's rule of thumb is one way of isolating extreme numbers for detailed study by the investigator. The method employs multiples of interquartile range to set up fences beyond which numbers are tagged as outside or far

6. See Good (1983) for an intriguing assessment of exploratory data analysis.

out. With k batches of numbers the rule of thumb is applied to the combined residuals of the batches. Data should be analyzed with and without extreme numbers (or with the extremes included vs. trimmed batches) to determine whether outliers influenced the conclusions. The possibility of re-expression as an aid in reducing or eliminating outliers was mentioned, as was Winsorizing (replacing the extremes at each end of a batch by the next-to-the-extremes) as a technique to remove outliers.

The status of spread as the fundamental concept in statistics was stressed again. Spread is essential in descriptive statistics and is the basis of inference. The general formula for a test statistic is the ratio of the spread between a statistic and a parameter (or the variability of means) to random variability. Suggestions were offered to increase the variability of the numerator and decrease the variability of the denominator in the general formula. Box's test for equality of variances was proposed as a form-robust test. The suggested routine for assessing variances for k separate batches with Box's test was demonstrated.

Re-expression of X was described as a possible way to achieve homogeneity of variance and normality and to diminish outliers. Reciprocal re-expression was advised for extreme positive skewness, logarithms for less extreme positive skewness, square roots for slight positive skewness, and arcsins for proportions. While re-expression is not advocated as often as it once was, it was urged that its possibilities be explored because calculators and computers greatly facilitate re-expression and the evaluation of its effects. The influence of re-expression on interactions was considered. If re-expression eliminates an interaction or makes it more interpretable, then re-expression could be valuable. If an interaction is invariant under various re-expressions, it suggests that the interaction is not an artifact of scale.

Power $(1 - \beta)$ is the probability of rejecting H_0 when H_0 is false. Power analysis involves: $1 - \beta$, power; γ, effect size; α, the criterion of significance; and N, batch size. The two major problems of power analysis are determining power and determining the batch size needed to attain a desired level of power. Power analysis can be performed before and after an experiment. An example of power analysis involved the means of two separate batches. Power analysis was demonstrated with known parameters and with Cohen's conventional effect sizes. Some relations among power, effect size, one- and two-tailed tests, and others were outlined.

Statistical packages for computers provide investigators with a powerful tool for data analysis. The output from computer analyses should be scrutinized with care and checked carefully, however. Doing data analysis on a computer does not guarantee its correctness.

Finally, the importance of a data-analysis orientation was reiterated. Data should be explored thoroughly before moving to inferential statistics. The results of a single study should be regarded as indications until they have been repeatedly replicated. Statistical analysis should be accompanied by the eyeball, common sense, and knowledge.

PROBLEMS

1. In the table below are the times for three rats to leave a start box and break a beam at the end of the start box.

   Rat 1: 54, 70, 61, 29, 38, 34, 28, 24, 26, 24, 34, 38, 36, 27, 25.
   Rat 2: 443, 163, 152, 50, 42, 36, 325, 37, 29, 23, 30, 28, 32, 28, 25.
   Rat 3: 126, 217, 105, 58, 60, 49, 52, 52, 43, 37, 29, 38, 40, 37, 36.

   Test rat 1's data for outside and far-out values. Repeat for rat 3.

2. Make a frequency distribution of all 45 scores. Convert the distribution to a histogram. Does the distribution appear to be normal? If not, how does it differ from a normal distribution?

3. Using your calculator, re-express the 45 starting times (Xs) into reciprocals ($1/X$) rounded to two decimal places. Make a frequency distribution and histogram of the re-expressed numbers. Did re-expression help to normalize the starting times?

4. Assume that the 45 scores in Problem 1 came from three groups with $N_G = 15$. Test the H_0: $\sigma_1^2 = \sigma_2^2 = \sigma_3^2$ with the F_{max} test and $\alpha = .05$.

5. Split the 15 scores in each group randomly into five subbatches of three scores. Test the H_0: $\sigma_1^2 = \sigma_2^2 = \sigma_3^2$ at $\alpha = .05$ with a variance test that is more form-robust than the F_{max} test.

6. What do you conclude regarding homogeneity of variance from the results of Problems 4 and 5?

7. A drug researcher predicts that a new drug will produce a *large effect* (a decrement) on blood pressure in hypertensive subjects. If 15 subjects are given the new drug and 15 controls receive a placebo, how much power would the researcher have to detect the large anticipated decrement with $\alpha = .05$ (one-tailed)?

8. If the drug researcher had done a two-tailed test with the same sample sizes, expected γ, and $\alpha = .05$, what power would the researcher have?

9. How many cases would the researcher need in each group in Problem 7 to achieve a power = .80?

10. If the researcher anticipated a medium decrement in blood pressure in Problem 7, what would the power be for α = .05 (one-tailed)?
11. How many cases would be needed per group to achieve a power = .80 in Problem 10?
12. An investigator predicts that $\mu_1 - \mu_2 = 10$ and $\sigma_1 = \sigma_2 = \sigma = 20$. In terms of Cohen's conventional effect sizes, what kind of an effect is the investigator predicting?

ADDENDUM: USING COMPUTERS IN STATISTICS

MINITAB is ideal for isolating outliers by Tukey's rule-of-thumb procedure. In the first example the reaction times for the three subjects in the chapter are put into memory, their means ($\overline{X}_j$) found, and their residuals ($X - \overline{X}_j$) computed and combined. The first box plot of these combined residuals (1) discloses one outside value, *, and two far out values, 0. In the second run, the reaction times have been re-expressed as logs to the base 10 before making a box plot of the residuals. The re-expression (2) reduced the outliers to two outside values, *, and one far out value 0. The final box plot is for a reciprocal re-expression of the reaction times. The box plot in this case is of the residuals of the combined re-expressed numbers (1000/X). This box plot reveals three outside values, *, in the left tail (3).

```
--SET IN C1
--1465 971 1041 921 836 839 856 889 822 1008 736 875 1119
--1106 855 1108 775 992 919 905 849 887 865 904
--SET IN C2
--712 761 814 652 866 1002 790 832 938 872 775 813 1081
--991 835 850 870 856 1423 786 870 1564
--SET IN C3
--1061 1031 819 1087 779 1056 996 1103 820 715 2041
--1240 869 1059 815 803 1046 959 843 902 861 995

--AVERAGE C1
    AVERAGE =         939.29
--AVERAGE C2
    AVERAGE =         906.95

--AVERAGE C3
    AVERAGE =         995.45
```

```
--SUBT 939.29 C1 PUT IN C4
--SUBT 906.95 C2 PUT IN C5
--SUBT 995.45 C3 PUT IN C6
--JOIN C4 TO C5 TO C6 PUT IN C7


--BOXPLOT DATA IN C7

                     -------
           -----I  +    I------        *    0          0           (1)
                     -------

              ---------+---------+---------+---------+----
ONE HORIZONTAL SPACE = 0.30E 02
FIRST TICK AT       0.

--LOGT C1 PUT IN C8
--LOGT C2 PUT IN C9
--LOGT C3 PUT IN C10
--AVERAGE C8
   AVERAGE =       2.9681
--AVERAGE C9
   AVERAGE =       2.9483
--AVERAGE C10
   AVERAGE =       2.9869
--SUBT 2.9681 C8 PUT IN C11
--SUBT 2.9483 C9 PUT IN C12
--SUBT 2.9869 C10 PUT IN C13
--JOIN C11 TO C12 TO C13 PUT IN C14
--BOXPLOT DATA IN C14

                     ----------
           --------I  +    I-------       *    *      0             (2)
                     ----------

              ---+---------+---------+---------+---------+--
ONE HORIZONTAL SPACE = 0.10E-01
FIRST TICK AT      -0.100

--DIVIDE 1000 BY C1 PUT IN C4
--DIVIDE 1000 BY C2 PUT IN C5
--DIVIDE 1000 BY C3 PUT IN C6
```

```
--AVERAGE C4
   AVERAGE =        1.0866
--AVERAGE C5
   AVERAGE =        1.1466
--AVERAGE C6
   AVERAGE =        1.0517
--SUBT 1.0866 C4 PUT IN C10
--SUBT 1.1466 C5 PUT IN C11
--SUBT 1.0517 C6 PUT IN C12
--JOIN C10 TO C11 TO C12 PUT IN C13
--BOXPLOT DATA IN C13


                                   ----------
            *  *   * ----------------I    +    I--------------          (3)
                                   ----------

            --------+---------+---------+---------+---------
ONE HORIZONTAL SPACE = 0.20E-01
FIRST TICK AT     -0.400
```

TABLES

A CRITICAL VALUES OF t

B CRITICAL VALUES OF THE PEARSON r

C CRITICAL VALUES OF r_s (SPEARMAN RANK-ORDER CORRELATION COEFFICIENT)

D PROPORTIONS OF AREA UNDER THE NORMAL CURVE

E TABLE FOR F_{max} PERCENTAGE POINTS OF THE RATIO, s^2_{max}/s^2_{min}

F TABLE OF F

G WILCOXON T VALUES FOR PAIRED REPLICATES

H WILCOXON T VALUES FOR UNPAIRED REPLICATES

I TABLE FOR THE KRUSKAL-WALLIS H TEST

J TABLE OF CRITICAL VALUES OF CHI-SQUARE

K CRITICAL χ^2_r VALUES FOR FRIEDMAN'S TEST

L POWER AS A FUNCTION OF δ AND SIGNIFICANCE CRITERION (α)

M δ AS A FUNCTION OF SIGNIFICANCE CRITERION (α) AND POWER

TABLES

TABLE A
Critical values of t

df	Alpha Levels for a Directional (One-tailed) Test					
	.10	.05	.025	.01	.005	.0005
1	3.078	6.314	12.706	31.821	63.657	636.619
2	1.886	2.920	4.303	6.965	9.925	31.598
3	1.638	2.353	3.182	4.541	5.841	12.941
4	1.533	2.132	2.776	3.747	4.604	8.610
5	1.476	2.015	2.571	3.365	4.032	6.859
6	1.440	1.943	2.447	3.143	3.707	5.959
7	1.415	1.895	2.365	2.998	3.499	5.405
8	1.397	1.860	2.306	2.896	3.355	5.041
9	1.383	1.833	2.262	2.821	3.250	4.781
10	1.372	1.812	2.228	2.764	3.169	4.587
11	1.363	1.796	2.201	2.718	3.106	4.437
12	1.356	1.782	2.179	2.681	3.055	4.318
13	1.350	1.771	2.160	2.650	3.012	4.221
14	1.345	1.761	2.145	2.624	2.977	4.140
15	1.341	1.753	2.131	2.602	2.947	4.073
16	1.337	1.746	2.120	2.583	2.921	4.015
17	1.333	1.740	2.110	2.567	2.898	3.965
18	1.330	1.734	2.101	2.552	2.878	3.922
19	1.328	1.729	2.093	2.539	2.861	3.883
20	1.325	1.725	2.086	2.528	2.845	3.850
21	1.323	1.721	2.080	2.518	2.831	3.819
22	1.321	1.717	2.074	2.508	2.819	3.792
23	1.319	1.714	2.069	2.500	2.807	3.767
24	1.318	1.711	2.064	2.492	2.797	3.745
25	1.316	1.708	2.060	2.485	2.787	3.725
26	1.315	1.706	2.056	2.479	2.779	3.707
27	1.314	1.703	2.052	2.473	2.771	3.690
28	1.313	1.701	2.048	2.467	2.763	3.674
29	1.311	1.699	2.045	2.462	2.756	3.659
30	1.310	1.697	2.042	2.457	2.750	3.646
40	1.303	1.684	2.021	2.423	2.704	3.551
60	1.296	1.671	2.000	2.390	2.660	3.460
120	1.289	1.658	1.980	2.358	2.617	3.373
∞	1.282	1.645	1.960	2.326	2.576	3.291
	.20	.10	.05	.02	.01	.001
	Alpha Levels for a Nondirectional (Two-tailed) Test					

Reprinted, with changes in notation, by permission of the publisher from E.S. Pearson and H.O. Hartley, *Biometrika Tables for Statisticians,* Volume 1, p. 146. Copyright 1966, Biometrika.

TABLE B
Critical values of the Pearson r

df (= N − 2; N = Number of Pairs)	Level of Significance for One-tailed Test			
	.05	.025	.01	.005
	Level of Significance for Two-tailed Test			
	.10	.05	.02	.01
1	.988	.997	.9995	.9999
2	.900	.950	.980	.990
3	.805	.878	.934	.959
4	.729	.811	.882	.917
5	.669	.754	.833	.874
6	.622	.707	.789	.834
7	.582	.666	.750	.798
8	.549	.632	.716	.765
9	.521	.602	.685	.735
10	.497	.576	.658	.708
11	.476	.553	.634	.684
12	.458	.532	.612	.661
13	.441	.514	.592	.641
14	.426	.497	.574	.623
15	.412	.482	.558	.606
16	.400	.468	.542	.590
17	.389	.456	.528	.575
18	.378	.444	.516	.561
19	.369	.433	.503	.549
20	.360	.423	.492	.537
21	.352	.413	.482	.526
22	.344	.404	.472	.515
23	.337	.396	.462	.505
24	.330	.388	.453	.496
25	.323	.381	.445	.487
26	.317	.374	.437	.479
27	.311	.367	.430	.471
28	.306	.361	.423	.463
29	.301	.355	.416	.456
30	.296	.349	.409	.449
35	.275	.325	.381	.418
40	.257	.304	.358	.393
45	.243	.288	.338	.372
50	.231	.273	.322	.354
60	.211	.250	.295	.325
70	.195	.232	.274	.302
80	.183	.217	.256	.283
90	.173	.205	.242	.267
100	.164	.195	.230	.254

Source: Table VII of Fisher and Yates. 1974. *Statistical tables for biological, agricultural and medical research*, 6th ed. Published by Longman Group Ltd., London (previously published by Oliver and Boyd, Ltd., Edinburgh), and by permission of the authors and publishers.

TABLE C
Critical values of r_s (Spearman rank-order correlation coefficient)*

	Level of significance for one-tailed test			
	.05	.025	.01	.005
	Level of significance for two-tailed test			
No. of pairs (N)	.10	.05	.02	.01
5	.900	1.000	1.000	—
6	.829	.886	.943	1.000
7	.714	.786	.893	.929
8	.643	.738	.833	.881
9	.600	.683	.783	.833
10	.564	.648	.746	.794
12	.506	.591	.712	.777
14	.456	.544	.645	.715
16	.425	.506	.601	.665
18	.399	.475	.564	.625
20	.377	.450	.534	.591
22	.359	.428	.508	.562
24	.343	.409	.485	.537
26	.329	.392	.465	.515
28	.317	.377	.448	.496
30	.306	.364	.432	.478

*From E. G. Olds (1938). *Annals of Mathematical Statistics, 9,* 133–148 and (1949). 20, 117–118. Reprinted by permission of the publisher.

TABLE D
Areas under the standard normal distribution

A	B	C	A	B	C	A	B	C
Z	0 to Z	beyond Z	Z	0 to Z	beyond Z	Z	0 to Z	beyond Z
0.00	.0000	.5000	0.55	.2088	.2912	1.10	.3643	.1357
0.01	.0040	.4960	0.56	.2123	.2877	1.11	.3665	.1335
0.02	.0080	.4920	0.57	.2157	.2843	1.12	.3686	.1314
0.03	.0120	.4880	0.58	.2190	.2810	1.13	.3708	.1292
0.04	.0160	.4840	0.59	.2224	.2776	1.14	.3729	.1271
0.05	.0199	.4801	0.60	.2257	.2743	1.15	.3749	.1251
0.06	.0239	.4761	0.61	.2291	.2709	1.16	.3770	.1230
0.07	.0279	.4721	0.62	.2324	.2676	1.17	.3790	.1210
0.08	.0319	.4681	0.63	.2357	.2643	1.18	.3810	.1190
0.09	.0359	.4641	0.64	.2389	.2611	1.19	.3830	.1170
0.10	.0398	.4602	0.65	.2422	.2578	1.20	.3849	.1151
0.11	.0438	.4562	0.66	.2454	.2546	1.21	.3869	.1131
0.12	.0478	.4522	0.67	.2486	.2514	1.22	.3888	.1112
0.13	.0517	.4483	0.68	.2517	.2483	1.23	.3907	.1093
0.14	.0557	.4443	0.69	.2549	.2451	1.24	.3925	.1075
0.15	.0596	.4404	0.70	.2580	.2420	1.25	.3944	.1056
0.16	.0636	.4364	0.71	.2611	.2389	1.26	.3962	.1038
0.17	.0675	.4325	0.72	.2642	.2358	1.27	.3980	.1020
0.18	.0714	.4286	0.73	.2673	.2327	1.28	.3997	.1003
0.19	.0753	.4247	0.74	.2704	.2296	1.29	.4015	.0985
0.20	.0793	.4207	0.75	.2734	.2266	1.30	.4032	.0968
0.21	.0832	.4168	0.76	.2764	.2236	1.31	.4049	.0951
0.22	.0871	.4129	0.77	.2794	.2206	1.32	.4066	.0934
0.23	.0910	.4090	0.78	.2823	.2177	1.33	.4082	.0918
0.24	.0948	.4052	0.79	.2852	.2148	1.34	.4099	.0901
0.25	.0987	.4013	0.80	.2881	.2119	1.35	.4115	.0885
0.26	.1026	.3974	0.81	.2910	.2090	1.36	.4131	.0869
0.27	.1064	.3936	0.82	.2939	.2061	1.37	.4147	.0853
0.28	.1103	.3897	0.83	.2967	.2033	1.38	.4162	.0838
0.29	.1141	.3859	0.84	.2995	.2005	1.39	.4177	.0823
0.30	.1179	.3821	0.85	.3023	.1977	1.40	.4192	.0808
0.31	.1217	.3783	0.86	.3051	.1949	1.41	.4207	.0793
0.32	.1255	.3745	0.87	.3078	.1922	1.42	.4222	.0778
0.33	.1293	.3707	0.88	.3106	.1894	1.43	.4236	.0764
0.34	.1331	.3669	0.89	.3133	.1867	1.44	.4251	.0749
0.35	.1368	.3632	0.90	.3159	.1841	1.45	.4265	.0735
0.36	.1406	.3594	0.91	.3186	.1814	1.46	.4279	.0721
0.37	.1443	.3557	0.92	.3212	.1788	1.47	.4292	.0708
0.38	.1480	.3520	0.93	.3238	.1762	1.48	.4306	.0694
0.39	.1517	.3483	0.94	.3264	.1736	1.49	.4319	.0681
0.40	.1554	.3446	0.95	.3289	.1711	1.50	.4332	.0668
0.41	.1591	.3409	0.96	.3315	.1685	1.51	.4345	.0655
0.42	.1628	.3372	0.97	.3340	.1660	1.52	.4357	.0643
0.43	.1664	.3336	0.98	.3365	.1635	1.53	.4370	.0630
0.44	.1700	.3300	0.99	.3389	.1611	1.54	.4382	.0618
0.45	.1736	.3264	1.00	.3413	.1587	1.55	.4394	.0606
0.46	.1772	.3228	1.01	.3438	.1562	1.56	.4406	.0594
0.47	.1808	.3192	1.02	.3461	.1539	1.57	.4418	.0582
0.48	.1844	.3156	1.03	.3485	.1515	1.58	.4429	.0571
0.49	.1879	.3121	1.04	.3508	.1492	1.59	.4441	.0559
0.50	.1915	.3085	1.05	.3531	.1469	1.60	.4452	.0548
0.51	.1950	.3050	1.06	.3554	.1446	1.61	.4463	.0537
0.52	.1985	.3015	1.07	.3577	.1423	1.62	.4474	.0526
0.53	.2019	.2981	1.08	.3599	.1401	1.63	.4484	.0516
0.54	.2054	.2946	1.09	.3621	.1379	1.64	.4495	.0505

Source: Table IIi of Fisher and Yates. 1974. *Statistical tables for biological, agricultural and medical research,* 6th ed. Published by Longman Group Ltd., London (previously published by Oliver and Boyd, Ltd., Edinburgh), and by permission of the authors and publishers.

TABLE D
Areas under the standard normal distribution (*continued*)

A	B	C	A	B	C	A	B	C
Z	0 to Z	beyond Z	Z	0 to Z	beyond Z	Z	0 to Z	beyond Z
1.65	.4505	.0495	2.22	.4868	.0132	2.79	.4974	.0026
1.66	.4515	.0485	2.23	.4871	.0129	2.80	.4974	.0026
1.67	.4525	.0475	2.24	.4875	.0125	2.81	.4975	.0025
1.68	.4535	.0465	2.25	.4878	.0122	2.82	.4976	.0024
1.69	.4545	.0455	2.26	.4881	.0119	2.83	.4977	.0023
1.70	.4554	.0446	2.27	.4884	.0116	2.84	.4977	.0023
1.71	.4564	.0436	2.28	.4887	.0113	2.85	.4978	.0022
1.72	.4573	.0427	2.29	.4890	.0110	2.86	.4979	.0021
1.73	.4582	.0418	2.30	.4893	.0107	2.87	.4979	.0021
1.74	.4591	.0409	2.31	.4896	.0104	2.88	.4980	.0020
1.75	.4599	.0401	2.32	.4898	.0102	2.89	.4981	.0019
1.76	.4608	.0392	2.33	.4901	.0099	2.90	.4981	.0019
1.77	.4616	.0384	2.34	.4904	.0096	2.91	.4982	.0018
1.78	.4625	.0375	2.35	.4906	.0094	2.92	.4982	.0018
1.79	.4633	.0367	2.36	.4909	.0091	2.93	.4983	.0017
1.80	.4641	.0359	2.37	.4911	.0089	2.94	.4984	.0016
1.81	.4649	.0351	2.38	.4913	.0087	2.95	.4984	.0016
1.82	.4656	.0344	2.39	.4916	.0084	2.96	.4985	.0015
1.83	.4664	.0336	2.40	.4918	.0082	2.97	.4985	.0015
1.84	.4671	.0329	2.41	.4920	.0080	2.98	.4986	.0014
1.85	.4678	.0322	2.42	.4922	.0078	2.99	.4986	.0014
1.86	.4686	.0314	2.43	.4925	.0075	3.00	.4987	.0013
1.87	.4693	.0307	2.44	.4927	.0073	3.01	.4987	.0013
1.88	.4699	.0301	2.45	.4929	.0071	3.02	.4987	.0013
1.89	.4706	.0294	2.46	.4931	.0069	3.03	.4988	.0012
1.90	.4713	.0287	2.47	.4932	.0068	3.04	.4988	.0012
1.91	.4719	.0281	2.48	.4934	.0066	3.05	.4989	.0011
1.92	.4726	.0274	2.49	.4936	.0064	3.06	.4989	.0011
1.93	.4732	.0268	2.50	.4938	.0062	3.07	.4989	.0011
1.94	.4738	.0262	2.51	.4940	.0060	3.08	.4990	.0010
1.95	.4744	.0256	2.52	.4941	.0059	3.09	.4990	.0010
1.96	.4750	.0250	2.53	.4943	.0057	3.10	.4990	.0010
1.97	.4756	.0244	2.54	.4945	.0055	3.11	.4991	.0009
1.98	.4761	.0239	2.55	.4946	.0054	3.12	.4991	.0009
1.99	.4767	.0233	2.56	.4948	.0052	3.13	.4991	.0009
2.00	.4772	.0228	2.57	.4949	.0051	3.14	.4992	.0008
2.01	.4778	.0222	2.58	.4951	.0049	3.15	.4992	.0008
2.02	.4783	.0217	2.59	.4952	.0048	3.16	.4992	.0008
2.03	.4788	.0212	2.60	.4953	.0047	3.17	.4992	.0008
2.04	.4793	.0207	2.61	.4955	.0045	3.18	.4993	.0007
2.05	.4798	.0202	2.62	.4956	.0044	3.19	.4993	.0007
2.06	.4803	.0197	2.63	.4957	.0043	3.20	.4993	.0007
2.07	.4808	.0192	2.64	.4959	.0041	3.21	.4993	.0007
2.08	.4812	.0188	2.65	.4960	.0040	3.22	.4994	.0006
2.09	.4817	.0183	2.66	.4961	.0039	3.23	.4994	.0006
2.10	.4821	.0179	2.67	.4962	.0038	3.24	.4994	.0006
2.11	.4826	.0174	2.68	.4963	.0037	3.25	.4994	.0006
2.12	.4830	.0170	2.69	.4964	.0036	3.30	.4995	.0005
2.13	.4834	.0166	2.70	.4965	.0035	3.35	.4996	.0004
2.14	.4838	.0162	2.71	.4966	.0034	3.40	.4997	.0003
2.15	.4842	.0158	2.72	.4967	.0033	3.45	.4997	.0003
2.16	.4846	.0154	2.73	.4968	.0032	3.50	.4998	.0002
2.17	.4850	.0150	2.74	.4969	.0031	3.60	.4998	.0002
2.18	.4854	.0146	2.75	.4970	.0030	3.70	.4999	.0001
2.19	.4857	.0143	2.76	.4971	.0029	3.80	.4999	.0001
2.20	.4861	.0139	2.77	.4972	.0028	3.90	.49995	.00005
2.21	.4864	.0136	2.78	.4973	.0027	4.00	.49997	.00003

TABLE E
Table for F_{max} percentage points of the ratio s^2_{max}/s^2_{min}—upper .05 points*

df \ k	2	3	4	5	6	7	8	9	10	11	12
2	39.0	87.5	142	202	266	333	403	475	550	626	704
3	15.4	27.8	39.2	50.7	62.0	72.9	83.5	93.9	104	114	124
4	9.60	15.5	20.6	25.2	29.5	33.6	37.5	41.1	44.6	48.0	51.4
5	7.15	10.8	13.7	16.3	18.7	20.8	22.9	24.7	26.5	28.2	29.9
6	5.82	8.38	10.4	12.1	13.7	15.0	16.3	17.5	18.6	19.7	20.7
7	4.99	6.94	8.44	9.70	10.8	11.8	12.7	13.5	14.3	15.1	15.8
8	4.43	6.00	7.18	8.12	9.03	9.78	10.5	11.1	11.7	12.2	12.7
9	4.03	5.34	6.31	7.11	7.80	8.41	8.95	9.45	9.91	10.3	10.7
10	3.72	4.85	5.67	6.34	6.92	7.42	7.87	8.28	8.66	9.01	9.34
12	3.28	4.16	4.79	5.30	5.72	6.09	6.42	6.72	7.00	7.25	7.48
15	2.86	3.54	4.01	4.37	4.68	4.95	5.19	5.40	5.59	5.77	5.93
20	2.46	2.95	3.29	3.54	3.76	3.94	4.10	4.24	4.37	4.49	4.59
30	2.07	2.40	2.61	2.78	2.91	3.02	3.12	3.21	3.29	3.36	3.39
60	1.67	1.85	1.96	2.04	2.11	2.17	2.22	2.26	2.30	2.33	2.36
∞	1.00	1.00	1.00	1.00	1.00	1.00	1.00	1.00	1.00	1.00	1.00

Upper .01 points

df \ k	2	3	4	5	6	7	8	9	10	11	12
2	199	448	729	1036	1362	1705	2063	2432	2813	3204	3605
3	47.5	85	120	151	184	21(6)	24(9)	28(1)	31(0)	33(7)	36(1)
4	23.2	37	49	59	69	79	89	97	106	113	120
5	14.9	22	28	33	38	42	46	50	54	57	60
6	11.1	15.5	19.1	22	25	27	30	32	34	36	37
7	8.89	12.1	14.5	16.5	18.4	20	22	23	24	26	27
8	7.50	9.9	11.7	13.2	14.5	15.8	16.9	17.9	18.9	19.8	21
9	6.54	8.5	9.9	11.1	12.1	13.1	13.9	14.7	15.3	16.0	16.6
10	5.85	7.4	8.6	9.6	10.4	11.1	11.8	12.4	12.9	13.4	13.9
12	4.91	6.1	6.9	7.6	8.2	8.7	9.1	9.5	9.9	10.2	10.6
15	4.07	4.9	5.5	6.0	6.4	6.7	7.1	7.3	7.5	7.8	8.0
20	3.32	3.8	4.3	4.6	4.9	5.1	5.3	5.5	5.6	5.8	5.9
30	2.63	3.0	3.3	3.4	3.6	3.7	3.8	3.9	4.0	4.1	4.2
60	1.96	2.2	2.3	2.4	2.4	2.5	2.5	2.6	2.6	2.7	2.7
∞	1.00	1.0	1.0	1.0	1.0	1.0	1.0	1.0	1.0	1.0	1.0

s^2_{max} is the largest and s^2_{min} the smallest in a set of k independent mean squares, each based on df degrees of freedom. Values in the column $k = 2$ and in the rows $df = 2$ and ∞ are exact. Elsewhere the third digit may be in error by a few units for the 5% points and several units for the 1% points. The third digit figures in brackets for $df = 3$ are the most uncertain.

*Reprinted, with changes in notation, by permission of the publisher from E. S. Pearson and H. O. Hartley, *Biometrika Tables for Statisticians*, Volume 1, p. 202. Copyright 1966, Biometrika.

TABLE F
Table of F of BETWEEN

Upper .10 points

df_2 \ df_1	1	2	3	4	5	6	7	8	9	10	12	15	20	24	30	40	60	120	∞
1	39.86	49.50	53.59	55.83	57.24	58.20	58.91	59.44	59.86	60.19	60.71	61.22	61.74	62.00	62.26	62.53	62.79	63.06	63.33
2	8.53	9.00	9.16	9.24	9.29	9.33	9.35	9.37	9.38	9.39	9.41	9.42	9.44	9.45	9.46	9.47	9.47	9.48	9.49
3	5.54	5.46	5.39	5.34	5.31	5.28	5.27	5.25	5.24	5.23	5.22	5.20	5.18	5.18	5.17	5.16	5.15	5.14	5.13
4	4.54	4.32	4.19	4.11	4.05	4.01	3.98	3.95	3.94	3.92	3.90	3.87	3.84	3.83	3.82	3.80	3.79	3.78	3.76
5	4.06	3.78	3.62	3.52	3.45	3.40	3.37	3.34	3.32	3.30	3.27	3.24	3.21	3.19	3.17	3.16	3.14	3.12	3.10
6	3.78	3.46	3.29	3.18	3.11	3.05	3.01	2.98	2.96	2.94	2.90	2.87	2.84	2.82	2.80	2.78	2.76	2.74	2.72
7	3.59	3.26	3.07	2.96	2.88	2.83	2.78	2.75	2.72	2.70	2.67	2.63	2.59	2.58	2.56	2.54	2.51	2.49	2.47
8	3.46	3.11	2.92	2.81	2.73	2.67	2.62	2.59	2.56	2.54	2.50	2.46	2.42	2.40	2.38	2.36	2.34	2.32	2.29
9	3.36	3.01	2.81	2.69	2.61	2.55	2.51	2.47	2.44	2.42	2.38	2.34	2.30	2.28	2.25	2.23	2.21	2.18	2.16
10	3.29	2.92	2.73	2.61	2.52	2.46	2.41	2.38	2.35	2.32	2.28	2.24	2.20	2.18	2.16	2.13	2.11	2.08	2.06
11	3.23	2.86	2.66	2.54	2.45	2.39	2.34	2.30	2.27	2.25	2.21	2.17	2.12	2.10	2.08	2.05	2.03	2.00	1.97
12	3.18	2.81	2.61	2.48	2.39	2.33	2.28	2.24	2.21	2.19	2.15	2.10	2.06	2.04	2.01	1.99	1.96	1.93	1.90
13	3.14	2.76	2.56	2.43	2.35	2.28	2.23	2.20	2.16	2.14	2.10	2.05	2.01	1.98	1.96	1.93	1.90	1.88	1.85
14	3.10	2.73	2.52	2.39	2.31	2.24	2.19	2.15	2.12	2.10	2.05	2.01	1.96	1.94	1.91	1.89	1.86	1.83	1.80
15	3.07	2.70	2.49	2.36	2.27	2.21	2.16	2.12	2.09	2.06	2.02	1.97	1.92	1.90	1.87	1.85	1.82	1.79	1.76
16	3.05	2.67	2.46	2.33	2.24	2.18	2.13	2.09	2.06	2.03	1.99	1.94	1.89	1.87	1.84	1.81	1.78	1.75	1.72
17	3.03	2.64	2.44	2.31	2.22	2.15	2.10	2.06	2.03	2.00	1.96	1.91	1.86	1.84	1.81	1.78	1.75	1.72	1.69
18	3.01	2.62	2.42	2.29	2.20	2.13	2.08	2.04	2.00	1.98	1.93	1.89	1.84	1.81	1.78	1.75	1.72	1.69	1.66
19	2.99	2.61	2.40	2.27	2.18	2.11	2.06	2.02	1.98	1.96	1.91	1.86	1.81	1.79	1.76	1.73	1.70	1.67	1.63
20	2.97	2.59	2.38	2.25	2.16	2.09	2.04	2.00	1.96	1.94	1.89	1.84	1.79	1.77	1.74	1.71	1.68	1.64	1.61
21	2.96	2.57	2.36	2.23	2.14	2.08	2.02	1.98	1.95	1.92	1.87	1.83	1.78	1.75	1.72	1.69	1.66	1.62	1.59
22	2.95	2.56	2.35	2.22	2.13	2.06	2.01	1.97	1.93	1.90	1.86	1.81	1.76	1.73	1.70	1.67	1.64	1.60	1.57
23	2.94	2.55	2.34	2.21	2.11	2.05	1.99	1.95	1.92	1.89	1.84	1.80	1.74	1.72	1.69	1.66	1.62	1.59	1.55
24	2.93	2.54	2.33	2.19	2.10	2.04	1.98	1.94	1.91	1.88	1.83	1.78	1.73	1.70	1.67	1.64	1.61	1.57	1.53
25	2.92	2.53	2.32	2.18	2.09	2.02	1.97	1.93	1.89	1.87	1.82	1.77	1.72	1.69	1.66	1.63	1.59	1.56	1.52
26	2.91	2.52	2.31	2.17	2.08	2.01	1.96	1.92	1.88	1.86	1.81	1.76	1.71	1.68	1.65	1.61	1.58	1.54	1.50
27	2.90	2.51	2.30	2.17	2.07	2.00	1.95	1.91	1.87	1.85	1.80	1.75	1.70	1.67	1.64	1.60	1.57	1.53	1.49
28	2.89	2.50	2.29	2.16	2.06	2.00	1.94	1.90	1.87	1.84	1.79	1.74	1.69	1.66	1.63	1.59	1.56	1.52	1.48
29	2.89	2.50	2.28	2.15	2.06	1.99	1.93	1.89	1.86	1.83	1.78	1.73	1.68	1.65	1.62	1.58	1.55	1.51	1.47
30	2.88	2.49	2.28	2.14	2.05	1.98	1.93	1.88	1.85	1.82	1.77	1.72	1.67	1.64	1.61	1.57	1.54	1.50	1.46
40	2.84	2.44	2.23	2.09	2.00	1.93	1.87	1.83	1.79	1.76	1.71	1.66	1.61	1.57	1.54	1.51	1.47	1.42	1.38
60	2.79	2.39	2.18	2.04	1.95	1.87	1.82	1.77	1.74	1.71	1.66	1.60	1.54	1.51	1.48	1.44	1.40	1.35	1.29
120	2.75	2.35	2.13	1.99	1.90	1.82	1.77	1.72	1.68	1.65	1.60	1.55	1.48	1.45	1.41	1.37	1.32	1.26	1.19
∞	2.71	2.30	2.08	1.94	1.85	1.77	1.72	1.67	1.63	1.60	1.55	1.49	1.42	1.38	1.34	1.30	1.24	1.17	1.00

Reprinted, with changes in notation, by permission of the publisher from E. S. Pearson and H. O. Hartley, *Biometrika Tables for Statisticians*, Vol. 1, pp 170–175. Copyright 1966, Biometrika.

TABLE F
Table of F (continued)

Upper .05 points

df_2 \ df_1	1	2	3	4	5	6	7	8	9	10	12	15	20	24	30	40	60	120	∞
1	161.4	199.5	215.7	224.6	230.2	234.0	236.8	238.9	240.5	241.9	243.9	245.9	248.0	249.1	250.1	251.1	252.2	253.3	254.3
2	18.51	19.00	19.16	19.25	19.30	19.33	19.35	19.37	19.38	19.40	19.41	19.43	19.45	19.45	19.46	19.47	19.48	19.49	19.50
3	10.13	9.55	9.28	9.12	9.01	8.94	8.89	8.85	8.81	8.79	8.74	8.70	8.66	8.64	8.62	8.59	8.57	8.55	8.53
4	7.71	6.94	6.59	6.39	6.26	6.16	6.09	6.04	6.00	5.96	5.91	5.86	5.80	5.77	5.75	5.72	5.69	5.66	5.63
5	6.61	5.79	5.41	5.19	5.05	4.95	4.88	4.82	4.77	4.74	4.68	4.62	4.56	4.53	4.50	4.46	4.43	4.40	4.36
6	5.99	5.14	4.76	4.53	4.39	4.28	4.21	4.15	4.10	4.06	4.00	3.94	3.87	3.84	3.81	3.77	3.74	3.70	3.67
7	5.59	4.74	4.35	4.12	3.97	3.87	3.79	3.73	3.68	3.64	3.57	3.51	3.44	3.41	3.38	3.34	3.30	3.27	3.23
8	5.32	4.46	4.07	3.84	3.69	3.58	3.50	3.44	3.39	3.35	3.28	3.22	3.15	3.12	3.08	3.04	3.01	2.97	2.93
9	5.12	4.26	3.86	3.63	3.48	3.37	3.29	3.23	3.18	3.14	3.07	3.01	2.94	2.90	2.86	2.83	2.79	2.75	2.71
10	4.96	4.10	3.71	3.48	3.33	3.22	3.14	3.07	3.02	2.98	2.91	2.85	2.77	2.74	2.70	2.66	2.62	2.58	2.54
11	4.84	3.98	3.59	3.36	3.20	3.09	3.01	2.95	2.90	2.85	2.79	2.72	2.65	2.61	2.57	2.53	2.49	2.45	2.40
12	4.75	3.89	3.49	3.26	3.11	3.00	2.91	2.85	2.80	2.75	2.69	2.62	2.54	2.51	2.47	2.43	2.38	2.34	2.30
13	4.67	3.81	3.41	3.18	3.03	2.92	2.83	2.77	2.71	2.67	2.60	2.53	2.46	2.42	2.38	2.34	2.30	2.25	2.21
14	4.60	3.74	3.34	3.11	2.96	2.85	2.76	2.70	2.65	2.60	2.53	2.46	2.39	2.35	2.31	2.27	2.22	2.18	2.13
15	4.54	3.68	3.29	3.06	2.90	2.79	2.71	2.64	2.59	2.54	2.48	2.40	2.33	2.29	2.25	2.20	2.16	2.11	2.07
16	4.49	3.63	3.24	3.01	2.85	2.74	2.66	2.59	2.54	2.49	2.42	2.35	2.28	2.24	2.19	2.15	2.11	2.06	2.01
17	4.45	3.59	3.20	2.96	2.81	2.70	2.61	2.55	2.49	2.45	2.38	2.31	2.23	2.19	2.15	2.10	2.06	2.01	1.96
18	4.41	3.55	3.16	2.93	2.77	2.66	2.58	2.51	2.46	2.41	2.34	2.27	2.19	2.15	2.11	2.06	2.02	1.97	1.92
19	4.38	3.52	3.13	2.90	2.74	2.63	2.54	2.48	2.42	2.38	2.31	2.23	2.16	2.11	2.07	2.03	1.98	1.93	1.88
20	4.35	3.49	3.10	2.87	2.71	2.60	2.51	2.45	2.39	2.35	2.28	2.20	2.12	2.08	2.04	1.99	1.95	1.90	1.84
21	4.32	3.47	3.07	2.84	2.68	2.57	2.49	2.42	2.37	2.32	2.25	2.18	2.10	2.05	2.01	1.96	1.92	1.87	1.81
22	4.30	3.44	3.05	2.82	2.66	2.55	2.46	2.40	2.34	2.30	2.23	2.15	2.07	2.03	1.98	1.94	1.89	1.84	1.78
23	4.28	3.42	3.03	2.80	2.64	2.53	2.44	2.37	2.32	2.27	2.20	2.13	2.05	2.01	1.96	1.91	1.86	1.81	1.76
24	4.26	3.40	3.01	2.78	2.62	2.51	2.42	2.36	2.30	2.25	2.18	2.11	2.03	1.98	1.94	1.89	1.84	1.79	1.73
25	4.24	3.39	2.99	2.76	2.60	2.49	2.40	2.34	2.28	2.24	2.16	2.09	2.01	1.96	1.92	1.87	1.82	1.77	1.71
26	4.23	3.37	2.98	2.74	2.59	2.47	2.39	2.32	2.27	2.22	2.15	2.07	1.99	1.95	1.90	1.85	1.80	1.75	1.69
27	4.21	3.35	2.96	2.73	2.57	2.46	2.37	2.31	2.25	2.20	2.13	2.06	1.97	1.93	1.88	1.84	1.79	1.73	1.67
28	4.20	3.34	2.95	2.71	2.56	2.45	2.36	2.29	2.24	2.19	2.12	2.04	1.96	1.91	1.87	1.82	1.77	1.71	1.65
29	4.18	3.33	2.93	2.70	2.55	2.43	2.35	2.28	2.22	2.18	2.10	2.03	1.94	1.90	1.85	1.81	1.75	1.70	1.64
30	4.17	3.32	2.92	2.69	2.53	2.42	2.33	2.27	2.21	2.16	2.09	2.01	1.93	1.89	1.84	1.79	1.74	1.68	1.62
40	4.08	3.23	2.84	2.61	2.45	2.34	2.25	2.18	2.12	2.08	2.00	1.92	1.84	1.79	1.74	1.69	1.64	1.58	1.51
60	4.00	3.15	2.76	2.53	2.37	2.25	2.17	2.10	2.04	1.99	1.92	1.84	1.75	1.70	1.65	1.59	1.53	1.47	1.39
120	3.92	3.07	2.68	2.45	2.29	2.17	2.09	2.02	1.96	1.91	1.83	1.75	1.66	1.61	1.55	1.50	1.43	1.35	1.25
∞	3.84	3.00	2.60	2.37	2.21	2.10	2.01	1.94	1.88	1.83	1.75	1.67	1.57	1.52	1.46	1.39	1.32	1.22	1.00

TABLE F
Table of F (continued)

Upper .025 points

df_2 \ df_1	1	2	3	4	5	6	7	8	9	10	12	15	20	24	30	40	60	120	∞
1	647.8	799.5	864.2	899.6	921.8	937.1	948.2	956.7	963.3	968.6	976.7	984.9	993.1	997.2	1001	1006	1010	1014	1018
2	38.51	39.00	39.17	39.25	39.30	39.33	39.36	39.37	39.39	39.40	39.41	39.43	39.45	39.46	39.46	39.47	39.48	39.49	39.50
3	17.44	16.04	15.44	15.10	14.88	14.73	14.62	14.54	14.47	14.42	14.34	14.25	14.17	14.12	14.08	14.04	13.99	13.95	13.90
4	12.22	10.65	9.98	9.60	9.36	9.20	9.07	8.98	8.90	8.84	8.75	8.66	8.56	8.51	8.46	8.41	8.36	8.31	8.26
5	10.01	8.43	7.76	7.39	7.15	6.98	6.85	6.76	6.68	6.62	6.52	6.43	6.33	6.28	6.23	6.18	6.12	6.07	6.02
6	8.81	7.26	6.60	6.23	5.99	5.82	5.70	5.60	5.52	5.46	5.37	5.27	5.17	5.12	5.07	5.01	4.96	4.90	4.85
7	8.07	6.54	5.89	5.52	5.29	5.12	4.99	4.90	4.82	4.76	4.67	4.57	4.47	4.42	4.36	4.31	4.25	4.20	4.14
8	7.57	6.06	5.42	5.05	4.82	4.65	4.53	4.43	4.36	4.30	4.20	4.10	4.00	3.95	3.89	3.84	3.78	3.73	3.67
9	7.21	5.71	5.08	4.72	4.48	4.32	4.20	4.10	4.03	3.96	3.87	3.77	3.67	3.61	3.56	3.51	3.45	3.39	3.33
10	6.94	5.46	4.83	4.47	4.24	4.07	3.95	3.85	3.78	3.72	3.62	3.52	3.42	3.37	3.31	3.26	3.20	3.14	3.08
11	6.72	5.26	4.63	4.28	4.04	3.88	3.76	3.66	3.59	3.53	3.43	3.33	3.23	3.17	3.12	3.06	3.00	2.94	2.88
12	6.55	5.10	4.47	4.12	3.89	3.73	3.61	3.51	3.44	3.37	3.28	3.18	3.07	3.02	2.96	2.91	2.85	2.79	2.72
13	6.41	4.97	4.35	4.00	3.77	3.60	3.48	3.39	3.31	3.25	3.15	3.05	2.95	2.89	2.84	2.78	2.72	2.66	2.60
14	6.30	4.86	4.24	3.89	3.66	3.50	3.38	3.29	3.21	3.15	3.05	2.95	2.84	2.79	2.73	2.67	2.61	2.55	2.49
15	6.20	4.77	4.15	3.80	3.58	3.41	3.29	3.20	3.12	3.06	2.96	2.86	2.76	2.70	2.64	2.59	2.52	2.46	2.40
16	6.12	4.69	4.08	3.73	3.50	3.34	3.22	3.12	3.05	2.99	2.89	2.79	2.68	2.63	2.57	2.51	2.45	2.38	2.32
17	6.04	4.62	4.01	3.66	3.44	3.28	3.16	3.06	2.98	2.92	2.82	2.72	2.62	2.56	2.50	2.44	2.38	2.32	2.25
18	5.98	4.56	3.95	3.61	3.38	3.22	3.10	3.01	2.93	2.87	2.77	2.67	2.56	2.50	2.44	2.38	2.32	2.26	2.19
19	5.92	4.51	3.90	3.56	3.33	3.17	3.05	2.96	2.88	2.82	2.72	2.62	2.51	2.45	2.39	2.33	2.27	2.20	2.13
20	5.87	4.46	3.86	3.51	3.29	3.13	3.01	2.91	2.84	2.77	2.68	2.57	2.46	2.41	2.35	2.29	2.22	2.16	2.09
21	5.83	4.42	3.82	3.48	3.25	3.09	2.97	2.87	2.80	2.73	2.64	2.53	2.42	2.37	2.31	2.25	2.18	2.11	2.04
22	5.79	4.38	3.78	3.44	3.22	3.05	2.93	2.84	2.76	2.70	2.60	2.50	2.39	2.33	2.27	2.21	2.14	2.08	2.00
23	5.75	4.35	3.75	3.41	3.18	3.02	2.90	2.81	2.73	2.67	2.57	2.47	2.36	2.30	2.24	2.18	2.11	2.04	1.97
24	5.72	4.32	3.72	3.38	3.15	2.99	2.87	2.78	2.70	2.64	2.54	2.44	2.33	2.27	2.21	2.15	2.08	2.01	1.94
25	5.69	4.29	3.69	3.35	3.13	2.97	2.85	2.75	2.68	2.61	2.51	2.41	2.30	2.24	2.18	2.12	2.05	1.98	1.91
26	5.66	4.27	3.67	3.33	3.10	2.94	2.82	2.73	2.65	2.59	2.49	2.39	2.28	2.22	2.16	2.09	2.03	1.95	1.88
27	5.63	4.24	3.65	3.31	3.08	2.92	2.80	2.71	2.63	2.57	2.47	2.36	2.25	2.19	2.13	2.07	2.00	1.93	1.85
28	5.61	4.22	3.63	3.29	3.06	2.90	2.78	2.69	2.61	2.55	2.45	2.34	2.23	2.17	2.11	2.05	1.98	1.91	1.83
29	5.59	4.20	3.61	3.27	3.04	2.88	2.76	2.67	2.59	2.53	2.43	2.32	2.21	2.15	2.09	2.03	1.96	1.89	1.81
30	5.57	4.18	3.59	3.25	3.03	2.87	2.75	2.65	2.57	2.51	2.41	2.31	2.20	2.14	2.07	2.01	1.94	1.87	1.79
40	5.42	4.05	3.46	3.13	2.90	2.74	2.62	2.53	2.45	2.39	2.29	2.18	2.07	2.01	1.94	1.88	1.80	1.72	1.64
60	5.29	3.93	3.34	3.01	2.79	2.63	2.51	2.41	2.33	2.27	2.17	2.06	1.94	1.88	1.82	1.74	1.67	1.58	1.48
120	5.15	3.80	3.23	2.89	2.67	2.52	2.39	2.30	2.22	2.16	2.05	1.94	1.82	1.76	1.69	1.61	1.53	1.43	1.31
∞	5.02	3.69	3.12	2.79	2.57	2.41	2.29	2.19	2.11	2.05	1.94	1.83	1.71	1.64	1.57	1.48	1.39	1.27	1.00

TABLE F
Table of F (continued)

Upper .01 points

df_2 \ df_1	1	2	3	4	5	6	7	8	9	10	12	15	20	24	30	40	60	120	∞
1	4052	4999·5	5403	5625	5764	5859	5928	5981	6022	6056	6106	6157	6209	6235	6261	6287	6313	6339	6366
2	98·50	99·00	99·17	99·25	99·30	99·33	99·36	99·37	99·39	99·40	99·42	99·43	99·45	99·46	99·47	99·47	99·48	99·49	99·50
3	34·12	30·82	29·46	28·71	28·24	27·91	27·67	27·49	27·35	27·23	27·05	26·87	26·69	26·60	26·50	26·41	26·32	26·22	26·13
4	21·20	18·00	16·69	15·98	15·52	15·21	14·98	14·80	14·66	14·55	14·37	14·20	14·02	13·93	13·84	13·75	13·65	13·56	13·46
5	16·26	13·27	12·06	11·39	10·97	10·67	10·46	10·29	10·16	10·05	9·89	9·72	9·55	9·47	9·38	9·29	9·20	9·11	9·02
6	13·75	10·92	9·78	9·15	8·75	8·47	8·26	8·10	7·98	7·87	7·72	7·56	7·40	7·31	7·23	7·14	7·06	6·97	6·88
7	12·25	9·55	8·45	7·85	7·46	7·19	6·99	6·84	6·72	6·62	6·47	6·31	6·16	6·07	5·99	5·91	5·82	5·74	5·65
8	11·26	8·65	7·59	7·01	6·63	6·37	6·18	6·03	5·91	5·81	5·67	5·52	5·36	5·28	5·20	5·12	5·03	4·95	4·86
9	10·56	8·02	6·99	6·42	6·06	5·80	5·61	5·47	5·35	5·26	5·11	4·96	4·81	4·73	4·65	4·57	4·48	4·40	4·31
10	10·04	7·56	6·55	5·99	5·64	5·39	5·20	5·06	4·94	4·85	4·71	4·56	4·41	4·33	4·25	4·17	4·08	4·00	3·91
11	9·65	7·21	6·22	5·67	5·32	5·07	4·89	4·74	4·63	4·54	4·40	4·25	4·10	4·02	3·94	3·86	3·78	3·69	3·60
12	9·33	6·93	5·95	5·41	5·06	4·82	4·64	4·50	4·39	4·30	4·16	4·01	3·86	3·78	3·70	3·62	3·54	3·45	3·36
13	9·07	6·70	5·74	5·21	4·86	4·62	4·44	4·30	4·19	4·10	3·96	3·82	3·66	3·59	3·51	3·43	3·34	3·25	3·17
14	8·86	6·51	5·56	5·04	4·69	4·46	4·28	4·14	4·03	3·94	3·80	3·66	3·51	3·43	3·35	3·27	3·18	3·09	3·00
15	8·68	6·36	5·42	4·89	4·56	4·32	4·14	4·00	3·89	3·80	3·67	3·52	3·37	3·29	3·21	3·13	3·05	2·96	2·87
16	8·53	6·23	5·29	4·77	4·44	4·20	4·03	3·89	3·78	3·69	3·55	3·41	3·26	3·18	3·10	3·02	2·93	2·84	2·75
17	8·40	6·11	5·18	4·67	4·34	4·10	3·93	3·79	3·68	3·59	3·46	3·31	3·16	3·08	3·00	2·92	2·83	2·75	2·65
18	8·29	6·01	5·09	4·58	4·25	4·01	3·84	3·71	3·60	3·51	3·37	3·23	3·08	3·00	2·92	2·84	2·75	2·66	2·57
19	8·18	5·93	5·01	4·50	4·17	3·94	3·77	3·63	3·52	3·43	3·30	3·15	3·00	2·92	2·84	2·76	2·67	2·58	2·49
20	8·10	5·85	4·94	4·43	4·10	3·87	3·70	3·56	3·46	3·37	3·23	3·09	2·94	2·86	2·78	2·69	2·61	2·52	2·42
21	8·02	5·78	4·87	4·37	4·04	3·81	3·64	3·51	3·40	3·31	3·17	3·03	2·88	2·80	2·72	2·64	2·55	2·46	2·36
22	7·95	5·72	4·82	4·31	3·99	3·76	3·59	3·45	3·35	3·26	3·12	2·98	2·83	2·75	2·67	2·58	2·50	2·40	2·31
23	7·88	5·66	4·76	4·26	3·94	3·71	3·54	3·41	3·30	3·21	3·07	2·93	2·78	2·70	2·62	2·54	2·45	2·35	2·26
24	7·82	5·61	4·72	4·22	3·90	3·67	3·50	3·36	3·26	3·17	3·03	2·89	2·74	2·66	2·58	2·49	2·40	2·31	2·21
25	7·77	5·57	4·68	4·18	3·85	3·63	3·46	3·32	3·22	3·13	2·99	2·85	2·70	2·62	2·54	2·45	2·36	2·27	2·17
26	7·72	5·53	4·64	4·14	3·82	3·59	3·42	3·29	3·18	3·09	2·96	2·81	2·66	2·58	2·50	2·42	2·33	2·23	2·13
27	7·68	5·49	4·60	4·11	3·78	3·56	3·39	3·26	3·15	3·06	2·93	2·78	2·63	2·55	2·47	2·38	2·29	2·20	2·10
28	7·64	5·45	4·57	4·07	3·75	3·53	3·36	3·23	3·12	3·03	2·90	2·75	2·60	2·52	2·44	2·35	2·26	2·17	2·06
29	7·60	5·42	4·54	4·04	3·73	3·50	3·33	3·20	3·09	3·00	2·87	2·73	2·57	2·49	2·41	2·33	2·23	2·14	2·03
30	7·56	5·39	4·51	4·02	3·70	3·47	3·30	3·17	3·07	2·98	2·84	2·70	2·55	2·47	2·39	2·30	2·21	2·11	2·01
40	7·31	5·18	4·31	3·83	3·51	3·29	3·12	2·99	2·89	2·80	2·66	2·52	2·37	2·29	2·20	2·11	2·02	1·92	1·80
60	7·08	4·98	4·13	3·65	3·34	3·12	2·95	2·82	2·72	2·63	2·50	2·35	2·20	2·12	2·03	1·94	1·84	1·73	1·60
120	6·85	4·79	3·95	3·48	3·17	2·96	2·79	2·66	2·56	2·47	2·34	2·19	2·03	1·95	1·86	1·76	1·66	1·53	1·38
∞	6·63	4·61	3·78	3·32	3·02	2·80	2·64	2·51	2·41	2·32	2·18	2·04	1·88	1·79	1·70	1·59	1·47	1·32	1·00

TABLE F (continued)
Table of F

Upper .005 points

df_2 \ df_1	1	2	3	4	5	6	7	8	9	10	12	15	20	24	30	40	60	120	∞
1	16211	20000	21615	22500	23056	23437	23715	23925	24091	24224	24426	24630	24836	24940	25044	25148	25253	25359	25465
2	198.5	199.0	199.2	199.2	199.3	199.3	199.4	199.4	199.4	199.4	199.4	199.4	199.4	199.5	199.5	199.5	199.5	199.5	199.5
3	55.55	49.80	47.47	46.19	45.39	44.84	44.43	44.13	43.88	43.69	43.39	43.08	42.78	42.62	42.47	42.31	42.15	41.99	41.83
4	31.33	26.28	24.26	23.15	22.46	21.97	21.62	21.35	21.14	20.97	20.70	20.44	20.17	20.03	19.89	19.75	19.61	19.47	19.32
5	22.78	18.31	16.53	15.56	14.94	14.51	14.20	13.96	13.77	13.62	13.38	13.15	12.90	12.78	12.66	12.53	12.40	12.27	12.14
6	18.63	14.54	12.92	12.03	11.46	11.07	10.79	10.57	10.39	10.25	10.03	9.81	9.59	9.47	9.36	9.24	9.12	9.00	8.88
7	16.24	12.40	10.88	10.05	9.52	9.16	8.89	8.68	8.51	8.38	8.18	7.97	7.75	7.65	7.53	7.42	7.31	7.19	7.08
8	14.69	11.04	9.60	8.81	8.30	7.95	7.69	7.50	7.34	7.21	7.01	6.81	6.61	6.50	6.40	6.29	6.18	6.06	5.95
9	13.61	10.11	8.72	7.96	7.47	7.13	6.88	6.69	6.54	6.42	6.23	6.03	5.83	5.73	5.62	5.52	5.41	5.30	5.19
10	12.83	9.43	8.08	7.34	6.87	6.54	6.30	6.12	5.97	5.85	5.66	5.47	5.27	5.17	5.07	4.97	4.86	4.75	4.64
11	12.23	8.91	7.60	6.88	6.42	6.10	5.86	5.68	5.54	5.42	5.24	5.05	4.86	4.76	4.65	4.55	4.44	4.34	4.23
12	11.75	8.51	7.23	6.52	6.07	5.76	5.52	5.35	5.20	5.09	4.91	4.72	4.53	4.43	4.33	4.23	4.12	4.01	3.90
13	11.37	8.19	6.93	6.23	5.79	5.48	5.25	5.08	4.94	4.82	4.64	4.46	4.27	4.17	4.07	3.97	3.87	3.76	3.65
14	11.06	7.92	6.68	6.00	5.56	5.26	5.03	4.86	4.72	4.60	4.43	4.25	4.06	3.96	3.86	3.76	3.66	3.55	3.44
15	10.80	7.70	6.48	5.80	5.37	5.07	4.85	4.67	4.54	4.42	4.25	4.07	3.88	3.79	3.69	3.58	3.48	3.37	3.26
16	10.58	7.51	6.30	5.64	5.21	4.91	4.69	4.52	4.38	4.27	4.10	3.92	3.73	3.64	3.54	3.44	3.33	3.22	3.11
17	10.38	7.35	6.16	5.50	5.07	4.78	4.56	4.39	4.25	4.14	3.97	3.79	3.61	3.51	3.41	3.31	3.21	3.10	2.98
18	10.22	7.21	6.03	5.37	4.96	4.66	4.44	4.28	4.14	4.03	3.86	3.68	3.50	3.40	3.30	3.20	3.10	2.99	2.87
19	10.07	7.09	5.92	5.27	4.85	4.56	4.34	4.18	4.04	3.93	3.76	3.59	3.40	3.31	3.21	3.11	3.00	2.89	2.78
20	9.94	6.99	5.82	5.17	4.76	4.47	4.26	4.09	3.96	3.85	3.68	3.50	3.32	3.22	3.12	3.02	2.92	2.81	2.69
21	9.83	6.89	5.73	5.09	4.68	4.39	4.18	4.01	3.88	3.77	3.60	3.43	3.24	3.15	3.05	2.95	2.84	2.73	2.61
22	9.73	6.81	5.65	5.02	4.61	4.32	4.11	3.94	3.81	3.70	3.54	3.36	3.18	3.08	2.98	2.88	2.77	2.66	2.55
23	9.63	6.73	5.58	4.95	4.54	4.26	4.05	3.88	3.75	3.64	3.47	3.30	3.12	3.02	2.92	2.82	2.71	2.60	2.48
24	9.55	6.66	5.52	4.89	4.49	4.20	3.99	3.83	3.69	3.59	3.42	3.25	3.06	2.97	2.87	2.77	2.66	2.55	2.43
25	9.48	6.60	5.46	4.84	4.43	4.15	3.94	3.78	3.64	3.54	3.37	3.20	3.01	2.92	2.82	2.72	2.61	2.50	2.38
26	9.41	6.54	5.41	4.79	4.38	4.10	3.89	3.73	3.60	3.49	3.33	3.15	2.97	2.87	2.77	2.67	2.56	2.45	2.33
27	9.34	6.49	5.36	4.74	4.34	4.06	3.85	3.69	3.56	3.45	3.28	3.11	2.93	2.83	2.73	2.63	2.52	2.41	2.29
28	9.28	6.44	5.32	4.70	4.30	4.02	3.81	3.65	3.52	3.41	3.25	3.07	2.89	2.79	2.69	2.59	2.48	2.37	2.25
29	9.23	6.40	5.28	4.66	4.26	3.98	3.77	3.61	3.48	3.38	3.21	3.04	2.86	2.76	2.66	2.56	2.45	2.33	2.21
30	9.18	6.35	5.24	4.62	4.23	3.95	3.74	3.58	3.45	3.34	3.18	3.01	2.82	2.73	2.63	2.52	2.42	2.30	2.18
40	8.83	6.07	4.98	4.37	3.99	3.71	3.51	3.35	3.22	3.12	2.95	2.78	2.60	2.50	2.40	2.30	2.18	2.06	1.93
60	8.49	5.79	4.73	4.14	3.76	3.49	3.29	3.13	3.01	2.90	2.74	2.57	2.39	2.29	2.19	2.08	1.96	1.83	1.69
120	8.18	5.54	4.50	3.92	3.55	3.28	3.09	2.93	2.81	2.71	2.54	2.37	2.19	2.09	1.98	1.87	1.75	1.61	1.43
∞	7.88	5.30	4.28	3.72	3.35	3.09	2.90	2.74	2.62	2.52	2.36	2.19	2.00	1.90	1.79	1.67	1.53	1.36	1.00

TABLE F
Table of F (continued)

Upper .001 points

df_2 \ df_1	1	2	3	4	5	6	7	8	9	10	12	15	20	24	30	40	60	120	∞
1	4053*	5000*	5404*	5625*	5764*	5859*	5929*	5981*	6023*	6056*	6107*	6158*	6209*	6235*	6261*	6287*	6313*	6340*	6366*
2	998.5	999.0	999.2	999.2	999.3	999.3	999.4	999.4	999.4	999.4	999.4	999.4	999.4	999.5	999.5	999.5	999.5	999.5	999.5
3	167.0	148.5	141.1	137.1	134.6	132.8	131.6	130.6	129.9	129.2	128.3	127.4	126.4	125.9	125.4	125.0	124.5	124.0	123.5
4	74.14	61.25	56.18	53.44	51.71	50.53	49.66	49.00	48.47	48.05	47.41	46.76	46.10	45.77	45.43	45.09	44.75	44.40	44.05
5	47.18	37.12	33.20	31.09	29.75	28.84	28.16	27.64	27.24	26.92	26.42	25.91	25.39	25.14	24.87	24.60	24.33	24.06	23.79
6	35.51	27.00	23.70	21.92	20.81	20.03	19.46	19.03	18.69	18.41	17.99	17.56	17.12	16.89	16.67	16.44	16.21	15.99	15.75
7	29.25	21.69	18.77	17.19	16.21	15.52	15.02	14.63	14.33	14.08	13.71	13.32	12.93	12.73	12.53	12.33	12.12	11.91	11.70
8	25.42	18.49	15.83	14.39	13.49	12.86	12.40	12.04	11.77	11.54	11.19	10.84	10.48	10.30	10.11	9.92	9.73	9.53	9.33
9	22.86	16.39	13.90	12.56	11.71	11.13	10.70	10.37	10.11	9.89	9.57	9.24	8.90	8.72	8.55	8.37	8.19	8.00	7.81
10	21.04	14.91	12.55	11.28	10.48	9.92	9.52	9.20	8.96	8.75	8.45	8.13	7.80	7.64	7.47	7.30	7.12	6.94	6.76
11	19.69	13.81	11.56	10.35	9.58	9.05	8.66	8.35	8.12	7.92	7.63	7.32	7.01	6.85	6.68	6.52	6.35	6.17	6.00
12	18.64	12.97	10.80	9.63	8.89	8.38	8.00	7.71	7.48	7.29	7.00	6.71	6.40	6.25	6.09	5.93	5.76	5.59	5.42
13	17.81	12.31	10.21	9.07	8.35	7.86	7.49	7.21	6.98	6.80	6.52	6.23	5.93	5.78	5.63	5.47	5.30	5.14	4.97
14	17.14	11.78	9.73	8.62	7.92	7.43	7.08	6.80	6.58	6.40	6.13	5.85	5.56	5.41	5.25	5.10	4.94	4.77	4.60
15	16.59	11.34	9.34	8.25	7.57	7.09	6.74	6.47	6.26	6.08	5.81	5.54	5.25	5.10	4.95	4.80	4.64	4.47	4.31
16	16.12	10.97	9.00	7.94	7.27	6.81	6.46	6.19	5.98	5.81	5.55	5.27	4.99	4.85	4.70	4.54	4.39	4.23	4.06
17	15.72	10.66	8.73	7.68	7.02	6.56	6.22	5.96	5.75	5.58	5.32	5.05	4.78	4.63	4.48	4.33	4.18	4.02	3.85
18	15.38	10.39	8.49	7.46	6.81	6.35	6.02	5.76	5.56	5.39	5.13	4.87	4.59	4.45	4.30	4.15	4.00	3.84	3.67
19	15.08	10.16	8.28	7.26	6.62	6.18	5.85	5.59	5.39	5.22	4.97	4.70	4.43	4.29	4.14	3.99	3.84	3.68	3.51
20	14.82	9.95	8.10	7.10	6.46	6.02	5.69	5.44	5.24	5.08	4.82	4.56	4.29	4.15	4.00	3.86	3.70	3.54	3.38
21	14.59	9.77	7.94	6.95	6.32	5.88	5.56	5.31	5.11	4.95	4.70	4.44	4.17	4.03	3.88	3.74	3.58	3.42	3.26
22	14.38	9.61	7.80	6.81	6.19	5.76	5.44	5.19	4.99	4.83	4.58	4.33	4.06	3.92	3.78	3.63	3.48	3.32	3.15
23	14.19	9.47	7.67	6.69	6.08	5.65	5.33	5.09	4.89	4.73	4.48	4.23	3.96	3.82	3.68	3.53	3.38	3.22	3.05
24	14.03	9.34	7.55	6.59	5.98	5.55	5.23	4.99	4.80	4.64	4.39	4.14	3.87	3.74	3.59	3.45	3.29	3.14	2.97
25	13.88	9.22	7.45	6.49	5.88	5.46	5.15	4.91	4.71	4.56	4.31	4.06	3.79	3.66	3.52	3.37	3.22	3.06	2.89
26	13.74	9.12	7.36	6.41	5.80	5.38	5.07	4.83	4.64	4.48	4.24	3.99	3.72	3.59	3.44	3.30	3.15	2.99	2.82
27	13.61	9.02	7.27	6.33	5.73	5.31	5.00	4.76	4.57	4.41	4.17	3.92	3.66	3.52	3.38	3.23	3.08	2.92	2.75
28	13.50	8.93	7.19	6.25	5.66	5.24	4.93	4.69	4.50	4.35	4.11	3.86	3.60	3.46	3.32	3.18	3.02	2.86	2.69
29	13.39	8.85	7.12	6.19	5.59	5.18	4.87	4.64	4.45	4.29	4.05	3.80	3.54	3.41	3.27	3.12	2.97	2.81	2.64
30	13.29	8.77	7.05	6.12	5.53	5.12	4.82	4.58	4.39	4.24	4.00	3.75	3.49	3.36	3.22	3.07	2.92	2.76	2.59
40	12.61	8.25	6.60	5.70	5.13	4.73	4.44	4.21	4.02	3.87	3.64	3.40	3.15	3.01	2.87	2.73	2.57	2.41	2.23
60	11.97	7.76	6.17	5.31	4.76	4.37	4.09	3.87	3.69	3.54	3.31	3.08	2.83	2.69	2.55	2.41	2.25	2.08	1.89
120	11.38	7.32	5.79	4.95	4.42	4.04	3.77	3.55	3.38	3.24	3.02	2.78	2.53	2.40	2.26	2.11	1.95	1.76	1.54
∞	10.83	6.91	5.42	4.62	4.10	3.74	3.47	3.27	3.10	2.96	2.74	2.51	2.27	2.13	1.99	1.84	1.66	1.45	1.00

* Multiply these entries by 100.

TABLE G*
Wilcoxon T values for paired replicates

Probabilities of a chance occurrence of a rank total of one sign, + or −, whichever is least, equal to, or less than T. T is given in the body of the table to the nearest whole number. N is the number of pairs.

N	.025 (one tail) .05 (two tail)	.01 .02	.005 .01
6	0	—	—
7	2	0	—
8	4	2	0
9	6	3	2
10	8	5	3
11	11	7	5
12	14	10	7
13	17	13	10
14	21	16	13
15	25	20	16
16	30	24	19
17	35	28	23
18	40	33	28
19	46	38	32
20	52	43	38
21	59	49	43
22	66	56	49
23	73	62	55
24	81	69	61
25	89	77	68

The values in the table were obtained by rounding off values given by Tukey in Memorandum Report 17, "The Simplest Signed Rank Tests," Statistical Research Group, Princeton University, 1949.

Source: From *Some rapid approximate statistical procedures*, Copyright © 1949, 1964, Lederle Laboratories Division of American Cyanamid Co., all rights reserved and reprinted with permission.

Probabilties of chance occurrence of a rank total equal to or less than T with N replicates per group. T is given in the body of the table to the nearest whole number.

N	.025 (one tail) .05 (two tail)	.01 .02	.005 .01
5	18	16	15
6	26	24	23
7	37	34	33
8	49	46	44
9	63	59	57
10	79	74	71
11	96	91	88
12	116	110	106
13	137	130	126
14	160	152	148
15	185	177	171
16	212	202	196
17	240	230	223
18	271	260	252
19	303	291	283
20	337	324	316

TABLE H*
Wilcoxon T values for unpaired replicates

Source: From *Some rapid approximate statistical procedures*, Copyright © 1949, 1964, Lederle Laboratories Division of American Cyanamid Co., all rights reserved and reprinted with permission.

TABLE I
Table for the Kruskal-Wallis H test

Sample sizes			H	p	Sample sizes			H	p
N_1	N_2	N_3			N_1	N_2	N_3		
2	1	1	2.7000	.500	4	3	2	6.4444	.008
								6.3000	.011
2	2	1	3.6000	.200				5.4444	.046
								5.4000	.051
2	2	2	4.5714	.067				4.5111	.098
			3.7143	.200				4.4444	.102
3	1	1	3.2000	.300					
					4	3	3	6.7455	.010
3	2	1	4.2857	.100				6.7091	.013
			3.8571	.133				5.7909	.046
								5.7273	.050
3	2	2	5.3572	.029				4.7091	.092
			4.7143	.048				4.7000	.101
			4.5000	.067					
			4.4643	.105	4	4	1	6.6667	.010
								6.1667	.022
3	3	1	5.1429	.043				4.9667	.048
			4.5714	.100				4.8667	.054
			4.0000	.129				4.1667	.082
								4.0667	.102
3	3	2	6.2500	.011					
			5.3611	.032	4	4	2	7.0364	.006
			5.1389	.061				6.8727	.011
			4.5556	.100				5.4545	.046
			4.2500	.121				5.2364	.052
								4.5545	.098
3	3	3	7.2000	.004				4.4455	.103
			6.4889	.011					
			5.6889	.029	4	4	3	7.1439	.010
			5.6000	.050				7.1364	.011
			5.0667	.086				5.5985	.049
			4.6222	.100				5.5758	.051
								4.5455	.099
4	1	1	3.5714	.200				4.4773	.102
4	2	1	4.8214	.057					
			4.5000	.076	4	4	4	7.6538	.008
			4.0179	.114				7.5385	.011
								5.6923	.049
4	2	2	6.0000	.014				5.6538	.054
			5.3333	.033				4.6539	.097
			5.1250	.052				4.5001	.104
			4.4583	.100					
			4.1667	.105	5	1	1	3.8571	.143
4	3	1	5.8333	.021	5	2	1	5.2500	.036
			5.2083	.050				5.0000	.048
			5.0000	.057				4.4500	.071
			4.0556	.093				4.2000	.095
			3.8889	.129				4.0500	.119

TABLE I
Table for the Kruskal-Wallis H test (continued)

Sample sizes			H	p	Sample sizes			H	p
N_1	N_2	N_3			N_1	N_2	N_3		
5	2	2	6.5333	.008				5.6308	.050
			6.1333	.013				4.5487	.099
			5.1600	.034				4.5231	.103
			5.0400	.056	5	4	4	7.7604	.009
			4.3733	.090				7.7440	.011
			4.2933	.122				5.6571	.049
5	3	1	6.4000	.012				5.6176	.050
			4.9600	.048				4.6187	.100
			4.8711	.052				4.5527	.102
			4.0178	.095	5	5	1	7.3091	.009
			3.8400	.123				6.8364	.011
5	3	2	6.9091	.009				5.1273	.046
			6.8218	.010				4.9091	.053
			5.2509	.049				4.1091	.086
			5.1055	.052				4.0364	.105
			4.6509	.091	5	5	2	7.3385	.010
			4.4945	.101				7.2692	.010
5	3	3	7.0788	.009				5.3385	.047
			6.9818	.011				5.2462	.051
			5.6485	.049				4.6231	.097
			5.5152	.051				4.5077	.100
			4.5333	.097	5	5	3	7.5780	.010
			4.4121	.109				7.5429	.010
5	4	1	6.9545	.008				5.7055	.046
			6.8400	.011				5.6264	.051
			4.9855	.044				4.5451	.100
			4.8600	.056				4.5363	.102
			3.9873	.098	5	5	4	7.8229	.010
			3.9600	.102				7.7914	.010
5	4	2	7.2045	.009				5.6657	.049
			7.1182	.010				5.6429	.050
			5.2727	.049				4.5229	.099
			5.2682	.050				4.5200	.101
			4.5409	.098	5	5	5	8.0000	.009
			4.5182	.101				7.9800	.010
5	4	3	7.4449	.010				5.7800	.049
			7.3949	.011				5.6600	.051
			5.6564	.049				4.5600	.100
								4.5000	.102

*Reprinted by permission of the American Statistical Association from W. H. Kruskal and W. A. Wallis. "Use of Ranks in One Criterion Variance Analysis," 447, pp. 614–617 and "Errata," 48, p. 910. *Journal of the American Statistical Association*, 1952, 1953.

TABLE J*
Table of critical values of chi-square

p df	0.250	0.100	0.050	0.025	0.010	0.005	0.001
1	1.32330	2.70554	3.84146	5.02389	6.63490	7.87944	10.828
2	2.77259	4.60517	5.99146	7.37776	9.21034	10.5966	13.816
3	4.10834	6.25139	7.81473	9.34840	11.3449	12.8382	16.266
4	5.38527	7.77944	9.48773	11.1433	13.2767	14.8603	18.467
5	6.62568	9.23636	11.0705	12.8325	15.0863	16.7496	20.515
6	7.84080	10.6446	12.5916	14.4494	16.8119	18.5476	22.458
7	9.03715	12.0170	14.0671	16.0128	18.4753	20.2777	24.322
8	10.2189	13.3616	15.5073	17.5345	20.0902	21.9550	26.125
9	11.3888	14.6837	16.9190	19.0228	21.6660	23.5894	27.877
10	12.5489	15.9872	18.3070	20.4832	23.2093	25.1882	29.588
11	13.7007	17.2750	19.6751	21.9200	24.7250	26.7568	31.264
12	14.8454	18.5493	21.0261	23.3367	26.2170	28.2995	32.909
13	15.9839	19.8119	22.3620	24.7356	27.6882	29.8195	34.528
14	17.1169	21.0641	23.6848	26.1189	29.1412	31.3194	36.123
15	18.2451	22.3071	24.9958	27.4884	30.5779	32.8013	37.697
16	19.3689	23.5418	26.2962	28.8454	31.9999	34.2672	39.252
17	20.4887	24.7690	27.5871	30.1910	33.4087	35.7185	40.790
18	21.6049	25.9894	28.8693	31.5264	34.8053	37.1565	42.312
19	22.7178	27.2036	30.1435	32.8523	36.1909	38.5823	43.820
20	23.8277	28.4120	31.4104	34.1696	37.5662	39.9968	45.315
21	24.9348	29.6151	32.6706	35.4789	38.9322	41.4011	46.797
22	26.0393	30.8133	33.9244	36.7807	40.2894	42.7957	48.268
23	27.1413	32.0069	35.1725	38.0756	41.6384	44.1813	49.728
24	28.2412	33.1962	36.4150	39.3641	42.9798	45.5585	51.179
25	29.3389	34.3816	37.6525	40.6465	44.3141	46.9279	52.618
26	30.4346	35.5632	38.8851	41.9232	45.6417	48.2899	54.052
27	31.5284	36.7412	40.1133	43.1945	46.9629	49.6449	55.476
28	32.6205	37.9159	41.3371	44.4608	48.2782	50.9934	56.892
29	33.7109	39.0875	42.5570	45.7223	49.5879	52.3356	58.301
30	34.7997	40.2560	43.7730	46.9792	50.8922	53.6720	59.703
40	45.6160	51.8051	55.7585	59.3417	63.6907	66.7660	73.402
50	56.3336	63.1671	67.5048	71.4202	76.1539	79.4900	86.661
60	66.9815	74.3970	79.0819	83.2977	88.3794	91.9517	99.607

*Reprinted, with changes in notation, by permission of the publisher from E. S. Pearson and H. O. Hartley, *Biometrika Tables for Statisticians*, Volume 1, p. 137. Copyright 1966, Biometrika.

TABLE K*

Critical χ_r^2 values for Friedman's test
$k = 3$ $N = 2, 3, 4, 5, 6, 7, 8, 9$

N = 2		N = 5		N = 7		N = 8		N = 9	
χ_r^2	p	χ_r^2	p	χ_r^2	p	χ_r^2	p	χ_r^2	p
0	1.000	.0	1.000	.000	1.000	.00	1.000	.000	1.000
1	.833	.4	.954	.286	.964	.25	.967	.222	.971
3	.500	1.2	.691	.857	.768	.75	.794	.667	.865
4	.167	1.6	.522	1.143	.620	1.00	.654	.889	.814
		2.8	.367	2.000	.486	1.75	.531	1.556	.569
		3.6	.182	2.571	.305	2.25	.355	2.000	.398
		4.8	.124	3.429	.237	3.00	.285	2.667	.328
		5.2	.093	3.714	.192	3.25	.236	2.889	.278
		6.4	.039	4.571	.112	4.00	.149	3.556	.187
		7.6	.024	5.429	.085	4.75	.120	4.222	.154
		8.4	.0085	6.000	.052	5.25	.079	4.667	.107
		10.0	.00077	7.143	.027	6.25	.047	5.556	.069
				7.714	.021	6.75	.038	6.000	.057
				8.000	.016	7.00	.030	6.222	.048
				8.857	.0084	7.75	.018	6.889	.031
				10.286	.0036	9.00	.0099	8.000	.019
				10.571	.0027	9.25	.0080	8.222	.016
				11.143	.0012	9.75	.0048	8.667	.010
				12.286	.00032	10.75	.0024	9.556	.0060
				14.000	.000021	12.00	.0011	10.667	.0035
						12.25	.00086	10.889	.0029
						13.00	.00026	11.556	.0013
						14.25	.000061	12.667	.00066
						16.00	.0000036	13.556	.00035
								14.000	.00020
								14.222	.000097
								14.889	.000054
								16.222	.000011
								18.000	.0000006

N = 3

χ_r^2	p
.000	1.000
.667	.944
2.000	.528
2.667	.361
4.667	.194
6.000	.028

N = 4

χ_r^2	p
.0	1.000
.5	.931
1.5	.653
2.0	.431
3.5	.273
4.5	.125
6.0	.069
6.5	.042
8.0	.0046

N = 6

χ_r^2	p
.00	1.000
.33	.956
1.00	.740
1.33	.570
2.33	.430
3.00	.252
4.00	.184
4.33	.142
5.33	.072
6.33	.052
7.00	.029
8.33	.012
9.00	.0081
9.33	.0055
10.33	.0017
12.00	.00013

TABLE K*
Critical χ_r^2 values for Friedman's test (*continued*)
$k = 4$ $N = 2, 3, 4$

$N = 2$		$N = 3$		$N = 4$			
χ_r^2	p	χ_r^2	p	χ_r^2	p	χ_r^2	p
.0	1.000	.2	1.000	.0	1.000	5.7	.141
.6	.958	.6	.958	.3	.992	6.0	.105
1.2	.834	1.0	.910	.6	.928	6.3	.094
1.8	.792	1.8	.727	.9	.900	6.6	.077
2.4	.625	2.2	.608	1.2	.800	6.9	.068
3.0	.542	2.6	.524	1.5	.754	7.2	.054
3.6	.458	3.4	.446	1.8	.677	7.5	.052
4.2	.375	3.8	.342	2.1	.649	7.8	.036
4.8	.208	4.2	.300	2.4	.524	8.1	.033
5.4	.167	5.0	.207	2.7	.508	8.4	.019
6.0	.042	5.4	.175	3.0	.432	8.7	.014
		5.8	.148	3.3	.389	9.3	.012
		6.6	.075	3.6	.355	9.6	.0069
		7.0	.054	3.9	.324	9.9	.0062
		7.4	.033	4.5	.242	10.2	.0027
		8.2	.017	4.8	.200	10.8	.0016
		9.0	.0017	5.1	.190	11.1	.00094
				5.4	.158	12.0	.000072

*Reprinted, with a change in notation, by permission of American Statistical Association from M. Friedman, "The Use of Ranks to Avoid the Assumption of Normality Implicit in the Analysis of Variance," *Journal of the American Statistical Association,* 32, pp. 688–689, 1937.

TABLE L
Power as a function of δ and significance criterion (α)

	One-tailed test (α)					One-tailed test (α)			
	.05	.025	.01	.005		.05	.025	.01	.005
	Two-tailed test (α)					Two-tailed test (α)			
δ	.10	.05	.02	.01	δ	.10	.05	.02	.01
0.0	.10*	.05*	.02	.01	2.5	.80	.71	.57	.47
0.1	.10*	.05*	.02	.01	2.6	.83	.74	.61	.51
0.2	.11*	.05	.02	.01	2.7	.85	.77	.65	.55
0.3	.12*	.06	.03	.01	2.8	.88	.80	.68	.59
0.4	.13*	.07	.03	.01	2.9	.90	.83	.72	.63
0.5	.14	.08	.03	.02	3.0	.91	.85	.75	.66
0.6	.16	.09	.04	.02	3.1	.93	.87	.78	.70
0.7	.18	.11	.05	.03	3.2	.94	.89	.81	.73
0.8	.21	.13	.06	.04	3.3	.96	.91	.83	.77
0.9	.23	.15	.08	.05	3.4	.96	.93	.86	.80
1.0	.26	.17	.09	.06	3.5	.97	.94	.88	.82
1.1	.30	.20	.11	.07	3.6	.97	.95	.90	.85
1.2	.33	.22	.13	.08	3.7	.98	.96	.92	.87
1.3	.37	.26	.15	.10	3.8	.98	.97	.93	.89
1.4	.40	.29	.18	.12	3.9	.99	.97	.94	.91
1.5	.44	.32	.20	.14	4.0	.99	.98	.95	.92
1.6	.48	.36	.23	.16	4.1	.99	.98	.96	.94
1.7	.52	.40	.27	.19	4.2	.99	.99	.97	.95
1.8	.56	.44	.30	.22	4.3	**	.99	.98	.96
1.9	.60	.48	.33	.25	4.4		.99	.98	.97
2.0	.64	.52	.37	.28	4.5		.99	.99	.97
2.1	.68	.56	.41	.32	4.6		**	.99	.98
2.2	.71	.59	.45	.35	4.7			.99	.98
2.3	.74	.63	.49	.39	4.8			.99	.99
2.4	.77	.67	.53	.43	4.9			.99	.99
					5.0			**	.99
					5.1				.99
					5.2				**

*Values inaccurate for *one-tailed* test by more than .01.
**The power at and below this point is greater than .995.

Note: From *Introductory Statistics for the Behavioral Sciences* (3rd ed., p. 363) by J. Welkowitz, R. B. Ewen, and J. Cohen, 1982, Orlando, FL: Academic Press. Copyright 1982 by Academic Press. Reprinted by permission.

TABLE M
δ as a function of significance criterion (α) and power

	One-tailed test (α)			
	.05	.025	.01	.005
	Two-tailed test (α)			
Power	.10	.05	.02	.01
.25	0.97	1.29	1.65	1.90
.50	1.64	1.96	2.33	2.58
.60	1.90	2.21	2.58	2.83
.67	2.08	2.39	2.76	3.01
.70	2.17	2.48	2.85	3.10
.75	2.32	2.63	3.00	3.25
.80	2.49	2.80	3.17	3.42
.85	2.68	3.00	3.36	3.61
.90	2.93	3.24	3.61	3.86
.95	3.29	3.60	3.97	4.22
.99	3.97	4.29	4.65	4.90
.999	4.37	5.05	5.42	5.67

Note: From *Introductory Statistics for the Behavioral Sciences* (3rd ed., p. 364) by J. Welkowitz, R. B. Ewen, and J. Cohen, 1982, Orlando, FL: Academic Press. Copyright 1982 by Academic Press. Reprinted by permission.

ANSWERS TO SELECTED PROBLEMS

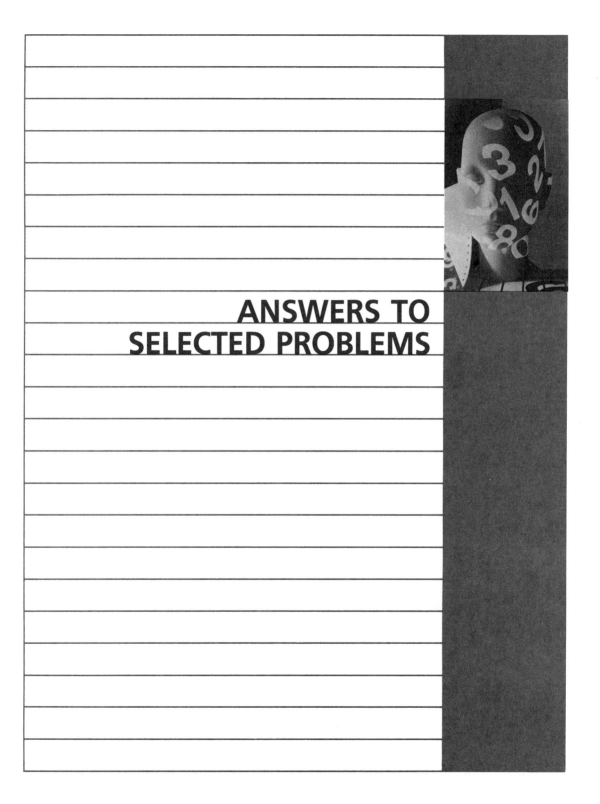

ANSWERS TO SELECTED PROBLEMS

CHAPTER 2

1.
1	23	39	59	86
2	23	40	63	90
6	24	44	64	91
9	33	47	68	92
10	33	47	77	92
17	35	53	81	93
18	37	56	82	93
22	39	57	85	97

2. a) $X_{low} = 1$, b) $X_{high} = 97$, c) range = 96, d) 47. No.

4.
Class	f	cf
180–194	4	71
165–179	1	67
150–164	3	66
135–149	4	63
120–134	8	59
105–119	5	51
90–104	17	46
75–89	14	29
60–74	10	15
45–59	3	5
30–44	2	2

7.

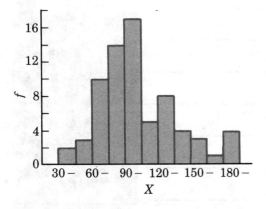
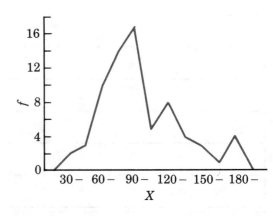

10. 50th percentile = 95.24. It is the point at or below which 50% of the scores fall.

CHAPTER 3

1. $\bar{X} = 84.95$. Yes, because 84.95 is near the center of the ordered batch.

3. $\overline{X} = 28.83$.
7. 50th percentile $\cong 27.5$.

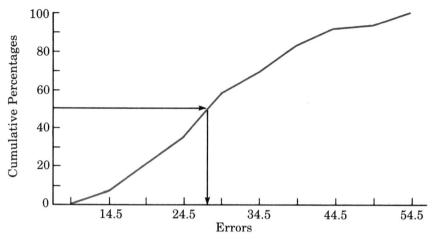

8. $\Sigma(X - \overline{X}) = 0.05$. $\Sigma(X - mdn) = -20$.

CHAPTER 4

1. a) $SS = 5872.95$, b) $s = 17.58$, c) $s^2 = 309.10$.
3. $sd = 17.14$.
5. $Q_1 = 21.75$, $Q_3 = 55.75$. $SIR = 17$.
6.

X_{low}	Q_1	Q_3	Q_1	X_{high}
11	21.75	43	55.75	64

9. $Z = 1.47$. $T = 64.7$. $SAT = 647$.
11. $X = 55.25$.

CHAPTER 5

1.

N_T	Coefficients	$\Sigma = T$
6	1 6 15 20 15 6 1	64
7	1 7 21 35 35 21 7 1	128
8	1 8 28 56 70 56 28 8 1	256

$p^8 + 8p^7q + 28p^6q^2 + 56p^5q^3 + 70p^4q^4 + 56p^3q^5 + 28p^2q^6 + 8pg^7 + q^8$
The exponent for the first term, p, is N_T. Thereafter, the exponent for p decreases by one and the exponent for q increases by one for each successive term.

3. $P(8H) = 1/256 = .0039$. $P(7H \text{ or more}) = 9/256 = .0352$. $P(8H \text{ or } 8T) = 2/256 = .0078$. $P(7H \text{ or more } or \text{ 7T or more}) = 18/256 = .0703$.
5. (1) H_0: $P(R) \leq .5$ (1) H_0: $P(R) \leq .5$
   (2) H_1: $P(R) > .5$ (2) H_1: $P(R) > .5$
   (3) $\alpha = .05$ (3) $\alpha = .05$
   (4) $P(7R, 2L) = .0898$ (4) $P(8R, 1L) = .0195$
   (5) $.0898 > .05$ (5) $.0195 < .05$
   (6) Do not reject H_0 (6) Reject H_0, accept H_1
7. (1) H_0: $P(L) \leq .5$
   (2) H_1: $P(L) > .5$
   (3) $\alpha = .05$
   (4) $P(Z = 1.84) = .0329$
   (5) $.0329 < .05$
   (6) Reject H_0, accept H_1
9. $P = .33$. $P = .0278$.

CHAPTER 6

3. $r = -.89$.
4. (1) H_0: $\rho \geq 0$
   (2) H_1: $\rho < 0$
   (3) $\alpha = .01$
   (4) $P(r = -.89) < .005$ $(df = 40)$
   (5) $.005 < .01$
   (6) Reject H_0, accept H_1
6. Do not reject H_0 with the two-tailed test $(P = .02)$. Reject H_0 with the one-tailed test $(P = .01)$.
8. $r_s = .12$. r_s is inappropriate because the relationship between X and Y is not linear.

CHAPTER 7

1. $r^2 (100) = 16.81$. $(1 - r^2)(100) = 83.19$.
3. $Y' = -0.98X + 13.69$.
5. $X' = -0.81Y + 12.01$.
6. $s^2_{y \cdot x} = 1.47$. The size of the $s^2_{y \cdot x}$ will increase because $(1 - r^2)$ will be larger.
8. No, because the same third variable (or variables) could be present on each occasion.

ANSWERS TO SELECTED PROBLEMS

CHAPTER 8

1. (1) H_0: $\mu_E \geq \mu_C$
   (2) H_1: $\mu_E < \mu_C$
   (3) $\alpha = .05$
   (4) $P(t = -1.95) < .05$ ($df = 22$)
   (5) $< .05 < .05$
   (6) Reject H_0, accept H_1
4. $Z = 2.65$. Although the sample $\overline{X}$ deviates significantly ($P = .008$) from μ, the student should generate a large number of samples, inspecting them for a systematic deviation from μ before rejecting the program.
6. $t = -4.02$, $df = 6$, $P < .005$. The women's average salary is significantly lower than the men's.
9. No. $t = -0.96$, $df = 9$, $P > .05$.
11. $28.23 \leq \mu \leq 30.63$. We can be 95% confident that μ is contained within these limits.

CHAPTER 9

1. $F = 6.34$, $df = 10, 10$, $P < .005$. The researcher's hypothesis is confirmed.
2. $F = 5.44$, $df = 7, 11$, $P < .02$. $\sigma_E^2 > \sigma_C^2$.
5. $F_{max} = 7.0$, $k = 4$, $df = 10$, $P < .05$. Reject H_0.
6.

s^2	8.76	12.13	29.62
4.23	2.07	2.87	7.00*
8.76		1.38	3.38
12.13			2.44

$F_{max\ .05} = 3.72$, $k = 2$, $df = 10$

*$P < .05$
Yes, because the overall test was significant.
9. $t = -0.36$, $df = 7$, $P > .05$. Do not reject H_0.

CHAPTER 10

1.

Source of Variation	SS	df	MS	F	$F_{.05}$
Groups	73.34	2	36.67	5.41 >	3.40
Within	162.74	24	6.78		
Total	236.08	26			

3. $100\eta^2 = 31.07$. $100\hat{\omega}^2 = 24.62$.
5. $F_{max} = 5.03$, $k = 3$, $df = 8$, $P > .05$. H_0 can not be rejected.

8.
Source of Variation	SS	df	MS	F	$F_{.05}$
Tests	95.25	2	47.625	29.31 > 3.74	
Cases	22.625	7	3.232		
Error	22.75	14	1.625		
Total	140.625	23			

Reject H_0.

9.
$\overline{X}$	5.00	7.25		LSD = 1.37
2.38	2.62*	4.87*		
5.00		2.25*		

*< .05

CHAPTER 11

2.
Source of Variation	SS	df	MS	F	$F_{.05}$
Dosages	19.44	2	9.72	7.36 > 3.23
Drugs	140.11	2	70.06	53.08 > 3.23
Dosages × Drugs	23.78	4	5.94	4.50 > 2.61
Within	59.50	45	1.32	
Total	242.83	53		

4. Dosage: $100\eta^2 = 8.0$. Drugs: $100\eta^2 = 57.7$.
Dosages × Drugs: $100\eta^2 = 9.8$.

6. The design is a 2 × 3 factorial but the researcher is analyzing it as a simple ANOVA. He is (a) losing information about the main effects, A and B, and their interactions; and (b) he is making some uninterpretable comparisons, such as $\overline{X}_{A_1B_3}$ versus $\overline{X}_{A_2B_1}$.

7. Yes, there is a 6/4 ratio of cases at each level of B.

Source of Variation	SS	df	MS	F	$F_{.05}$
A	33.80	1	33.80	17.16 > 4.26
B	94.47	2	47.24	23.98 > 3.40
A × B	8.03	2	4.02	2.04 < 3.40
Within	47.17	24	1.97	
Total	183.47	29		

Both A and B are significant.

9.

Source of Variation	SS	df	MS	F	$F_{.05}$
Drugs	1.37	1	1.37	1.12 <	4.08
Length	135.03	2	67.52	55.34 >	3.23
Drugs × Length	9.03	2	4.52	3.70 >	3.23
Error	43.88	36	1.22		

The N_Gs are not proportional. Length and Drugs × Length are significant sources of variation. $N_h = 6.5452$.

CHAPTER 12

1. $T = 54$, N per batch $= 9$, $P < .005$. The study supports the scientist's hypothesis.
3. $T = 3$, N pairs $= 8$, $P < .025$. The scientist's hypothesis was confirmed.
5. $H = 7.76$. $H \cong \chi^2$, $df = 2$, $P < .025$. The overall test of the ms is significant.
7. $\chi_r^2 = 13.56$, $k = 3$, $N = 8$, $P < .00026$. The null hypothesis may be rejected.

CHAPTER 13

1. $\chi^2 = 14.40$, $df = 1$, $P < .001$. Yes, drinking experience and driving performance are associated—inexperienced drinkers drive more poorly than moderate drinkers.
3. $\chi^2 = 43.91$, $df = 2$, $P < .001$. Yes, there is a relationship between age of the delegate and type of solution.
5. $\chi^2 = 4.5$, $df = 1$, $P < .05$. A larger proportion of the subjects passed task II.
6. $Q = 94.31$. $Q \cong \chi^2$, $df = 4$, $P < .001$. The H_0 can be rejected.

CHAPTER 14

1. Rat 1: 70 and 61 are outside numbers. ($IR = 12$).
   Rat 3: 217 is far out; 126 and 105 are outside numbers. ($IR = 23$).

2.

Classes	f	cf
400–449	1	45
350–399	0	44
300–349	1	44
250–299	0	43
200–249	1	43
150–199	2	42
100–149	2	40
50–99	8	38
0–49	30	30

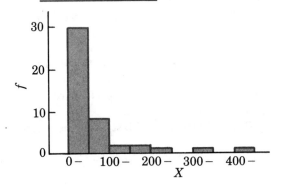

The distribution is markedly and positively skewed.

4. $F_{max} = 79.84$, $k = 3$, $df = 14$, $P < .01$. The null hypothesis may be rejected.

7. Power $(1 - \beta) \cong .71$.

9. N per group $= 19.38$ or approximately 19 cases per group are needed to achieve a power $= .80$.

12. $\gamma = .5$. A $\gamma = .5$ is a medium effect size in Cohen's conception of effect sizes.

REFERENCES

REFERENCES

Andrews, D.F., Bickel, P.J., Hampel, F.R., Huber, P.J., Rogers, W.H., & Tukey, J.W. (1972). *Robust estimates of location*. Princeton, NJ: Princeton University Press.

Arbuthnott, J. (1710). An argument for divine providence taken from the constant regularity observed in the births of both sexes. *Philosophical Transactions, 27,* 186–190.

Barnett, V. (1983). Discussion. *Technometrics, 25,* 150–152.

Barnett, V., & Lewis, T. (1978). *Outliers in statistical data*. New York: John Wiley.

Bartlett, M.S. (1937). Properties of sufficiency and statistical tests. *Proceedings of the Royal Society. Series A, 160,* 268–282.

Beckman, R.J., & Cook, R.D. (1983). Outlier. . . . . .s. *Technometrics, 25,* 119–149.

Bernhardson, C.S. (1975). Type I error rates when multiple comparison procedures follow a significant F test of ANOVA. *Biometrics, 31,* 229–232.

Blalock, H.M. Jr. (1979). *Social statistics*. New York: McGraw-Hill.

Boneau, C.A. (1960). The effect of violations of assumptions underlying the t test. *Psychological Bulletin, 57,* 47–64.

Box, G.E.P. (1953). Non-normality and tests on variances. *Biometrika, 40,* 318–335.

Box, G.E.P., Hunter, W.G., & Hunter, J.S. (1978). *Statistics for experimenters*. New York: Wiley.

Bradley, J.V. (1968). *Distribution-free statistical tests*. Englewood Cliffs, NJ: Prentice-Hall.

Bradley, J.V. (1980). Nonrobustness in classical tests on means and variances: A large-scale sampling study. *Bulletin of the Psychonomic Society, 15,* 275–278.

Bresnahan, J.L., & Shapiro, M.M. (1966). A general equation and technique for exact partitioning of chi-square contingency tables. *Psychological Bulletin, 66,* 252–262.

Camilli, G., & Hopkins, K.D. (1978). Applicability of chi-square to 2 × 2 contingency tables with small expected cell frequencies. *Psychological Bulletin, 85,* 163–167.

Camilli, G., & Hopkins, K.D. (1979). Testing for association in 2 × 2 tables with very small sample sizes. *Psychological Bulletin, 86,* 1011–1014.

Campbell, N.R. (1938) *Symposium: Measurement and its importance for philosophy*. Aristotelian Society. Suppl. Vol. 17. London: Harrison.

Carmer, S.G., & Swanson, M.R. (1973). An evaluation of ten pairwise multiple comparison procedures by Monte Carlo methods. *Journal of the American Statistical Association, 68,* 66–74.

Castellan, N.J. Jr. (1965). On the partitioning of contingency tables. *Psychological Bulletin, 64,* 330–338.

Church, J.D., & Wike, E.L. (1976). The robustness of homogeneity of variance tests for asymmetric distributions: A Monte Carlo study. *Bulletin of the Psychonomic Society, 7,* 417–420.

Church, J.D., & Wike, E.L. (1979). A Monte Carlo study of nonparametric multiple-comparison tests for a two-way layout. *Bulletin of the Psychonomic Society, 14,* 95–98.

Cohen, J. (1965). Some statistical issues in psychological research. In B.B. Wolman (Ed.), *Handbook of clinical psychology.* New York: McGraw-Hill.

Cohen, J. (1969). *Statistical power analysis for the behavioral sciences.* New York: Academic Press.

Cohen, J., & Cohen, P. (1975). *Applied multiple regression/correlation analysis for the behavioral sciences.* Hillsdale, NJ: Lawrence Erlbaum.

Conover, W.J. (1971). *Practical nonparametric statistics.* New York: Wiley.

Conover, W.J., Johnson, M.E., & Johnson, M.M. (1981). A comparative study of tests of homogeneity of variances, with applications to the outer continental shelf bidding data. *Technometrics, 23,* 351–361.

Cook, T.D., & Campbell, D.T. (1979). *Quasi-experimentation: Design & analysis issues for field settings.* Chicago: Rand McNally.

Daniel, W.W. (1978). *Applied nonparametric statistics.* Boston: Houghton Mifflin.

Delucchi, K.L. (1983). The use and misuse of chi-square: Lewis and Burke revisited. *Psychological Bulletin, 94,* 166–176.

Dixon, W.J. (Ed.), Brown, M.B., Engelman, L., Frane, J.W., Hill, M.A., Jennrich, R.I., & Toporek, J.D. (1983). *BMDP statistical software.* Berkeley, CA: University of California Press.

Dixon, W.J., & Massey, F.J. Jr. (1951). *Introduction to statistical analysis.* New York: McGraw-Hill.

Draper, N., & Smith, H. (1966). *Applied regression analysis.* New York: Wiley.

Ferguson, G.A. (1981). *Statistical analysis in psychology and education* (5th ed.). New York: McGraw-Hill.

Fleiss, J.L. (1973). *Statistical methods for rates and proportions.* New York: Wiley.

Friedman, M. (1937). The use of ranks to avoid the assumption of normality implicit in the analysis of variance. *Journal of the American Statistical Association, 32,* 675–701.

Gaito, J. (1980). Measurement scales and statistics: Resurgence of an old misconception. *Psychological Bulletin, 87,* 564–567.

Glass, G.V., Peckham, P.D., & Sanders, J.R. (1972). Consequences of failure to meet assumptions underlying the fixed-effects analysis of variance and covariance. *Review of Educational Research, 42,* 237–288.

REFERENCES

Good, I.J. (1983). The philosophy of exploratory data analysis. *Philosophy of Science, 50,* 283–295.

Greenhouse, S.W., & Geisser, S. (1959). On methods in the analysis of profile data. *Psychometrika, 24,* 95–111.

Guilford, J.P. (1956). *Fundamental statistics in psychology and education* (3rd ed.). New York: McGraw-Hill.

Haertel, R.J., & Lane, D.M. (1979). A FORTRAN IV program for testing differences in variance using a jackknife procedure. *Behavior Research Methods & Instrumentation, 11,* 74.

Harlow, H.F. (1962). Fundamental principles for preparing psychology journal articles. *Journal of Comparative and Physiological Psychology, 55,* 893–896.

Hartley, H.O. (1950). The maximum F-ratio as a shortcut test for heterogeneity of variance. *Biometrika, 37,* 308–312.

Hays, W.L. (1963). *Statistics for psychologists.* New York: Holt, Rinehart and Winston.

Hilgard, E.R. (1955). Discussion of probabilistic functionalism. *Psychological Review, 62,* 226–228.

Hinkle, D.E., Wiersma, W., & Jurs, S.G. (1979). *Applied statistics for the behavioral sciences.* Chicago: Rand McNally.

Hollander, M., & Wolfe, D.A. (1973). *Nonparametric statistical methods.* New York: Wiley.

Hopkins, K.D., & Glass, G.V. (1978). *Basic statistics for the behavioral sciences.* Englewood Cliffs, NJ: Prentice-Hall.

Howell, D.C. (1982). *Statistical methods for psychology.* Boston: Duxbury Press.

Huynh, H., & Feldt, L.S. (1970). Conditions under which mean square ratios in repeated measurements designs have exact F-distributions. *Journal of the American Statistical Association, 65,* 582–589.

Huynh, H., & Mandeville, G.K. (1979). Validity conditions in repeated measures designs. *Psychological Bulletin, 86,* 964–973.

Inman, R.L., & Conover, W.J. (1983). *A modern approach to statistics.* New York: Wiley.

Kendall, M.G. (1948). *Rank correlation methods.* London: Charles Griffin.

Keppel, G. (1982). *Design and analysis: A researcher's handbook* (2nd ed.). Englewood Cliffs, NJ: Prentice-Hall.

King, R.S., & Julstrom, B. (1982). *Applied statistics using the computer.* Sherman Oaks, CA: Alfred, 1982.

Kirk, R.E. (1978). *Introductory statistics.* Monterey, CA: Brooks/Cole.

Kruskal, W.H., & Wallis, W.A. (1952). Use of ranks in one-criterion variance analysis. *Journal of the American Statistical Association, 47,* 583–621.

Layard, M.W.J. (1973). Robust large-sample tests for homogeneity of variances. *Journal of the American Statistical Association, 68,* 195–198.

Lilliefors, H.W. (1967). On the Kolmogorov-Smirnov test for normality with mean and variance unknown. *Journal of the American Statistical Association, 62,* 399–402.

Lindman, H.R. (1974). *Analysis of variance in complex experimental designs.* San Francisco: Freeman.

Lindquist, E.F. (1953). *Design and analysis of experiments in psychology and education.* Boston: Houghton Mifflin.

McNemar, Q. (1947). Note on the sampling error of the difference between correlated proportions or percentages. *Psychometrika, 12,* 153–157.

Miller, R.G. Jr. (1966). *Simultaneous statistical inference.* New York: McGraw-Hill.

Miller, R.G. Jr. (1968). Jackknifing variances. *Annals of Mathematical Statistics, 39,* 567–582.

Mosteller, F., Fienberg, S.E., & Rourke, R.E. (1983). *Beginning statistics with data analysis.* Reading, MA: Addison-Wesley.

Mosteller, F., & Tukey, J.W. (1968). Data analysis, including statistics. In G.W. Lindzey & E. Aronson (Eds.), *The handbook of social psychology* (2nd ed.) Vol. II. Reading, MA: Addison-Wesley.

Mosteller, F., & Tukey, J.W. (1977). *Data analysis and regression: A second course in statistics.* Reading, MA: Addison-Wesley.

O'Brien, R.G. (1981). A simple test of variance effects in experimental designs. *Psychological Bulletin, 89,* 570–574.

Quenouille, M.H. (1959). *Rapid statistical calculations.* New York: Hafner.

Rogan, J.C., Keselman, H.J., & Mendoza, J. (1979). Analysis of repeated measurements. *British Journal of Mathematical and Statistical Psychology, 32,* 269–286.

Runyon, R.P., & Haber, A. (1984). *Fundamentals of behavioral statistics* (5th ed.). Reading, MA: Addison-Wesley.

Ryan, T.A. Jr., Joiner, B.L., & Ryan, B.F. (1981). *Minitab reference manual.* Lawrence, KS: University of Kansas, Academic Computation Center.

Satterwaite, F.E. (1946). An approximate distribution of estimates of variance components. *Biometrics Bulletin, 2,* 110–114.

Siegel, S. (1956). *Nonparametric statistics for the behavioral sciences.* New York: McGraw-Hill.

Smith, J.E.K. (1976). Data transformations in analysis of variance. *Journal of Verbal Learning and Verbal Behavior, 15,* 339–346.

Snedecor, G.W., & Cochran, W.G. (1967). *Statistical Methods* (6th ed.). Ames, IA: Iowa State University Press.

Stavig, G.R. (1978). The median controversy. *Perceptual and Motor Skills, 47,* 1177–1178.

Stevens, S.S. (1951). Mathematics, measurement, and psychophysics. In S.S. Stevens (Ed.), *Handbook of experimental psychology.* New York: Wiley.

REFERENCES

Student. (1927). Errors of routine analysis. *Biometrika, 19,* 151–164.

Teel, L.S. (1981). *Personality dimensions and psychophysiological indices of ischemic pain tolerance.* Unpublished doctoral dissertation, University of Kansas, Lawrence.

Tsai, W.S., Duran, B.S., & Lewis, T.O. (1975). Small-sample behavior of some multisample nonparametric tests for scale. *Journal of the American Statistical Association, 70,* 791–796.

Tufte, E.R. (1983). *The visual display of quantitative information.* Cheshire, CT: Graphics Press.

Tukey, J.W. (1962). The future of data analysis. *Annals of Mathematical Statistics, 33,* 1–67.

Tukey, J.W. (1969). Analyzing data: Sanctification or detective work. *American Psychologist, 24,* 83–91.

Tukey, J.W. (1977). *Exploratory data analysis.* Reading, MA: Addison-Wesley.

User's guide SPSS^x. (1983). New York: McGraw-Hill.

Velleman, P., & Hoaglin, D. (1981). *ABC's of EDA.* Boston: Duxbury Press.

Walker, H.M., & Lev, J. (1953). *Statistical inference.* New York: Holt.

Welkowitz, J., Ewen, R.B., & Cohen, J. (1982). *Introductory statistics for the behavioral sciences* (3rd Ed.). New York: Academic Press.

Wike, E.L. (1971). *Data analysis: A statistical primer for psychology students.* Chicago, IL: Aldine.

Wike, E.L., & Church, J.D. (1978). A Monte Carlo investigation of four nonparametric multiple-comparison tests for k independent groups. *Bulletin of the Psychonomic Society, 11,* 25–28.

Wike, E.L., & Church, J.D. (1982a). Nonrobustness in F tests: 1. A replication and extension of Bradley's study. *Bulletin of the Psychonomic Society, 20,* 165–167.

Wike, E.L., & Church, J.D. (1982b). Nonrobustness in F tests: 2. Further extensions of Bradley's study. *Bulletin of the Psychonomic Society, 20,* 168–170.

Wilcoxon, F., Katti, S.K., & Wilcox, R.A. (1963). *Critical values and probability levels for the Wilcoxon rank sum test and the Wilcoxon signed ranks test.* Pearl River, NY: American Cyanamid Company and The Florida State University.

Winer, B.J. (1971). *Statistical principles in experimental design* (2nd ed.). New York: McGraw-Hill.

GLOSSARY OF KEY TERMS

additive rule of probabilities The rule for determining the probability of either event A occurring or event B occurring is $P(A) + P(B) - P(AB)$, where $P(AB)$ is the probability of both events occurring.

algebraic logic A type of logic used in some calculators. The calculator will have an $\boxed{=}$ key and entries follow the normal sequence as they appear in an equation or formula.

alphanumeric A computer term for letters, numbers, characters, or combinations thereof that represent category membership. Alphanumeric is a contraction of alphabetic and numeric.

alternative hypothesis (H_1) A statistical hypothesis that specifies a value or set of values for the population parameter different from that asserted in the null hypothesis (H_0). H_1 conforms to the research hypothesis.

analysis of variance (ANOVA) The partitioning of the total sum of squares into its components and testing null hypotheses regarding the components. The H_0 tests permit decisions regarding the means of two or more populations.

analysis of variance summary table A table displaying the results of an ANOVA in terms of the SSs, dfs, MSs, and F ratios for the various effects.

applied statistics Use of the logic and techniques of theoretical statistics to make sense of numbers. Applied statistics includes descriptive and inferential statistics.

arithmetic mean ($\overline{X}$) The sum of the numbers in a batch divided by the batch count. A measure of the center of a batch.

assumption of homogeneity of variance The assumption that the numbers in separate batches are random samples from populations with equal variances.

assumption of independence The assumption that the numbers in separate batches are random samples from populations in which the numbers are uncorrelated.

assumption of normality The assumption that the numbers in separate batches are random samples from normally distributed populations.

ballantine (Venn diagram) A diagram that pictures the relationship between two variables. It consists of two circles that display the proportions of explained and unexplained variance in X and Y.

Bartlett's test A test of equality of variances for two or more separate populations. Unlike the $F_{\max}$ test, Bartlett's does not require equal batch sizes.

batch mode A computer procedure in which a program and data are input together. The program is then executed immediately or at a later time.

between-design A factorial design with nested subjects (separate batches).

between-groups sum of squares (SS_G) The variability of the batch means about the grand mean weighted by the number of cases in the batches.

Big T-for-twins test (Wilcoxon's signed-ranks test) A nonparametric test for the medians of two related populations.

Big T-for-two test (Wilcoxon's rank-sum test) A nonparametric test for the medians of two separate populations.

Big T test (Wilcoxon's signed-ranks test) A nonparametric test for the median of a single population.

bimodal A distribution with two nonadjacent modes.

binomial distribution A theoretical distribution consisting of a family of distributions for the outcomes of an experiment in which either of two outcomes may occur on a given trial. The distributions are symmetrical when $p = .5$ and become more normal as the number of trials (N_T) increases.

binomial expansion A method of obtaining a binomial distribution. The quantity, $(p + q)$, is raised to the N_T power, where p is the probability of one outcome, $q = 1 - p$ is the probability of the other outcome, and N_T is the number of trials.

bivariate data An arrangement of data in which X and Y values are available for each case.

bivariate normal distribution An extension of the assumption of normality to the joint distribution of two variables in a scatter plot in which the third dimension is the frequency of the $X - Y$ dots.

box-and-whisker plot A graph of Tukey's five-number summary of a batch.

Box's test for variances A robust test for homogeneity of variance. The test is an F test of the logged subbatch variances in the k batches.

center A value that is representative of a batch.

chi-square (χ^2) distribution A theoretical distribution consisting of a family of curves varying with the degrees of freedom (df). Such distributions are positively skewed for small df but gradually approach normality with larger df.

chi-square test for goodness of fit A test of the degree that the observed frequencies deviate from expected frequencies in two or more bins.

chi-square test of independence A test of association between two factors in a contingency table.

class interval size (i) In a frequency distribution the width of a class, that is, the difference between the upper real limit (URL) and the lower real limit (LRL).

Cochran's Q test A test of proportions in k related populations.

compound symmetry The assumption for a repeated measures ANOVA that the variances and covariances for the levels of the repeated factor are equal.

computational formula A convenient formula for calculating a statistic or parameter.

confidence interval The values falling between two interval estimates. The investigator can have $100(1 - \alpha)$ confidence that the limits contain the parameter.

confidence limits Interval estimates about which the investigator can have $100(1 - \alpha)$ confidence that the limits contain the parameter.

conservative F test An F test for a repeated factor in which the degrees of freedom are reduced by the value for epsilon (ε).

contingency table A matrix of categories resulting from combining all classes of a factor with all classes of another factor. Each category contains the frequency of cases.

continuous measurement Measurement in which the value for a variable can take on an infinite number of values.

conventional effect sizes Sets of small, medium, and large effect sizes for various statistics proposed by Cohen and used in power analysis.

correct decisions Not rejecting the null hypothesis (H_0) when H_0 is true or rejecting H_0 when H_0 is false.

correction for continuity A correction that is applied when a test statistic, which can only have discrete values, is evaluated by a continuous theoretical distribution.

Cramér's coefficient (V) An index of association for a contingency table that has more than one degree of freedom.

criterion A variable that an investigator wants to predict.

criterion of significance (α) The probability of rejecting the null hypothesis (H_0) when H_0 is true.

critical value of a test statistic The value of a test statistic for the α level. A statistical decision regarding the null hypothesis (H_0) can be made by comparing the actual value of the test statistic with the critical value.

crossed subjects An experimental design in which the cases undergo all treatments or treatment-combinations (related batches).

cross-over interaction An interaction in which the direction of the differences in the means for a factor is reversed at the lowest and highest level of a second factor.

cumulative frequency (cf) The number of values or cases falling within a class and all classes below it.

cumulative frequency distribution A display of the cumulative frequencies for a set of classes.

cumulative percentage curve (ogive) A graph constructed by plotting the cumulative percentages for a set of classes at the upper real limits of the classes and connecting the plotted points by straight lines.

curvilinear relationship A relationship in which the line fitting a set of X and Y points in a scatter plot is a curved line.

degrees of freedom (df) The number of values in a set less the number of restrictions on the set.

delta (δ) A measure relating effect size (γ) and batch size (N). Delta is used in power analysis.

descriptive statistics (exploratory data analysis) Characterizing and summarizing batches of numbers by statistics and pictures.

deviation from the mean The difference between a number and the mean of a batch.

direct relationship A relationship in which increases in X are associated with increases in Y. X and Y are then said to be positively correlated.

distribution-free or nonparametric statistics A class of statistical tests whose assumptions are less restrictive than the assumptions for parametric statistical tests like t, F, r, and so forth.

divergent interaction An interaction in which the differences in the means for a factor increase over the levels of a second factor.

effect size The extent to which the null hypothesis (H_0) is false.

epsilon (ϵ) A measure of the degree of heterogeneity of the variance-covariance matrix in a repeated measures ANOVA.

error in prediction The difference between an actual value and a predicted value for a given case.

eta squared (η^2) The ratio of the sum of squares for an effect to the total sum of squares. Eta squared is a measure of the proportion of variability due to the effect. It is also called correlation ratio.

expected frequency (f_e) The expected count for a bin or category based upon marginal counts or a theory.

explained variance The proportion (r^2) of the total variance in Y (or X) due to X (or Y).

factorial design An experimental plan in which all levels of a factor are combined with all levels of another factor (or factors).

far out number A number that falls beyond a far fence. Tukey defines a far fence as $3(IR)$ from the nearest Q value.

F distribution A theoretical distribution for the ratio of two variances. It consists of a family of positively skewed distributions that varies as a function of the degrees of freedom for the variances.

Fisher's LSD test A method for comparing means after a significant F test. The differences in the means are compared with LSD, a least significant difference value, to test significance.

Fisher's sign test A nonparametric test for the median of a single population or the medians of two related populations.

five-number summary A description of a batch, proposed by Tukey, in terms of the lowest value, Q_1, Q_2, Q_3, and the highest value.

fixed A computer term for real numbers or numbers resulting from continuous measurement. In practice, counts are often treated as fixed.

☐ GLOSSARY OF KEY TERMS

fixed-effects model A model for expected mean squares in ANOVA when the levels of a factor are arbitrarily selected.

frequency (f) The number of values or cases falling within a class in a frequency distribution.

frequency distribution A display of the frequencies falling in a set of classes.

frequency polygon A graph formed by plotting the frequencies for a set of classes at the midpoints of the classes and connecting the plotted points by straight lines.

Friedman's χ_r^2 test A nonparametric test for the medians of k related populations.

F test for variances A test for equality of the variances for two separate populations.

F_{max} test A test for the equality of the variances for two or more separate populations.

gamma (γ) Effect size in the population standardized.

harmonic mean A special mean that is the reciprocal of the mean for the reciprocals of the values. For example, N_h is employed in unweighted means analysis and power analysis.

histogram A graph in which bars depict the frequencies for a set of classes.

indication A provisional conclusion resulting from the exploratory and confirmatory analysis of a single experiment.

inferential statistics (confirmatory data analysis) Making statements about populations on the basis of random samples.

integers Whole numbers (1, 2, . . . ,) used in counts or to perform repetitive operations (loops) in computing.

interaction The combined effects of two or more factors upon a dependent variable.

interaction analysis Investigating an interaction by the determination and testing of simple main effects.

interaction degrees of freedom The degrees of freedom for any interaction is the product of the degrees of freedom for the interacting factors.

interactive mode A computer procedure in which commands are executed upon input and the results are immediately output.

interquartile range (IR) The distance between the 75th percentile (Q_3) and the 25th percentile (Q_1). A measure of spread for a batch.

interval estimation Specifying the limits within which a parameter might fall on the basis of information from a batch (or batches).

interval measurement A measurement scale with an arbitrary zero point and equal units.

inverse relationship A relationship in which increases in X are associated with decreases in Y. X and Y are then said to be negatively correlated.

Kendall's tau coefficient (r_k) A correlation coefficient for bivariate data in the form of ranks.

Kruskal-Wallis H test A nonparametric test for the medians of k separate populations.

kurtosis The degree of peakedness of a distribution. A normal distribution is termed mesokurtic. One that is too peaked is termed leptokurtic. One that is too flat is termed platykurtic.

Lilliefors test A graphical procedure for the assessment of normality.

linear relationship A relationship in which the line fitting a set of X and Y points in a scatter plot is a straight line.

line of best fit A regression line in which the squared errors in prediction are minimal; the line thus conforms to the least-squares criterion.

main effect A factor (an independent variable) in a factorial design.

McNemar's test for correlated frequencies A test of proportions in two related populations.

mean square A variance estimate obtained by dividing a sum of squares by its degrees of freedom.

median (*Mdn*) The $(N + 1)/2$th number in an ordered batch. If N is odd, the median is the middle number. If N is even, the median is halfway between the two middle numbers. Also the 50th percentile in a frequency distribution. The median is a measure of a center of a batch.

midpoint (X') The center of a class.

midrank method The assignment of average ranks to tied values when converting numbers to ranks.

mode The most frequently occurring number in a batch and a crude measure of the center of a batch.

moment (*M*) An average of the deviations from the mean raised to a certain power. Combinations of moments provide quantitative indices of skewness and kurtosis.

Monte Carlo study A simulated experiment in which repeated random samples result in approximate solutions to problems. Today, these studies are done on computers.

multiplicative rule of probabilities The rule for determining the probability of event A occurring and then event B occurring and then . . . is: $P(A) \times P(B) \times \ldots$.

nested subjects An experimental design in which the cases are randomly assigned to the treatments or treatment-combinations (separate batches).

nominal measurement A measurement scale consisting of a set of nonordered categories.

nonpooled error term A measure of random variability for the t-for-two test when the assumption of homogeneity of variance is not tenable.

nonrejection region Values of a test statistic that do not permit rejection of the null hypothesis.

normal distribution A theoretical distribution consisting of a family of symmetrical, bell-shaped distributions and having a known mathematical function.

null hypothesis (H_0) A statistical hypothesis that specifies a population parameter. Usually H_0 is the negation of the research hypothesis.

observed frequency (f_o) The obtained count for a bin or category.

omega squared ($\hat{\omega}^2$) A population estimate of the proportion of variability due to an effect in ANOVA.

one-tailed (one-sided) test An hypothesis test in which the alternative hypothesis (H_1) specifies a parametric value in a particular direction and one tail of the theoretical distribution is used to evaluate the null hypothesis (H_0).

ordinal measurement A 1, 2, . . ., N measurement scale in which the integers signify only greater or less.

orthogonal comparisons A set of $k - 1$ independent comparisons of k statistics.

outlier A number (or numbers) far from the bulk of the batch that is discrepant to an investigator.

outside number A number that falls beyond a near fence. Tukey defines the near fence as 1.5 (IR) from the nearest Q value.

pairwise tests The $k(k - 1)/2$ comparisons that result when each batch is compared to every other batch.

parameter A number describing and summarizing a population.

Pascal's triangle A procedure for determining the coefficients for the binomial expansion.

Pearson product-moment r A correlation coefficient indicating the strength and direction of relationship between two variables. This coefficient is an estimate of the population parameter, ρ.

percentile A number at or below which a specified percentage of the numbers in a batch fall.

phi coefficient (ϕ) A Pearson product-moment correlation coefficient for a 2 × 2 contingency table.

point biserial correlation coefficient (r_{pb}) A Pearson product-moment r for a data arrangement in which one variable is dichotomous (e.g., male vs. female) and the other variable consists of numbers. r_{pb}^2 is the proportion of accounted-for variance in the numbers due to the dichotomous variable.

point estimation Guessing a single number for a population parameter (e.g., μ) from a statistic (e.g., $\overline{X}$).

pooled error term A measure of random variability for a t-for-two test when the assumption of homogeneity of variance is tenable.

population A complete aggregation of cases, "things," or numbers.

power efficiency The efficiency of a test (e.g., a nonparametric test) relative to a parametric test. Power efficiency is measured in terms of the number of cases required by the tests to achieve the same power.

predictor A variable that an investigator employs to predict the criterion.

probability of an event (P) The number of ways an event may occur (W) in relation to the total number of mutually exclusive and equally likely possible outcomes (T).

proportionality When the ratio of cases is the same in corresponding treatment-combinations of a factorial design.

protected test Performing multiple comparisons of the statistics from batches only when an overall test is significant.

protected t test A modified t test for comparing means after a significant F test.

Q_2 The median obtained from an ordered batch or the 50th percentile calculated from a frequency distribution.

random-effects model A model for the expected mean squares in ANOVA when the levels of a factor are randomly selected from a population of levels.

random sample A sample so chosen that each element in the population has an equal likelihood of being included.

range A measure of spread consisting of the difference between the largest and smallest numbers in a batch.

ratio measurement A measurement scale characterized by a true zero point, equal units, and equality of ratios.

rational formula A logical formula for a statistic or a parameter.

real limits The actual boundaries of a class in a frequency distribution. The lower real limit (LRL) is half a unit below the stated lower limit (LL) and the upper real limit (URL) is half a unit above the stated upper limit (UL).

rectangular (uniform) distribution A distribution in which each value occurs with the same frequency.

re-expression The transformation of the numbers (Xs) obtained in an experiment to a different scale (e.g., $\log_{10} X$). Re-expression is usually done to achieve normality and equality of variances.

regression The part of correlational analysis concerned with prediction.

regression coefficient (b_y or b_x) A "weight" in a regression equation that makes the regression line a line of best fit.

regression equation The equation for a regression line.

regression line A line of best fit for a set of bivariate data.

rejection region Values of a test statistic that lead to rejection of the null hypothesis.

related batches (matched groups) Batches resulting from pairs formed on the basis of scores from a matching variable or cases tested twice.

research hypothesis An hypothesis that an investigator wishes to test in an experiment.

residuals analysis The assessment of normality by examining plots of residuals $(X - \overline{X}_j)$.

restriction of range A limitation in the spread of values for a variable. Restriction of range in one or both variables reduces the size of the correlation coefficient.

reverse (RPN) logic A type of logic used in some calculators. The calculator has an $\boxed{\text{enter}}$ key, does not have an $\boxed{=}$ key, and the entries do not follow the normal sequence as they appear in an equation or formula.

robust statistics Statistics that are relatively unperturbed by departures from normality.

robust test A statistical test that is little affected by departures from the assumptions.

sample (batch) A subset of a population.

sampling distribution A frequency distribution of a statistic based upon a large number of random samples of size N.

sampling distribution of the mean A frequency distribution of means obtained from a large number of random samples of size N.

SAT score A type of standard score with a mean of 500 and a standard deviation of 100.

Satterwaite degrees of freedom Approximate degrees of freedom for a t-for-two test when the assumption of homogeneity of variance is not tenable.

scatter plot A graph of a set of bivariate data.

semi-interquartile range (SIR) One-half of the interquartile range (IR). A measure of spread for a batch.

separate batches (independent groups) Batches in which inclusion depends on random sampling or random assignment.

shape The form of a frequency distribution generally with reference to a normal distribution.

simple main effect The variability of the means for all levels of a factor at one level of another factor.

skewness The degree of symmetry of a distribution. A positively skewed distribution has a tail on the right; a negatively skewed distribution has a tail on the left.

slope The ratio of rise to run for a regression line. The slope is b_y or b_x.

sorting Ordering the numbers in a batch from the smallest number to the largest.

Spearman's rho coefficient (r_s) A Pearson product-moment r for bivariate data in the form of ranks.

spread The degree of scatter of the numbers in a batch about its center.

standard deviation (s) An estimate of the population standard deviation (σ) based upon a batch. The positive square root of the squared deviations from the mean of the batch divided by $N - 1$.

standard deviation (σ) The standard deviation of a population. The positive square root of the squared deviations from the population mean (μ) divided by the population count (N_p). A measure of spread for a population.

standard deviation (sd) The actual standard deviation of a batch. The positive square root of the squared deviations from the mean of the batch divided by N. A measure of spread for a batch.

standard error The standard deviation of a sampling distribution for a statistic. Since the sampling distribution results from random samples, the standard error is a measure of random variability for the statistic.

standard error of a mean ($\sigma_{\bar{x}}$ or $s_{\bar{x}}$) The standard deviation of the sampling distribution of the mean.

standard error of estimate The square root of the variance of estimate. A measure of predictive accuracy.

standard (unit) normal distribution A normal distribution with a mean of 0 and a standard deviation of 1.

standard score (Z) A deviation from the mean divided by the standard deviation. Z scores have a mean of 0 and standard deviation of 1.

statistic A number describing and summarizing a batch of numbers.

statistical power The probability $(1 - \beta)$ of rejecting the null hypothesis (H_0) when H_0 is false.

stem-and-leaf display A display, devised by Tukey, of the actual values (leaves) falling within a set of classes (stems).

stated limits The listed boundaries for a class in a frequency distribution. LL is the stated lower limit and UL is the stated upper limit.

subject mortality The loss of subjects in an experiment. Subject mortality can result in disproportionality.

summation operator (Σ) A symbol directing the addition of numbers. The subscripts and superscripts specify the limits of the values to be summed and the letter(s) following the operator denotes the variable(s).

sum of squares (SS) The sum of the squared deviations from the mean of a batch. A measure of spread of a batch.

t distribution A theoretical distribution consisting of a family of symmetrical, bell-shaped, platykurtic distributions that become increasingly normal as the degrees of freedom increase.

☐ **GLOSSARY OF KEY TERMS**

test statistic   A statistic based upon a batch (or batches) that is employed to evaluate a statistical hypothesis.

theoretical distribution   A distribution such as a binomial or normal one whose mathematical function and properties are well known. A theoretical distribution is used to determine the probability of an obtained test statistic.

theoretical statistics   The branch of statistics concerned with abstract problems such as mathematical derivations and proofs.

third variable problem   In correlation when an observed r between two variables is due to the presence of a third variable (or variables).

total sum of squares (SS_T)   The variability of all numbers in a set about the grand mean.

t-for-twins test (t test for correlated means)   A test of the means for two related populations.

t-for-two test (t test for independent means)   A test of the means for two separate populations.

treatments by subjects (repeated measures) analysis of variance   An ANOVA for testing the means of k related populations.

trimean   A robust center, devised by Tukey, that is based upon a weighted combination of quartile values.

trimmed mean   A robust center that is the mean of a batch after a specified percentage of numbers has been deleted from each tail.

T score   A standard score with a mean of 50 and a standard deviation of 10.

t test for a single mean   A test of the mean of a single population when σ is unknown and N is any size.

t test for variances   A test for equality of variances for two related populations.

two-number summary   A description of a batch in terms of the mean and standard deviation (or variance).

two-tailed (two-sided) test   An hypothesis test in which the alternative hypothesis (H_1) specifies a parameter that departs from the null hypothesis (H_0) in both directions and both tails of the theoretical distribution are used to evaluate H_0.

Type 1 error of inference   The rejection of the null hypothesis (H_0) when H_0 is true. The probability of committing a Type 1 error is α.

Type 2 error of inference   The failure to reject the null hypothesis (H_0) when H_0 is false. The probability of a Type 2 error is β.

Tukey's rule of thumb   A procedure for designating certain numbers as possible outliers.

unbiased estimator   A statistic that provides an estimate of a parameter that is systematically neither too large nor too small.

unexplained variance The proportion $(1 - r^2)$ of the total variance in Y (or X) due to unknown causes.

unweighted means analysis A method of analyzing a factorial design when disproportionality is present.

variable Something that varies, meaning it takes on different values.

variance A squared standard deviation: s^2, $(sd)^2$, and σ^2. Variance is a measure of spread.

variance interpretation of r The analysis of total variance of a variable into explained and unexplained portions.

variance of estimate The variability of the actual scores about the predicted scores. A measure of predictive accuracy.

Winsorizing The replacement of the extreme numbers at each end of an ordered batch by the next-to-the extreme numbers. A method of coping with outliers.

within-design A factorial design with crossed subjects.

within-groups sum of squares (SS_W) The variability of the numbers in a batch about the mean of the batch, pooled over all batches.

Yates' correction for continuity An adjustment in the discrepancies between observed and expected frequencies that was designed to improve the determination of probabilities for discrete χ^2 values by continuous χ^2 distributions.

***Y* intercept** The Y value at which $X = 0$.

***Z* test for a single mean** A test of the mean of a single population when σ is known and N is large.

***Z* version of the Big *T*-for-two test** A large sample version of Wilcoxon's rank-sum test.

***Z* version of the Big *T* test** A large sample version of Wilcoxon's signed-ranks test.

INDEX

Alternative hypothesis, 84, 93–95
   null and, 84
   research hypothesis and, 85, 93
Analysis of variance, k separate batches, 214–31
   assumptions, 226–29
   correction term, 221
   degrees of freedom, 216–17, 221
   eta squared, 225
   F ratio, 217–18, 221
   mean squares, 217–21
   omega squared, 225–26
   statistical models, 217–218
   summary table, 221–22
   sums of squares, 215–16, 219–21
Arithmetic mean, 10, 38–43
   deviations from, 47
   frequency distribution and, 51–53
   median and, 47–48
   outliers and, 47, 65
   population parameter, 43
   reliability, 47

Ballantine (Venn) diagram. *See* Variance interpretation
Barnett, V., 49, 318

Bartlett's test, k separate variances, 203–5
   F_{max} test and, 203
   pairwise tests and, 205
Beckman, R.J., 49, 318
Bernhardson, C.S., 223
Binomial distribution, 87–97
   binomial expansion, 90–92
   mean, 97–98
   normal distribution and, 90, 97–99
   properties, 89–90
   standard deviation, 97–98
   table, 90, 96–97
BMDP, 48, 69, 237, 320, 331
Boneau, C.A., 199–200
Box, G.E.P., 196, 227, 322, 327
Box-and-whisker plot, 68, 320
Box's test, k separate variances, 322–25
   power, 325
   rationale, 322
Bradley, J.V., 227–28, 277, 290
Bresnahan, J.L., 372

Calculations, checking, 10–11
   errors, 10–11
   omitting formulas, 10
   questioning answers, 10

Calculators, algebraic and reverse logic, 41–42
   dangers, 11–12
   selection, 11
   use, 11
Camilli, G., 303
Campbell, D.T., 108, 144
Campbell, N.R., 4–6
Carmer, S.G., 223
Castellan, N.J. Jr., 306
Centers, 18, 22, 25, 38–49, 51–53
Central limit theorem, 163n
Chi-square distribution, 204
   Bartlett's test, 204–5
   Cochran's Q test, 309
   contingency tables, 302, 304
   degrees of freedom, 204, 285–86, 289, 299, 304, 307, 309
   Friedman's χ_r^2 test, 289
   Kruskal-Wallis H test, 286–87
   McNemar's χ^2 test, 307
   one-and two-tailed tests, 300, 302
   single batch, 299
Chi-square tests. See Frequency analysis
Church, J.D., 195, 203, 227–28, 288, 290, 325
Cochran, W.G., 206
Cohen, J., 5, 71, 133, 143–44, 174, 225, 291, 321, 328–31
Cohen, P., 5, 225
Computers, 331–32
   advantages, 331–32
   dangers, 11–12, 332
   data analysis and, 5–6, 9, 11, 331
   treatment of numbers, 6
Conover, W.J., 124, 194, 196, 277, 290, 307
Cook, R.D., 49, 318
Cook, T.D., 108, 144
Correct decisions in inference, 85–87
Correction for continuity, 98
   Yates', 303
Criterion of significance, 85
Cumulative frequency distribution, 19–20
Cumulative percentage polygon (ogive), 45–46
   median, 45–46
   percentiles, 46

Daniel, W.W., 277, 280, 287, 290
Data-analytic orientation, 8–9, 332–33
   indications, 9, 333
   "praying over" data, 8, 332–333
   replications, 9, 120–22, 333
Decision-reality matrix, 85, 327
Degrees of freedom, 120, 169–70
Delucchi, K.L., 306
Distribution-free statistics. See Nonparametric statistics
Distribution shape, 22–25
   discrete, 89, 98
   kurtosis, 23, 69
   normal, 22, 73
   rectangular (uniform), 47
   skewness, 23, 69
   symmetrical, 24
   unimodal and bimodal, 39
Dixon, W.J., 48, 67, 69, 237, 320, 331
Draper, N., 113

Effect size. See Power analysis
Estimation, 181–84
   confidence interval, 182
   confidence limits, 182–83
   hypothesis testing and, 181, 183–84
   interval, 181–82
   point, 181
Ewen, R.B., 71, 291, 328–29

F distribution, 197
   analysis of variance, 221
   degrees of freedom, 198
   testing two variances, 198
Feldt, L.S., 236
Ferguson, G.A., 59, 97, 170
Fienberg, S.E., 319–20
Firm knowledge, Fisher, 120–1
Fisher, R.A., 120, 180, 215, 223, 237, 280, 284
Fisher's sign test, 280–81, 285–86
   assumptions, 281, 286
   binomial test, 280–81, 285–86
Fleiss, J.L., 306, 309
Formulas, rational and computational, 59, 116
Frequency analysis, 298–311
   assumptions, 300, 303, 306, 308, 311
   Cochran's Q test, 309–10
   Cramér's V, 305
   degrees of freedom, 299, 302, 307

(Frequency analysis, *continued*)
  goodness of fit and independence, 299, 301, 303
  k related batches, 308–11
  k separate batches, 303–6
  McNemar's χ^2 test, 307–8
  null hypotheses, 299, 301, 307, 309
  observed and expected frequencies, 299, 301–2, 304
  one-and two-tailed tests, 300–302
  partitioning tables, 305–6
  phi coefficient, 302–3
  protected pairwise tests, 310
  sign test, 307–8
  single batch, 298–300
  two related batches, 306–8
  two separate batches, 300–3
  Yates' correction for continuity, 303
Frequency distribution, 18–20
  class midpoint, 19
  class size, 18, 20
  number of classes, 18, 20
  real limits of classes, 19
  stated limits of classes, 19
Frequency polygon, 21–27
  histogram and, 22
  information from, 22–27
Friedman's χ_r^2 test, k related batches, 288–90
  assumptions, 290
  null hypothesis, 289
  pairwise tests, 290
  ranking, 288–89
F test, two separate variances, 196–99
  degrees of freedom, 197–98
  Lilliefors test, 195–96, 201
  normality and, 195, 201
  power, 195
  re-expression, 195, 201
  residuals analysis, 195–96, 201
  type 1 error rate, 195

Geisser, S., 336–37
Glass, G.V., 227, 326
Greenhouse, S.W., 236–37
Guilford, J.P., 22

Haber, A., 142

Harlow, H.F., 264
Harmonic mean, 267, 330
Hartley's F_{max} test, 198–99, 201–3
  features, 199
  Lillefors test, 203
  normality, 203
  pairwise tests, 203
  power, 195
  residuals analysis, 203
  type 1 error rates, 195
Hays, W.L., 118, 225–26, 261
Hilgard, E.R., 144
Hinges, 320n
Hinkle, D.E., 72, 100, 194, 223
Histogram, 21–22
  information from, 22–27
Hoaglin, D., 320
Hollander, M., 71, 96, 277, 281, 284n, 291
Hopkins, K.D., 303
Howell, D.C., 192
Hunter, J.S., 196, 327
Hunter, W.G., 196, 327
Huynh, H., 236
Hypothesis testing, 84–87, 99–100
  steps in, 94–95, 120

Inman, R.L., 124, 196
Interquartile range, 62, 319–20

Johnson, M.E., 194
Johnson, M.M., 194
Joiner, B.L., 33, 68, 96, 320
Julstrom, B., 332
Jurs, S.G., 72, 100, 194, 223

Kendall, M.G., 122, 125
King, R.S., 332
Kirk, R.E., 5
Kruskal-Wallis H test, k separate batches, 286–88
  assumptions, 288
  null hypothesis, 287
  pairwise tests, 286, 288
  ranking, 286–87

Lev, J., 206
Levels of measurement, 4–6
  arithmetic operations, 5

(Levels of Measurement, *continued*)
    nominal, ordinal, interval, and ratio, 5
    permissible statistics, 5
    statistical assumptions, 5
Lewis, T., 49
Linear regression, 132, 135–44
    approximate slope, 137–38
    criterion and predictor, 135, 140
    equations, 135, 138–41
    errors in prediction, 139, 141–43
    least squares criterion, 135
    lines of best fit, 135
    prediction, 132, 135, 138–41
    regression coefficients, 135–38, 140–41
    regression lines, 135, 138–41
    size of r, 140, 142–44
    Y intercept, 135–36

Mandeville, G.K., 236
Massey, F.J. Jr., 67
Mean. *See* Arithmetic mean
Median, 38–39, 42–46
    cumulative percentage polygon, 45–46
    frequency distribution, 44–45
    indeterminate number, 47
    inference, 47, 276–91
    outliers, 47
    sorting-and-counting method, 43–44
Miller, R.G. Jr., 181, 195, 306
MINITAB, 33, 68, 80–81, 96, 103–4, 128–29, 148–50, 187–88, 210–11, 241–42, 272–73, 286, 293–95, 313–14, 320, 331, 336–38
Mode, 38–39
    problems with, 46–47
Moments about the mean, 69
    kurtosis, 69
    skewness, 69
Monte Carlo studies, optimal statistics, 9, 288, 290
    violations of assumptions, 9, 195, 199–200, 203, 227–28
Mosteller, F., 57, 65–66, 319–20, 326

Nonparametric statistics, 125, 276–91, 298–311
    advantages and disadvantages, 276–77
    defined, 276
    power efficiency, 277

Normal distribution, 73
    binomial distribution and, 90, 97–99
    critical Z values, 165
    equations, 73
    inference, 76
    Pearson correlation, 118, 145
    ranks tests, 279–80, 283–86
    shape, 69
    standard (unit) normal, 73–74
    Z scores, 73–76
Null hypothesis, 84–87, 94–95
    alternative hypothesis and, 84–93
Numbers, kinds of, 4–7
    computers, 5–6
    frequencies, 6–7
    impact of, 4, 152–59
    organization, 16–21
    pictures, 21–27, 68, 108–13, 254–57, 261–62
    ranks, 6–7, 122–26, 277–90
    real, 6–7
    signs, 280, 285

O'Brien, R.G., 325
One-tailed test, 94–97, 121, 178
    directional prediction, 94, 167
    experimental replication, 94
    power, 94, 331
    probability, 94, 167
    steps, 94
Outliers, 26–27, 318–20
    detection, 22, 27, 318–20
    effects, 27, 65–66, 112–13, 125
    interpretation, 26–27, 318
    Pearson correlation, 112–13
    re-expression, 320
    research, 49, 318
    Spearman rho, 124
    treatment of, 27, 319–20
    Tukey's rule of thumb, 319–20
    Winsorizing, 27, 320

Pascal's triangle, 91, 281
Pearson product-moment correlation, 113–122
    bivariate normality, 206
    causality, 144–46
    covariance, 236
    direction in causality, 144n

(Pearson correlation, *continued*)
  limits, 115
  outliers, 112–13, 125
  point-biserial correlation, 174–75
  population rho, 113
  reporting probabilities, 121–22
  restriction of range, 144–46
  sign, 115, 132
  size, 115, 132
  Spearman rho, 122, 125
  standard scores, 113–16, 133
  third variable problem, 144, 146
Peckham, P.D., 227, 326
Percentiles, 27–31
  advantage and disadvantage, 72
  compared-to-what problem, 27, 30, 70
  critical class, 28
  quartiles, 62–65, 68
Pie diagram, 226–27, 251–52
Point-biserial correlation, 174–76
  Pearson correlation, 174–75
  t and, 175–76
Population, 7
Power analysis, 327–31
  batch size, 145, 327
  conventional effect size, 330
  criterion of significance, 327, 331
  delta, 329
  effect size, 145, 327–29, 331
  factors, 327–28
  nonparametric tests, 145
  one- and two-tailed tests, 94, 331
  power defined, 86, 327
  problems, 328
Probability, additive rule, 93
  cumulative, 96–97
  multiplicative rule, 93
  one- and two-tailed, 94, 167
  reporting probabilities, 121–22
  test statistic, 85, 89
Protected tests, frequencies, 310
  means (LSD), 222–25, 234–35, 253, 258–59
  medians, 288, 290
  tables, 224n
  variances, 203, 205

Quenouille, M.H., 137

Range, 18, 20, 57, 65–67
  batch size, 66–67
  Pearson correlation, 144–46
  relation to outliers, 67
  unreliability, 66
Re-expression, 195, 325–27
  calculators and computers, 326–27
  interactions, 327
  potential benefits, 325
  rejection of, 326
  types, 325–26
Regression. *See* Linear regression
Relationship, 108–27
  contingency tables, 305, 309–10
  curvilinear, 110–13, 125–26
  direct, 108, 113
  direction, 108–9, 113, 123
  inverse, 108–10, 113
  linear, 111, 113
  strength, 110–13, 123
Repeated measures analysis of variance, 231–37
  assumptions, 235–36
  conservative F tests, 236–37
  degrees of freedom, 234, 236–37
  F ratio, 234
  Greenhouse-Geisser method, 236–37
  interaction, 233–34
  mean squares, 234
  summary table, 234
  sums of squares, 234
  violations of assumptions, 236
Research hypotheses. *See* Alternative hypothesis
Robust statistics, 48–49
  tests, 193–95, 227, 236, 322, 325
Rourke, R.E., 319–20
Runyon, R.P., 142
Ryan, B.F., 33, 68, 96, 320
Ryan, T.A. Jr., 33, 68, 96, 320

Sample, defined, 7
  random, 170
Sampling distribution, 65
  difference in related means, 177
  difference in separate means, 170–1, 193
  mean, 66, 163, 167–68

(Sampling distribution, *continued*)
   Pearson correlation, 118–19, 145
   semi-interquartile range, 66
   Spearman rho, 122, 125
   standard deviation, 65
Sanders, J.R., 227, 326
Scatter plot, 108–13, 117–18, 135–41
   information from, 113
Semi-interquartile range, 57, 62–65
   cumulative percentage polygon, 64–65
   frequency distribution, 63–64
   inference, 66
   outliers, 66
   sorting-and-counting method, 63
Shapiro, M.M., 306
Siegel, S., 5, 193, 277
Smith, H., 113
Snedecor, G.W., 206
Sorting, defined, 17
   information from, 18, 117
Spearman's rho correlation, 122–26, 276n
   Kendall's tau, 122
   limits, 123
   outliers, 124
   sampling distribution, 122, 125
   sign, 123
   size, 123
"Spotty" data, 66
Spreads, 22, 25–26, 56–67, 192–207, 214, 321–25
   nonparametric tests, 290
SPSS^x, 69, 331–32
Standard deviation, 56
   actual, 58
   estimate of population, 57–58, 214
   frequency distribution, 62, 79–80
   outliers, 65–66
   population, 58
   reliability, 65–66
Standard error, 119
   difference in related means, 179
   difference in separate means, 170–74
   expected rank sum, 280, 283
   mean, 163
   of estimate, 142–43
   Pearson correlation, 119, 145–46
Standard scores, characteristics of, 70–73
   compared-to-what problem and, 70

   constants, 71
   Pearson correlation, 113–16
   SAT scores, 72
   T scores, 72
Statistical assumptions, 5–6
   bivariate normality, 206
   compound symmetry, 235–36
   continuous measurement, 281
   homogeneity of variance, 171
   identity of shape, 284
   independence, 171
   linearity, 171
   normality, 171
   random sampling, 171
   symmetry, 276
Statistics, applied and theoretical, 7
   descriptive and inferential, 7
   exploratory data analysis and confirmatory, 8
   tasks, 7–8
Stem-and-leaf display, 20–21, 67
   advantages, 27
Stevens, S.S., 5–6
Subject mortality, 264
Summarizing batches, 67–69
   box-and-whisker plot, 68
   five-number summary, 67–68
   two-number summary, 67–69
Sum of squares, 57, 214
   between groups, 215–16, 219–20
   cases, 233
   disproportionality, 265–67
   interaction, 233–34, 251
   main effects, 251
   tests, 233
   total, 215, 219
   within groups, 216, 221
Summation operator, sigma, 40
Swanson, M.R., 223

t distribution, 119
   degrees of freedom, 120
   heterogeneity of variance, 193, 199–200
   normal distribution, 119
   Pearson correlation, 119–20
   related means, 178
   related variances, 205

(*t* distribution, *continued*)
   separate means, 171
   single means, 168
   Spearman rho, 125
   unequal batch sizes, 200
t for twins (*t* test, related means), 162–76
   assumptions, 180
   degrees of freedom, 179–80
   difference scores, 177–79
   Pearson correlation, 177, 180
   related defined, 176
   separate batches, 180–81
   t test for single mean, 177
t for two (*t* test for separate means), 162, 170–76, 199–201
   assumptions, 171, 193
   Behrens-Fisher problem, 200
   degrees of freedom, 171, 173, 200–1
   error terms, 171–74, 200
   F test, 229–30
   point-biserial correlation, 174–76
   Satterwaite degrees of freedom, 200–1
   separate defined, 170
t test, single mean, 162, 167–69
   assumptions, 276–81
   batch size, 162
   degrees of freedom, 169
   standard deviation, 162
t test, two related variances, 205–6
   assumptions, 206
   degrees of freedom, 205
   Pearson correlation, 205
Teel, L.S., 205
Test statistic, critical values, 121, 165
   general formula for, 100, 119, 163, 321
   probability, 85, 88, 94–95
Trimean, 48
Trimmed mean, 49, 320
Tufte, E.R., 109n
Tukey, J.W., 6, 8–9, 20–21, 48, 57, 62, 65–68, 120, 223, 319–20, 326
Two-factor analysis of variance, 246–68
   advantages, 247–48
   assumptions, 262, 267
   definition, 246
   degrees of freedom, 251–52, 268
   disproportionality, 264–68
   eta squared, 261–63
   extensions, 263–64
   F ratios, 252, 256–59, 265, 268
   interaction, 248, 253–54, 263–64, 267
   interaction analysis, 254–61, 267
   main effects analysis, 253, 260–61
   mean squares, 252, 268
   nested and crossed subjects, 246–47
   null hypothesis, 253
   omega squared, 261–63
   possible outcomes, 253–54
   simple main effects, 255–60
   summary table, 252, 258–59, 268
   sums of squares, 250–51, 265–67
   unweighted means analysis, 265–68
Two-tailed test, 94–96, 118, 125, 162–67
   power, 94, 331
   steps, 95
Type 1 error of inference, 85–87, 120
   consequences, 86–87
   type 2 error, 86
Type 2 error of inference, 85–87
   consequences, 86–87
   type 1 error, 86

Variable, 108n
   bivariate data, 107
   dependent, 180, 246
   independent, 246
   matching, 176, 180
Variance, actual, 59
   estimate of population, 59, 65
   mean square, 214–15
   of estimate, 142–43
   outliers, 65–66
   population, 58
   reasons for testing, 192–94, 207
   reliability, 65, 322
   treatments, 192–93, 321
Variance interpretation, 133
   analysis of variance, 225–26, 261–63
   Ballantine (Venn) diagram, 133–34
   explained and unexplained variance, 132–34
   nonparametric tests and, 290–91
   Pearson correlation, 132–33
   t for separate means, 174–76

(Variance interpretation, *continued*)
   variance of estimate, 143
Velleman, P., 376

Walker, H.M., 206
Welkowitz, J., 71, 291, 328–29
Wiersma, W., 72, 100, 194, 223
Wike, E.L., 12, 195, 203, 227–28, 288, 290, 325
Wilcoxon's rank-sum test (Big T for two), 281–84
   assumptions, 284
   Mann-Whitney test, 283
   ranking, 282
   Z version, 283
Wilcoxon's signed-ranks test (Big T and Big T for twins), 277–80, 284–85
   assumptions, 281, 286
   difference scores, 275, 284
   ranking, 278
   Z version, 279–80
Wolfe, D.A., 71, 96, 277, 281, 284n, 291

Z test, single mean, 162–67
   batch size, 162
   critical values, 165
   standard deviation, 162

ABOUT THE AUTHOR

Edward L. Wike is professor of psychology at the University of Kansas, Lawrence. During more than three decades of teaching and research at the university, Dr. Wike has written dozens of articles and reviews in the areas of motivation and statistics. He is the author of *Secondary Reinforcement: Selected Experiments* and *Data Analysis: A Primer for College Students*. Dr. Wike, who teaches courses in methodology and statistics, was named Outstanding Educator of America in 1975. A fellow of the American Psychological Association, Dr. Wike received a Ph.D. in experimental psychology from the University of California at Los Angeles.

k SEPARATE MEANS

$$SS_T = \Sigma X^2 - \frac{(\Sigma X)^2}{N} \tag{10.1}$$

With unequal N_Gs,

$$SS_G = \frac{(\Sigma X_1)^2}{N_1} + \ldots + \frac{(\Sigma X_k)^2}{N_k} - \frac{(\Sigma X)^2}{N} \tag{10.1}$$

With equal N_Gs,

$$SS_G = \frac{(\Sigma X_1)^2 + \ldots + (\Sigma X_k)^2}{N_G} - \frac{(\Sigma X)^2}{N} \tag{10.1}$$

$$SS_W = \Sigma X_1^2 - \frac{(\Sigma X_1)^2}{N_1} + \ldots + \Sigma X_k^2 - \frac{(\Sigma X_k)^2}{N_k} \tag{10.1}$$

$$SS_W = SS_T - SS_G \tag{10.1}$$

$$df_T = df_G + df_W \tag{10.1}$$

where $df_T = N - 1$, $df_G = k - 1$, and $df_W = N - k$ (10.1)

$$MS_G = \frac{SS_G}{df_G} \quad \text{and} \quad MS_W = \frac{SS_W}{df_W}$$

$$F = \frac{MS_G}{MS_W} \text{ for } df_1 = df_G = k - 1 \text{ and } df_2 = df_W = N - k \tag{10.2}$$

$$LSD = t_\alpha \sqrt{\frac{MS_W(2)}{N_G}} \tag{10.3}$$

k RELATED MEANS

$$SS_S = \frac{(\Sigma X_{S1})^2 + \ldots + (\Sigma X_{Sc})^2}{N_T} - \frac{(\Sigma X)^2}{N} \tag{10.6}$$

$$SS_E = SS_{G \times S} = SS_T - SS_G - SS_S \tag{10.6}$$

where $df_S = c - 1$, $df_E = (k - 1)(c - 1)$, and $MS_E = \frac{SS_E}{df_E}$ (10.6)

$$F = \frac{MS_G}{MS_E} \text{ for } df_1 = df_G = k - 1 \text{ and } df_2 = df_E = (k - 1)(c - 1) \tag{10.6}$$

$$LSD = t_\alpha \sqrt{\frac{MS_E(2)}{N_G}} \tag{10.6}$$

conservative $df_G = df_G \epsilon = (k - 1)\epsilon$ (10.7)

conservative $df_E = df_E \epsilon = (k - 1)(c - 1)\epsilon$ (10.7)

and $\epsilon = \frac{1}{k - 1}$ (10.7)

FACTORIAL EXPERIMENT

$$SS_A = \frac{(\Sigma X_{A_1})^2 + \ldots + (\Sigma X_{A_l})^2}{N_A} - \frac{(\Sigma X)^2}{N} \tag{11.2}$$